Synthesis, Crystal Structures and Hirshfeld Surface Analysis of Coordination Compounds (Volume II)

Synthesis, Crystal Structures and Hirshfeld Surface Analysis of Coordination Compounds (Volume II)

Editor

Waldemar Maniukiewicz

MDPI • Basel • Beijing • Wuhan • Barcelona • Belgrade • Manchester • Tokyo • Cluj • Tianjin

Editor
Waldemar Maniukiewicz
Lodz University of
Technology
Poland

Editorial Office
MDPI
St. Alban-Anlage 66
4052 Basel, Switzerland

This is a reprint of articles from the Special Issue published online in the open access journal *Crystals* (ISSN 2073-4352) (available at: https://www.mdpi.com/journal/crystals/special_issues/Coordination_Compounds_VolumeII).

For citation purposes, cite each article independently as indicated on the article page online and as indicated below:

LastName, A.A.; LastName, B.B.; LastName, C.C. Article Title. *Journal Name* **Year**, *Volume Number*, Page Range.

ISBN 978-3-0365-7928-3 (Hbk)
ISBN 978-3-0365-7929-0 (PDF)

Contents

crystals

MDPI

Article

Surface and Void Space Analysis of the Crystal Structures of Two Lithium Bis(pentafluoroethanesulfonyl)imide Salts

Anne Collart [1], Matthias Zeller [2] and Patrick C. Hillesheim [1,3,*]

[1] Department of Chemistry and Physics, Ave Maria University, Ave Maria, FL 34142, USA; anne.collart@my.avemaria.edu
[2] Department of Chemistry, Purdue University, West Lafayette, IN 47907, USA; zeller4@purdue.edu
[3] Department of Chemistry and Physics, Florida Gulf Coast University, Fort Myers, FL 33965, USA
* Correspondence: patrick.hillesheim@avemaria.edu

Abstract: Analysis of two crystal structures of lithium bis(pentafluoroethanesulfonyl)imide is presented. Two orientations of the anion, that is a cis and trans orientation, are observed. Both structures exhibit unique interactions leading to the formation of discrete fluorous domains in the solid-state. A notable difference in the F···F interactions is seen when contrasting the two orientations wherein the trans geometry has a higher percentage of fluorine interactions than the cis orientation. The inclusion of water molecules in one of the structures also leads to the formation of a polar domain formed through a series of cyclical hydrogen bonding rings. The two structures allow for a detailed examination of the bond distances and angles involved in the formation of the two structures. Analysis of the void space in the two structures leads to the observation that the trans conformation exhibits notably higher void space as compared with the cis orientation. Hirshfeld surface analysis is used to help rationalize the interactions leading to unique changes in geometries and structure.

Keywords: crystal structure; alkali salts; surface analysis; ionic liquids

Citation: Collart, A.; Zeller, M.; Hillesheim, P.C. Surface and Void Space Analysis of the Crystal Structures of Two Lithium Bis(pentafluoroethanesulfonyl)imide Salts. *Crystals* **2022**, *12*, 701. https://doi.org/10.3390/cryst12050701

Academic Editor: Waldemar Maniukiewicz

Received: 28 March 2022
Accepted: 11 May 2022
Published: 15 May 2022

1. Introduction

An ionic liquid (IL) is a class of soft material that consists of a charge disperse organic cation and anion. It follows that the choice for which specific charge disperse anion to use is of vital importance for the rational design of ILs [1]. The field of ionic liquids has seen prolific development since the introduction of modern air and moisture stable anions [2,3]. The most common anions studied today typically contain fluorine groups which help disperse charges while also repelling other anions [4,5]. The three most common anions used in the synthesis of ILs are $[PF_6]^-$, $[BF_4]$, and $[TF_2N]^-$ [6]. ILs incorporating the tetrafluoroborate [2] and hexafluorophosphate [7] anions were reported first, followed by the introduction of the $[TF_2N]^-$ anion (see Figure 1) [8]. To date, the $[TF_2N]^-$ anion remains one of the most important anions given the favorable physicochemical properties it tends to impart on IL systems. However, other perfluoroalkyl-based anions capable of forming ILs have also been reported such as FAP [9], NCyF [10], and $B(hfip)_4$ (Figure 1) [11].

With approximately 1000 reported crystal structures [12], $[TF_2N]^-$ has found applications across a variety of fields. For example, $[TF_2N]^-$ has been used in catalyst design [13], in the formation of conductive materials [14], and in separations applications [15]. The crystal structures of the alkali salts of $[TF_2N]^-$ were reported by Xue et al. [16] In their studies, it was noted that the alkali salts have unique solvent behavior based on the nature of the alkali metal (i.e., Li vs. Rb). The inclusion of water and crystallization solvents was noted for several samples in the series examined. Further, it was found that the $[TF_2N]^-$ anion could exist in either the cis or trans isomers, simply defined by the orientation of the CF_3 groups with respect to the central S—N—S plane [17]. The trans isomer is energetically lower than the cis when examining the anion by itself [18]. In the presence of a metal ion, however, the anion often adopts the cis conformation to allow for the formation of chelate rings from the sulfonyl oxygens. This appears to be the preferred orientation when interacting with metal cations, simply based on the reported structures in the CSD showing a preference for the cis isomer. It should be noted that there are reported examples wherein the trans conformation is seen binding to a metal also [19]. Both cis and trans conformations of $[TF_2N]^-$ have also been observed for several crystal structures of ILs [20,21]. The conformational flexibility of $[TF_2N]^-$ along with its charge dispersion and favorable thermal properties is what makes the anion particularly attractive for IL applications [1].

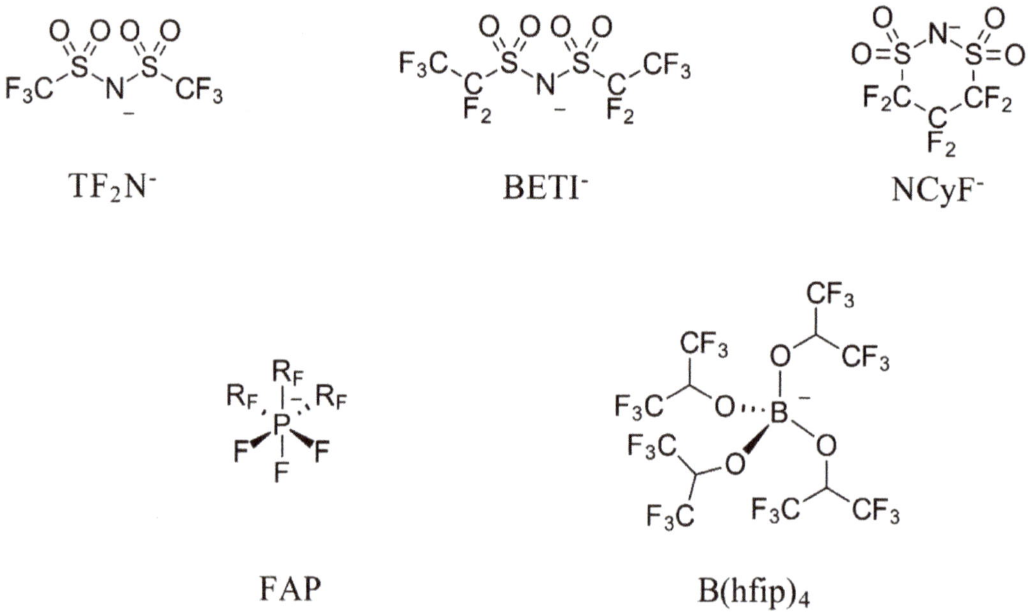

Figure 1. Common perfluorinated anions used in the development of ionic liquids and the abbreviations used herein.

In contrast to the many structures incorporating $[TF_2N]^-$, few $[BETI]^-$ structures that have been reported. Two reported structures are organometallic, with a cobaltocenium cation balancing the $[BETI]^-$ anion [22]. Three additional structures are classified into the realm of ILs given the use of quaternary ammonium cations [23–25]. A report by Henderson and Frech described the structures and thermal behavior of several glyme-coordinated $[BETI]^-$ species [26]. While structural reports on $[BETI]^-$ are limited, there are numerous reports detailing the use and application of $[BETI]^-$ anions, establishing this anion's diverse application across many fields of study. For example, $[BETI]^-$ has been used to form fluorescent materials [27]. Further, $[BETI]^-$ has been used for carbon capture systems, helping to improve solubility when compared with other structurally related perfluorinated anions in the study [28]. Systems incorporating the $[BETI]^-$ anion have also been applied for extractions, specifically in the separation of rare-earth metals [29].

As expected, based on the structural and compositional similarities, $[BETI]^-$ shares several parallels with $[TF_2N]^-$. For example, both anions impart high thermal stability for ILs while simultaneously depressing the melting point [30]. Structurally, the $[BETI]^-$ anion can adopt both a cis or trans conformation as well [23]. The energetics of the conformers were examined by the Davis group and found that similar to $[TF_2N]^-$, there is an energetic preference for the trans conformation [31]. Given the increased ethyl chain length it is also likely that there would exist additional conformers arising from unique rotations of the ethyl chains in combination with the cis/trans isomers. A more in-depth computational and experimental investigation of $[BETI]^-$ would certainly be of benefit to the IL community, especially in light of the anion crisis [32].

Herein we present the crystal structure of lithium bis(pentafluoroethanesulfonyl)imide and diaqua lithium bis(pentafluoroethanesulfonyl)imide. Given that the lithium salts are fundamental components in the synthesis of ILs, and that the metathesis of ILs is typically carried out in an aqueous solution, understanding the structure of the alkali salts of $[BETI]^-$ is of importance to the broader scientific community. Further, these two crystal structures represent some of the few current examples of simple metal complexes with this anion. The high-resolution structures are also of importance for the development of theoretical models to understand the electronic structure of the anion allowing for the development of novel materials [33]. Moreover, the $[BETI]^-$ anion is observed in a cis and trans orientations, allowing for a more in-depth analysis of variations brought about by these unique conformers.

2. Materials and Methods

2.1. Chemicals

Lithium bis(pentafluoroethanesulfonyl)imide was purchased from TCI Chemicals and used as is without further purification.

2.2. Single Crystal Diffraction

Single crystals suitable for X-ray diffraction were grown from a saturated aqueous solution of the anhydrous LiBETI salt. The solution was placed in a 10 mL scintillation vial and loosely capped to allow for evaporation. Slow evaporation over 10 months yielded both sets of crystals in a single vial.

Single crystals of both samples were coated with a trace of Fomblin oil and were transferred to the goniometer head of a Bruker Quest diffractometer. Data collection of [Li{BETI}] used Cu Kα wavelength (λ = 1.54178 Å) on an instrument with kappa geometry, an I-μ-S microsource X-ray tube, laterally graded multilayer (Goebel) mirror single crystal for monochromatization, and a PhotonIII_C14 area detector while [Li{BETI}$(H_2O)_2$] was collected using Mo Kα wavelength (λ = 0.71073 Å) on an instrument with a fixed chi angle, a sealed tube fine focus X-ray tube, single crystal curved graphite incident beam monochromator and a PhotonII area detector. Both instruments are equipped with an Oxford Cryosystems low temperature device and examination and data collection were performed at 150 K. Data were collected, reflections were indexed and processed, and the files scaled and corrected for absorption using APEX3 [34] and SADABS [35]. The space groups were assigned and the structures were solved by direct methods using XPREP within the SHELXTL [36] suite of programs and refined by full matrix least squares against F^2 with all reflections using Shelxl2018 [37] and the graphical interfaces Shelxle [38] and Olex2 [39]. Water H atoms were located as electron density and were refined. O-H bond distances were restrained to 0.84(2) Å. $U_{iso}(H)$ values were set to a 1.5 times $U_{eq}(O)$.

Complete crystallographic data, in CIF format, have been deposited with the Cambridge Crystallographic Data Centre. Table 1 contains the relevant crystallographic data for both compounds. CCDC 2126084 and 2126085 contains the supplementary crystallographic data for this paper. These data can be obtained free of charge from The Cambridge Crystallographic Data Centre via www.ccdc.cam.ac.uk/data_request/cif.

Table 1. Crystallographic data and refinement details for [Li{BETI}] and Li{BETI}$(H_2O)_2$.

	Li{BETI}	**Li{BETI}$(H_2O)_2$**
Chemical formula	$C_4F_{10}LiNO_4S_2$	$C_4H_4F_{10}LiNO_6S_2$
M_r	387.11	423.14
Crystal system, space group	Monoclinic, $C2/c$	Triclinic, $P\bar{1}$
Temperature (K)	150	150
a, b, c (Å)	42.733 (5), 5.4927 (7), 10.4751 (14)	6.4581 (3), 8.0725 (4), 13.0495 (5)
α (°)	90	95.160 (3)
β (°)	99.632 (11)	103.406 (3)
γ (°)	90	91.403 (3)
V (Å^3)	2424.0 (5)	658.37 (5)
Z	8	2
Radiation type	Cu $K\alpha$	Mo $K\alpha$
μ (mm^{-1})	5.50	0.56
Crystal size (mm)	0.21 × 0.15 × 0.02	0.20 × 0.05 × 0.03
	Data collection	
Diffractometer	Bruker AXS D8 Quest diffractometer with PhotonIII_C14 charge-integrating and photon counting pixel array detector	Bruker AXS D8 Quest diffractometer with PhotonII charge-integrating pixel array detector (CPAD)
Absorption correction	Multi-scan *SADABS* 2016/2	Multi-scan *SADABS* 2016/2
T_{min}, T_{max}	0.503, 0.754	0.676, 0.747
No. of measured, independent and observed [$I > 2\sigma(I)$] reflections	7332, 2443, 1991	21544, 4985, 3337
R_{int}	0.097	0.066
$(\sin\theta/\lambda)_{max}$ (Å^{-1})	0.638	0.770
	Refinement	
$R[F^2 > 2\sigma(F^2)]$, $wR(F^2)$, S	0.042, 0.112, 1.05	0.042, 0.091, 1.01
No. of reflections	2443	4985
No. of parameters	199	229
$\Delta\rho_{max}$, $\Delta\rho_{min}$ (e Å^{-3})	0.36, −0.38	0.49, −0.54

2.3. *Software*

Hirshfeld surfaces, images, and fingerprint plots were calculated and produced using CrystalExplorer17 [40]. Discussion about the analysis of fingerprints and surfaces can be found in the appropriate references [41,42]. Images and analysis of the structures was accomplished using Olex2 [39]. Void space images and calculations were accomplished using Mercury [43]. Void spaces were calculated with a 0.8 Å probe radius.

3. Results and Discussion

[Li{BETI}] crystallizes in the C2/c space group with one molecule in the asymmetric unit. [Li{BETI}$(H_2O)_2$] crystallizes in the $P\bar{1}$ space group, also with one molecule in the asymmetric unit. The asymmetric units of both structures are shown in Figure 2. In both structures, the sulfonyl oxygen atoms are binding to the metal center forming a six-membered chelate ring. Despite forming a similar chelate structure, there are several notable differences in the structures. For example, in [Li{BETI}], the sulfonyl groups make shorter bonds to the lithium with S—O···Li distances of 1.927(5) and 1.917(5). In [Li{BETI}(H_2O)$_2$], however, the distances are slightly longer with S—O···Li distances of 2.094(4) and 2.163(3). One noteworthy distinction between [Li{BETI}] and [Li{BETI}(H_2O)$_2$] is that [Li{BETI}(H_2O)$_2$] is the first example of the [BETI]$^-$ anion in the cis geometry wherein the anion is not disordered [23]. As discussed above, the [TF_2N]$^-$ anion exists both in the cis and trans geometry, with there appearing to be an approximately equal distribution of both isomers in the reported crystalline structures. Currently, however, [Li{BETI}] has only been reported in the lower energy trans conformation.

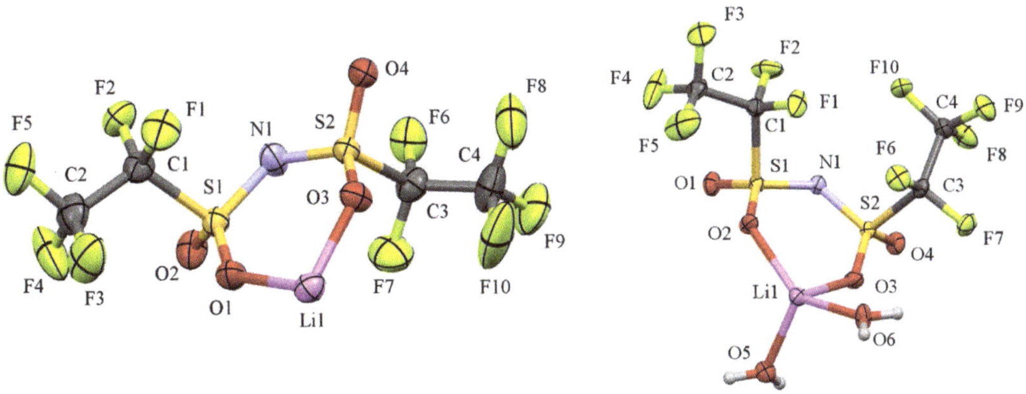

Figure 2. Asymmetric units of [Li{BETI}] (**left**) and [Li{BETI}(H_2O)$_2$] (**right**) shown with 50% probability ellipsoids. Gray, carbon; green, fluorine; red, oxygen; pink, lithium; yellow, sulfur; blue, nitrogen.

Several potential factors influence the distances and interactions in the two molecules. Predominantly, the coordination geometry of the lithium is a factor. In [Li{BETI}], the lithium exists in a pseudo-tetrahedral geometry with four oxygen atoms making bonds to the metal. Two of the oxygens are from a single bidentate [BETI]$^-$ anion, and the remaining oxygens are from symmetry adjacent anions, effectively making an infinite chain linked through the lithium ion (see Figure 3). In [Li{BETI}(H_2O)$_2$], the lithium atom exists in a distorted trigonal bipyramidal geometry with two water molecules, one bidentate [BETI]$^-$ anion, and a single sulfonyl oxygen from an adjacent anion making up the coordination sphere (see Figure 3). The hydrated structure, thus, forms a near-linear arrangement of the lithium ions. The bound water molecules help arrange the layers of the structure by forming reciprocal hydrogen bonds to symmetry adjacent water molecules. The unique coordination in addition to the cis/trans geometries lead to distinctive arrangements of the structures.

The long-range ordering of both structures does share similarities, as expected (see Figure 4). For instance, both show distinctive fluorous layers wherein the perfluoroalkyl groups are in close contact with each other. Further, there exists a distinct layer wherein the lithium ions and sulfonyl groups exist. The inclusion of the serendipitous water molecules in [Li{BETI}(H_2O)$_2$], in addition to the structural changes associated with the coordination of the lithium ion, influences the void space in the lattice (see Figure 5). Examining Figure 5 one can readily observe the differences in the two structures. [Li{BETI}] has significantly higher void space in the lattice (5.8% of volume, 140.59 Å^3) as compared with [Li{BETI}(H_2O)$_2$] (1.1% of volume, 7.03 Å^3). [Li{BETI}] has void space near the lithium ion as well as in the fluorous domain whereas the voids in [Li{BETI}(H_2O)$_2$] are in the hydrophilic pocket formed by the waters and sulfonyl oxygen atoms. The differences in the void space is also observed through examining the varied interactions in the crystalline structures (*vide infra*).

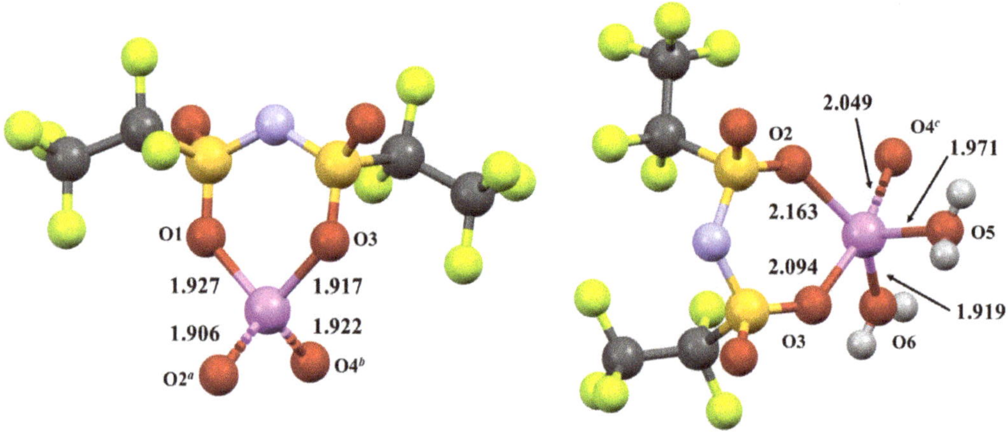

Figure 3. Images depicting the lithium cation coordination geometry for [Li{BETI}] (**left**) and [Li{BETI}$(H_2O)_2$] (**right**). Bond distances are shown in Å. Dashed bonds indicate symmetry adjacent atoms. a = +x, −y, $\frac{1}{2}$ + z; b = +x, −1 + y, +z; c = −1 + x, +y, +z.

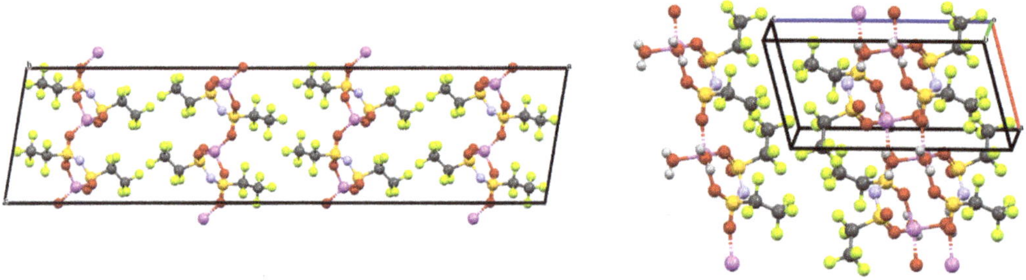

Figure 4. Packing diagrams for [Li{BETI}] (**left**) and [Li{BETI}$(H_2O)_2$] (**right**). Alternating layers of fluorine and oxygen atoms are prominent features of both structures. Unit cells are shown for references.

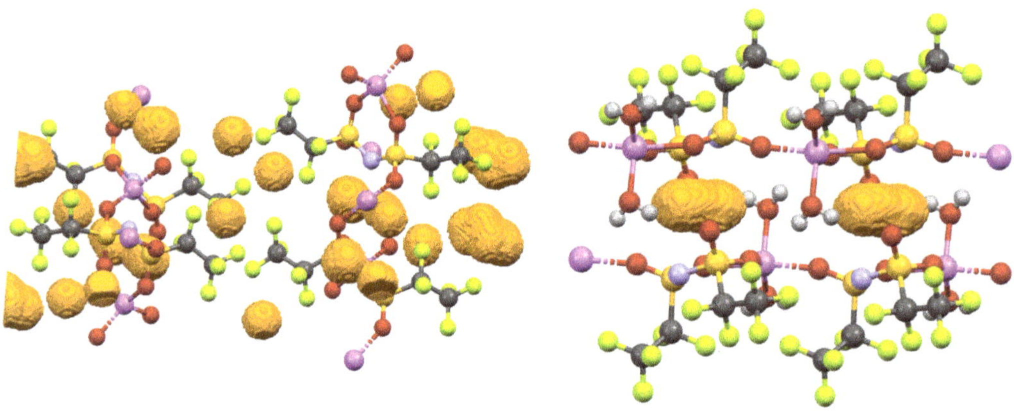

Figure 5. Visualization of the void space (yellow spheres) in [Li{BETI}] (**left**) and [Li{BETI}$(H_2O)_2$] (**right**).

To better understand the interactions and structures of the two compounds, Hirshfeld surface analysis was performed [44]. The surfaces for both structures are shown in Figures 6 and 7, respectively, and the fingerprint plots [41] in Figure 8. Several key similarities are observed in the fingerprints of the structures. A diagonal region of red interactions is observed in both fingerprints, indicative of a large number of interactions at similar distances [45]. These interactions are primarily from F···F interactions. Further, there are regions of disperse spots in both fingerprints, especially at longer distances ($d_i \approx d_e \approx 2.5$ Å). These disperse spots can be indicative of inefficient crystal packing. While not directly quantifiable, the increased void space in [Li{BETI}] is observed as the larger regions of disperse interactions seen in the fingerprint for [Li{BETI}] when contrasted with [Li{BETI}(H_2O)$_2$]. Finally, a set of pincer-like features are seen in both structures, with [Li{BETI}] showing more pronounced features. These shapes arise from the fact that the structures are polymeric, forming interactions linking multiple asymmetric units through O···Li bonds which form the pincer shapes. These bonds manifest as the indicated red indentations observed on the d_{norm} surfaces (Figures 6 and 7).

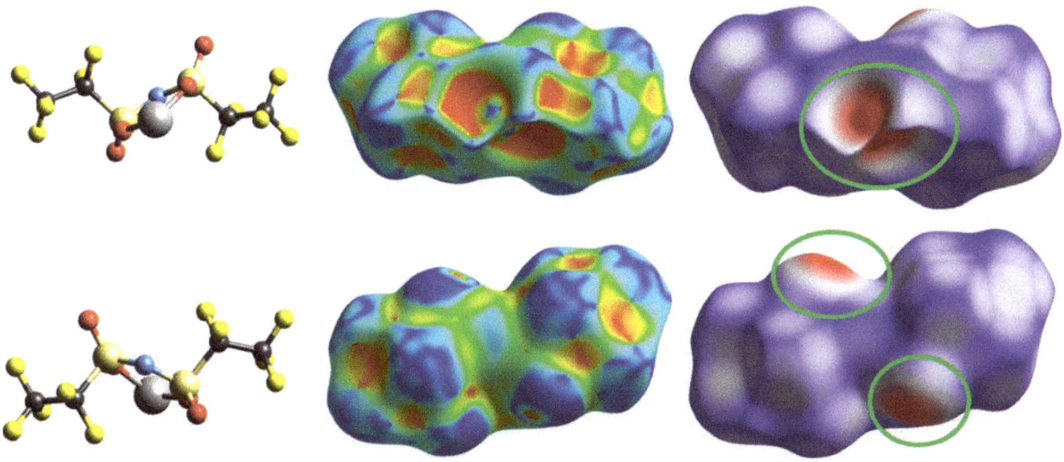

Figure 6. Depiction of the [Li{BETI}] structure (**left**) and the Hirshfeld surfaces for [Li{BETI}] mapped with the shape index (**middle**) and d_{norm} (**right**). Green circles indicate the connection points for the oxygen atoms to adjacent lithium ions.

Several key differences can also be observed when looking at the fingerprints (Figure 8). Looking at the shape of the fingerprints at higher d_i/d_e values, [Li{BETI}] shows a flared tail while [Li{BETI}(H_2O)$_2$] tapers to a blunted end. As discussed, this corroborates the void space analysis and the differences between the two structures. Looking closely at the F···F interactions in both fingerprints (the diagonal red line, *vide supra*), [Li{BETI}(H_2O)$_2$] displays shorter interactions as observed by the starting distances of the red line in [Li{BETI}(H_2O)$_2$] ($d_i \approx d_e \approx 1.5$ Å) vs. [Li{BETI}] ($d_i \approx d_e \approx 1.7$ Å). Additionally, the shape of this red region is unique between the two molecules with [Li{BETI}(H_2O)$_2$] having a more distinct, narrow red line while the interactions in **1** manifest as a more disperse set. This observation points to the unique sets of interactions that are present in both structures.

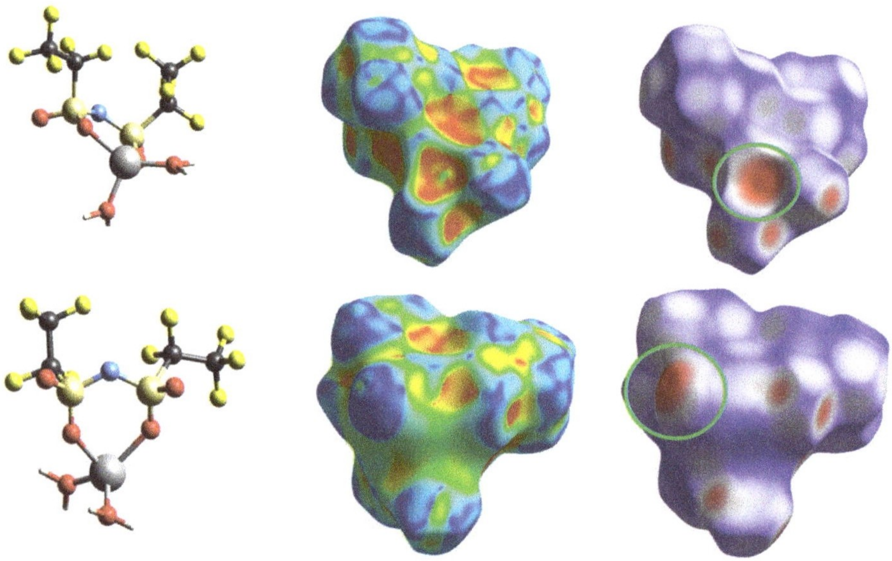

Figure 7. Depiction of the [Li{BETI}(H_2O)$_2$] structure (**left**) and the Hirshfeld surfaces for [Li{BETI}(H_2O)$_2$] mapped with the shape index (**middle**) and d_{norm} (**right**). Green circles indicate the points wherein the oxygen atoms coordinate to the lithium ions in adjacent molecules.

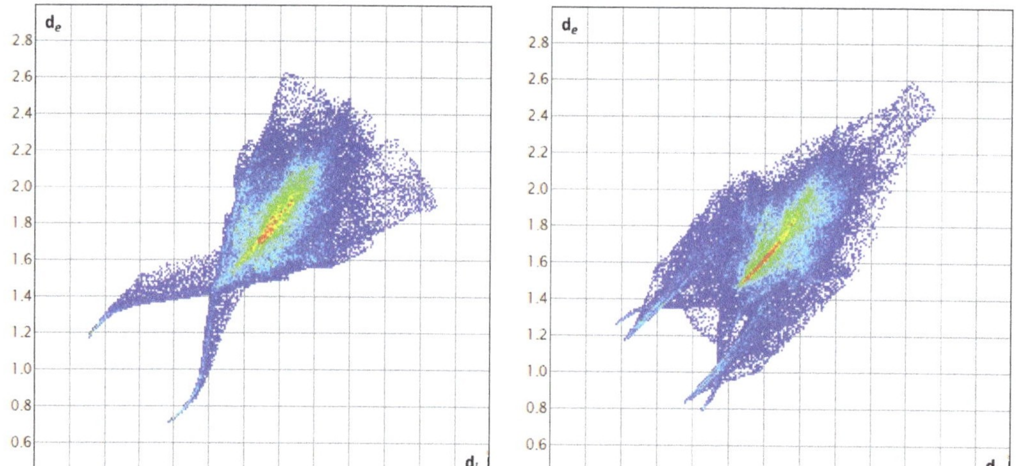

Figure 8. The fingerprint plots for [Li{BETI}] (**left**) and [Li{BETI}(H_2O)$_2$] (**right**).

The total percentage of F···F interactions is higher in [Li{BETI}] (60.5%) than in [Li{BETI}(H_2O)$_2$] (46.3%). This may seem to suggest that the F···F interactions are, in some way, favorable. However, this change in percentage is simply due to the positioning of the perfluoroethyl chain in [Li{BETI}(H_2O)$_2$]. The cis orientation places atoms F1, F6, F10 in close proximity intramolecularly, effectively preventing any intermolecular interactions from these three positions. The trans structure, however, does not show any steric 'blocking' of fluorine atoms, leading to an increased percentage of F···F interactions overall. Further, there are no O—H···F interactions observed arising from the water molecules in [Li{BETI}(H_2O)$_2$], implying that the change in F···F interactions is not influenced by the presence of water and the potential hydrogen bonding. It should be stated that reports have shown F···F interactions to be destabilizing or, at the least, do not stabilize packing [46] though some reports seem to contradict this [47].

Given the packing of the anions into discrete fluorous domains, along with the results of the surface analysis, we sought for evidence of any F···F halogen bonds [48–50]. While the existence of halogen bonding between fluorine atoms remains debated [51], there are a number of shorter contacts between fluorine atoms which may be evidence of type II halogen bonding. For example, the terminal CF_3 group in [Li{BETI}] shows close contacts with a symmetry adjacent CF_3 group (i = 1 − x, −y, 1 − z). Specifically, F4 is interacting with $F4^i$ at a distance of 3.045(6) Å (∠C2—F4···$F4^i$ = 93.3(3)°). F6 also exhibits F···F interactions at similar distances to those seen with F4 (d(F6···$F8^j$) = 3.087(4) Å, ∠C3—F6···$F8^j$ = 126.6(2), j = x, 1 − y, −1/2 + z). These close interactions are seen on the shape index surface as the indicated red/yellow indentations (see Figures 6 and 7). Aside from these interactions, however, we do not observe any defined halogen bonding. For clarity, the F···F interactions for both molecules are visualized in Figure 9, allow for comparison between the two structures.

Table 2 shows a complete listing of the total interaction percentages arising from specific elements in both structures. As expected, the fluorine atoms dominate the intermolecular interactions (by percentage) followed by the oxygen. The roles of the oxygen interactions are unique, however, when contrasting structures **1** and **2**. Given the presence of the water molecules in **2**, hydrogen bonding is observed, linking discrete asymmetric units together. A depiction of a portion of the hydrogen bonding is shown in Figure 10. Further, the hydrogen bonding is observed as the reciprocal set of spikes in the fingerprint plot ($d_i \approx 1.1$ Å, $d_e \approx 0.9$ Å) and at the appropriate reciprocal distances, that is $d_i \approx 0.9$ Å, $d_e \approx 1.1$ Å. Overall the hydrogen bonding in [Li{BETI}$(H_2O)_2$] is quite complex, with several unique set of hydrogen bonding rings being observed [52]. Two $R_2^2(8)$ rings are formed from the reciprocal interactions between O2 and O5 and O5 and O6. These are highlighted as green and yellow, respectively, in Figure 9. With respect to the O2 and O5 hydrogen interactions, the distance is 2.32 Å (d(O2···$H5B^m$, m = −x, 1 − y, 1 − z). The O5 and O6 hydrogen interactions distance is 2.22 Å (d(O5···$H6B^n$, n = −x, 2 − y, 1 − z). An additional $R_3^3(8)$ ring is observed involving the two water molecules (O5 and O6) and two sulfonyl oxygens O1 and O2. This is shown as the purple highlighted ring in Figure 9. The distances between O5 and O2 as well as O5 and O6 are given previously, 2.32 Å and 2.22 Å, respectively. The O1···H6A distance is 2.16 Å d(O1···$H6A^o$, o = +x, 1 + y, +z). The arrangement of these rings and formation of the hydrogen bonding network help to form the void space discussed within the structure of [Li{BETI}$(H_2O)_2$].

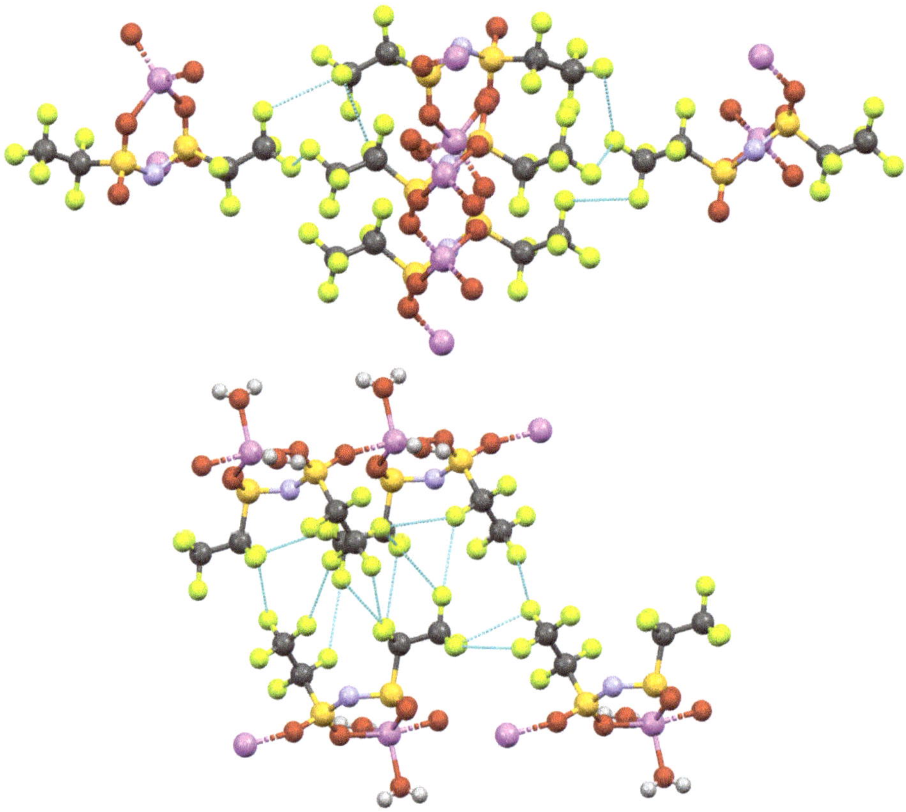

Figure 9. Depiction of the F···F interactions (light blue lines) in [Li{BETI}] (**top**) and [Li{BETI}(H_2O)$_2$] (**bottom**).

Table 2. Calculated interactions of atoms in both [Li{BETI}] and [Li{BETI}(H_2O)$_2$] based on Hirshfeld surface calculations.

	[Li{BETI}]	**[Li{BETI}(H_2O)$_2$]**
F···All	69.5%	54.9%
O···All	19.9%	21.2%
N···All	2.5%	2.4%

Given the absence of water molecules, the oxygen atoms in **1** show a markedly different set of interactions. Predominantly there are no overtly stabilizing non-covalent interactions [53,54] arising from the oxygen atoms. All four of the oxygen atoms are binding to a lithium ion, with O1 and O3 acting as the bidentate binding points, while O2 and O4 act as the bridging points to the adjacent lithium ions. The majority of the ~20% of the O interactions in [Li{BETI}] arise from long-distance O···F interactions. The distances and angles examined do not seem to indicate any form of chalcogen bonds being formed [55]. Additional interactions between O and N and O and O complete the noteworthy interactions observed in [Li{BETI}]. From our perspective, neither the O···N nor O···O interactions are stabilizing but are simply an artifact of the packing from the crystal based on the distances and angles observed.

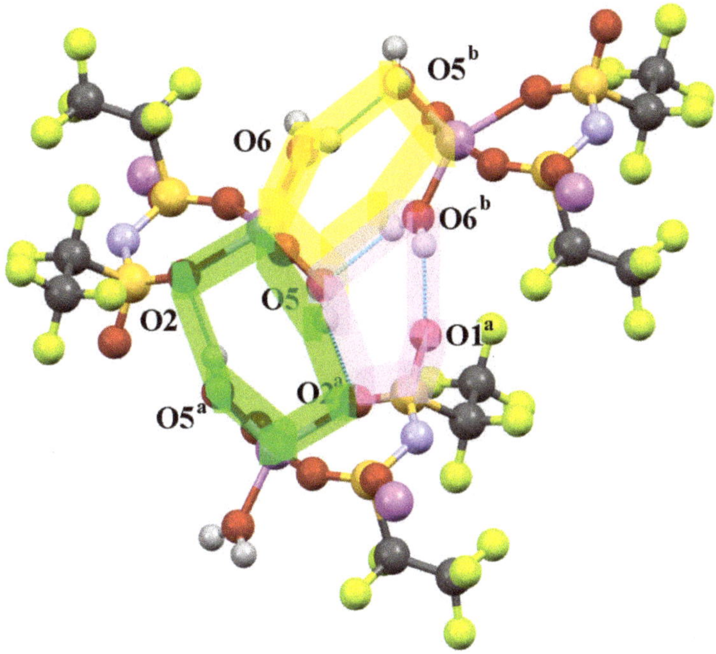

Figure 10. Depiction of hydrogen bonding ring interactions in [Li{BETI}$(H_2O)_2$]. Different highlights are used to distinguish different hydrogen bonding rings which are formed. Green = $R_2^2(8)$; purple $R_3^3(8)$; yellow = $R_2^2(8)$. a = −x, 1 − y, 1 − z; b = −x, 2 − y, 1 − z.

4. Conclusions

The two crystal structures of lithium bis(pentafluoroethanesulfonyl)imide and lithium bis(pentafluoroethanesulfonyl)imide diaqua are reported. These structures represent two new samples for the sparse but growing set of structural data for the $[BETI]^-$ anion. The two structures clearly show the cis and trans geometries of the anion, allowing for the detailed examination of distances and angles arising from this change in structure. For example, the cis geometry displays unique intramolecular F···F interactions due to the proximity of the perfluoroalkyl chains. Thus, the diaqua structure shows a lower F···F interaction percentage due to the arrangements of the perfluoroethyl chains hindering intermolecular interactions. In contrast, the trans geometry allows for increased interactions with the perfluoroalkyl chains. Overall, F···F close-contacts making up the majority of the interactions present in the solid-state.

[Li{BETI}$(H_2O)_2$], however, shows significantly lower void space in the crystal structure due, in part, to shorter F···F interactions and hydrogen bonding arising from the water molecules. The hydrogen bonding in the molecule is quite unique, with three distinctive reciprocal hydrogen bonding rings formed. Given the linear nature of hydrogen bonds, a void pocket is formed within the region wherein the bound water molecules are located.

Author Contributions: A.C.: writing—original draft, review and editing, formal analysis; M.Z.: data curation, investigation, writing—review and editing, formal analysis; P.C.H.: conceptualization, methodology, validation, visualization, formal analysis, supervision, project administration, writing—original draft, review and editing. All authors have read and agreed to the published version of the manuscript.

Funding: Parts of this material is based upon work supported by the National Science Foundation through the Major Research Instrumentation Program under Grants No. CHE–1625543. The APC was funded by Ave Maria University Department of Chemistry and Physics.

Institutional Review Board Statement: Not applicable.

Informed Consent Statement: Not applicable.

Data Availability Statement: Not applicable.

Acknowledgments: This work was supported by Ave Maria University Department of Chemistry and Physics and Florida Gulf Coast University Department of Chemistry and Physics.

Conflicts of Interest: The authors declare no conflict of interest.

References

1. MacFarlane, D.R.; Kar, M.; Pringle, J.M. *Fundamentals of Ionic Liquids*; Wiley-VCH Verlag GmbH & Co. KGaA: Weinheim, Germany, 2017. [CrossRef]
2. Wilkes, J.S.; Zaworotko, M.J. Air and Water Stable 1-Ethyl-3-Methylimidazolium Based Ionic Liquids. *J. Chem. Soc. Chem. Commun.* **1992**, *13*, 965–967. [CrossRef]
3. Deetlefs, M.; Fanselow, M.; Seddon, K.R. Ionic Liquids: The View from Mount Improbable. *RSC Adv.* **2016**, *6*, 4280–4288. [CrossRef]
4. Dupont, J. On the Solid, Liquid and Solution Structural Organization of Imidazolium Ionic Liquids. *J. Braz. Chem. Soc.* **2004**, *15*, 341–350. [CrossRef]
5. Wang, Y.; Parvis, F.; Hossain, I.; Ma, K.; Jarošová, R.; Swain, G.M.; Blanchard, G.J. Local and Long-Range Organization in Room Temperature Ionic Liquids. *Langmuir* **2021**, *37*, 605–615. [CrossRef] [PubMed]
6. Welton, T. Ionic Liquids: A Brief History. *Biophys. Rev.* **2018**, *10*, 691–706. [CrossRef] [PubMed]
7. Fuller, J.; Carlin, R.T.; De Long, H.C.; Haworth, D. Structure of 1-Ethyl-3-Methylimidazolium Hexafluorophosphate: Model for Room Temperature Molten Salts. *J. Chem. Soc. Chem. Commun.* **1994**, *3*, 299. [CrossRef]
8. Bonhôte, P.; Dias, A.-P.; Papageorgiou, N.; Kalyanasundaram, K.; Grätzel, M. Hydrophobic, Highly Conductive Ambient-Temperature Molten Salts. *Inorg. Chem.* **1996**, *35*, 1168–1178. [CrossRef] [PubMed]
9. Ignat'ev, N.V.; Welz-Biermann, U.; Kucheryna, A.; Bissky, G.; Willner, H. New Ionic Liquids with Tris(Perfluoroalkyl)Trifluorophosphate (FAP) Anions. *J. Fluor. Chem.* **2005**, *126*, 1150–1159. [CrossRef]
10. Philippi, F.; Welton, T. Targeted Modifications in Ionic Liquids–from Understanding to Design. *Phys. Chem. Chem. Phys.* **2021**, *23*, 6993–7021. [CrossRef]
11. Kaliner, M.; Rupp, A.; Krossing, I.; Strassner, T. Tunable Aryl Alkyl Ionic Liquids with Weakly Coordinating Tetrakis((1,1,1,3,3,3-Hexafluoropropan-2-Yl)Oxy)Borate [B(Hfip)4] Anions. *Chem.–Eur. J.* **2016**, *22*, 10044–10049. [CrossRef]
12. Groom, C.R.; Bruno, I.J.; Lightfoot, M.P.; Ward, S.C. The Cambridge Structural Database. *Acta Crystallogr. Sect. B Struct. Sci. Cryst. Eng. Mater.* **2016**, *72*, 171–179. [CrossRef] [PubMed]
13. Hashmi, A.S.K.; Braun, I.; Rudolph, M.; Rominger, F. The Role of Gold Acetylides as a Selectivity Trigger and the Importance of Gem-Diaurated Species in the Gold-Catalyzed Hydroarylating-Aromatization of Arene-Diynes. *Organometallics* **2012**, *31*, 644–661. [CrossRef]
14. Cheng, Y.; Yang, J.; Hung, J.-H.; Patra, T.K.; Simmons, D.S. Design Rules for Highly Conductive Polymeric Ionic Liquids from Molecular Dynamics Simulations. *Macromolecules* **2018**, *51*, 6630–6644. [CrossRef]
15. Fan, F.-L.; Qin, Z.; Cao, S.-W.; Tan, C.-M.; Huang, Q.-G.; Chen, D.-S.; Wang, J.-R.; Yin, X.-J.; Xu, C.; Feng, X.-G. Highly Efficient and Selective Dissolution Separation of Fission Products by an Ionic Liquid [Hbet][Tf2N]: A New Approach to Spent Nuclear Fuel Recycling. *Inorg. Chem.* **2019**, *58*, 603–609. [CrossRef]
16. Xue, L.; Padgett, C.W.; DesMarteau, D.D.; Pennington, W.T. Synthesis and Structures of Alkali Metal Salts of Bis[(Trifluoromethyl)Sulfonyl]Imide. *Solid State Sci.* **2002**, *4*, 1535–1545. [CrossRef]
17. Nockemann, P.; Thijs, B.; Pittois, S.; Thoen, J.; Glorieux, C.; Van Hecke, K.; Van Meervelt, L.; Kirchner, B.; Binnemans, K. Task-Specific Ionic Liquid for Solubilizing Metal Oxides. *J. Phys. Chem. B* **2006**, *110*, 20978–20992. [CrossRef]
18. Fujii, K.; Fujimori, T.; Takamuku, T.; Kanzaki, R.; Umebayashi, Y.; Ishiguro, S. Conformational Equilibrium of Bis(Trifluoromethanesulfonyl) Imide Anion of a Room-Temperature Ionic Liquid: Raman Spectroscopic Study and DFT Calculations. *J. Phys. Chem. B* **2006**, *110*, 8179–8183. [CrossRef]
19. McOwen, D.W.; Seo, D.M.; Borodin, O.; Vatamanu, J.; Boyle, P.D.; Henderson, W.A. Concentrated Electrolytes: Decrypting Electrolyte Properties and Reassessing Al Corrosion Mechanisms. *Energy Environ. Sci.* **2014**, *7*, 416–426. [CrossRef]
20. Holbrey, J.D.; Reichert, W.M.; Rogers, R.D. Crystal Structures of Imidazolium Bis(Trifluoromethanesulfonyl)Imide 'Ionic Liquid' Salts: The First Organic Salt with a Cis-TFSI Anion Conformation. *Dalton Trans.* **2004**, *15*, 2267–2271. [CrossRef]
21. Dean, P.M.; Pringle, J.M.; Forsyth, C.M.; Scott, J.L.; MacFarlane, D.R. Interactions in Bisamide Ionic Liquids—Insights from a Hirshfeld Surface Analysis of Their Crystalline States. *New J. Chem.* **2008**, *32*, 2121. [CrossRef]
22. Mochida, T.; Funasako, Y.; Inagaki, T.; Li, M.-J.; Asahara, K.; Kuwahara, D. Crystal Structures and Phase-Transition Dynamics of Cobaltocenium Salts with Bis(Perfluoroalkylsulfonyl)Amide Anions: Remarkable Odd-Even Effect of the Fluorocarbon Chains in the Anion. *Chem.-Eur. J.* **2013**, *19*, 6257–6264. [CrossRef] [PubMed]
23. Traver, J.; Chenard, E.; Zeller, M.; Guillet, G.L.; Lynch, W.E.; Hillesheim, P.C. Directing Cation-Cation Interactions in Thiamine Compounds: Analysis of a Series of Organic Salts Based on Vitamin B1. *J. Mol. Struct.* **2021**, *1232*, 130046. [CrossRef]

24. Siegel, D.J.; Anderson, G.I.; Cyr, N.; Lambrecht, D.S.; Zeller, M.; Hillesheim, P.C.; Mirjafari, A. Molecular Design Principles of Ionic Liquids with a Sulfonyl Fluoride Moiety. *New J. Chem.* **2021**, *45*, 2443–2452. [CrossRef]
25. Yoshida, Y.; Saito, G. Ionic Liquids Based on Diethylmethyl(2-Methoxyethyl)Ammonium Cations and Bis(Perfluoroalkanesulfonyl)Amide Anions: Influence of Anion Structure on Liquid Properties. *Phys. Chem. Chem. Phys.* **2011**, *13*, 20302–20310. [CrossRef] [PubMed]
26. Henderson, W.A.; McKenna, F.; Khan, M.A.; Brooks, N.R.; Young, V.G.; Frech, R. Glyme–Lithium Bis(Trifluoromethanesulfonyl)Imide and Glyme–Lithium Bis(Perfluoroethanesulfonyl)Imide Phase Behavior and Solvate Structures. *Chem. Mater.* **2005**, *17*, 2284–2289. [CrossRef]
27. Siraj, N.; Hasan, F.; Das, S.; Kiruri, L.W.; Steege Gall, K.E.; Baker, G.A.; Warner, I.M. Carbazole-Derived Group of Uniform Materials Based on Organic Salts: Solid State Fluorescent Analogues of Ionic Liquids for Potential Applications in Organic-Based Blue Light-Emitting Diodes. *J. Phys. Chem. C* **2014**, *118*, 2312–2320. [CrossRef]
28. Yim, J.-H.; Oh, B.-K.; Lim, J.S. Solubility Measurement and Correlation of CO_2 in Bis(Pentafluoroethylsulfonyl)Imide ([BETI]) Anion-Based Ionic Liquids: [EMIM][BETI], [BMIM][BETI], [HMIM][BETI]. *J. Chem. Eng. Data* **2020**, *65*, 4378–4386. [CrossRef]
29. Sun, X.; Luo, H.; Dai, S. Mechanistic Investigation of Solvent Extraction Based on Anion-Functionalized Ionic Liquids for Selective Separation of Rare-Earth Ions. *Dalton Trans.* **2013**, *42*, 8270–8275. [CrossRef]
30. Tokuda, H.; Hayamizu, K.; Ishii, K.; Susan, A.B.H.; Watanabe, M. Physicochemical Properties and Structures of Room Temperature Ionic Liquids. 1. Variation of Anionic Species. *J. Phys. Chem. B* **2004**, *108*, 16593–16600. [CrossRef]
31. Siu, B.; Cassity, C.G.; Benchea, A.; Hamby, T.; Hendrich, J.; Strickland, K.J.; Wierzbicki, A.; Sykora, R.E.; Salter, E.A.; O'Brien, R.A.; et al. Thermally Robust: Triarylsulfonium Ionic Liquids Stable in Air for 90 Days at 300 °C. *RSC Adv.* **2017**, *7*, 7623–7630. [CrossRef]
32. Benchea, A.; Siu, B.; Soltani, M.; McCants, J.H.; Salter, E.A.; Wierzbicki, A.; West, K.N.; Davis, J.H.J. An Evaluation of Anion Suitability for Use in Ionic Liquids with Long-Term, High-Temperature Thermal Stability. *New J. Chem.* **2017**, *41*, 7844–7848. [CrossRef]
33. Pulido, A.; Chen, L.; Kaczorowski, T.; Holden, D.; Little, M.A.; Chong, S.Y.; Slater, B.J.; McMahon, D.P.; Bonillo, B.; Stackhouse, C.J.; et al. Functional Materials Discovery Using Energy–Structure–Function Maps. *Nature* **2017**, *543*, 657–664. [CrossRef] [PubMed]
34. Bruker. *Apex3 V2019.1-0, SAINT V8.40A*; Bruker AXS Inc.: Madison, WI, USA, 2019.
35. Krause, L.; Herbst-Irmer, R.; Sheldrick, G.M.; Stalke, D. Comparison of Silver and Molybdenum Microfocus X-ray Sources for Single-Crystal Structure Determination. *J. Appl. Crystallogr.* **2015**, *48*, 3–10. [CrossRef] [PubMed]
36. Bruker. *SHELXTL Suite of Programs, Version 6.14, 2000–2003, Bruker Advanced X-ray Solutions*; Bruker AXS Inc.: Madison, WI, USA, 2000–2003.
37. Sheldrick, G.M. A Short History of *SHELX*. *Acta Crystallogr. A* **2008**, *64*, 112–122. [CrossRef]
38. Hübschle, C.B.; Sheldrick, G.M.; Dittrich, B. *ShelXle*: A Qt Graphical User Interface for *SHELXL*. *J. Appl. Crystallogr.* **2011**, *44*, 1281–1284. [CrossRef]
39. Dolomanov, O.V.; Bourhis, L.J.; Gildea, R.J.; Howard, J.A.K.; Puschmann, H. *OLEX2*: A Complete Structure Solution, Refinement and Analysis Program. *J. Appl. Crystallogr.* **2009**, *42*, 339–341. [CrossRef]
40. Spackman, P.R.; Turner, M.J.; McKinnon, J.J.; Wolff, S.K.; Grimwood, D.J.; Jayatilaka, D.; Spackman, M.A. *CrystalExplorer*: A Program for Hirshfeld Surface Analysis, Visualization and Quantitative Analysis of Molecular Crystals. *J. Appl. Crystallogr.* **2021**, *54*, 1006–1011. [CrossRef]
41. Spackman, M.A.; McKinnon, J.J. Fingerprinting Intermolecular Interactions in Molecular Crystals. *CrystEngComm* **2002**, *4*, 378–392. [CrossRef]
42. Spackman, M.A.; McKinnon, J.J.; Jayatilaka, D. Electrostatic Potentials Mapped on Hirshfeld Surfaces Provide Direct Insight into Intermolecular Interactions in Crystals. *CrystEngComm* **2008**, *10*, 377–388. [CrossRef]
43. Macrae, C.F.; Bruno, I.J.; Chisholm, J.A.; Edgington, P.R.; McCabe, P.; Pidcock, E.; Rodriguez-Monge, L.; Taylor, R.; van de Streek, J.; Wood, P.A. *Mercury CSD 2.0*—New Features for the Visualization and Investigation of Crystal Structures. *J. Appl. Crystallogr.* **2008**, *41*, 466–470. [CrossRef]
44. Spackman, M.A.; Jayatilaka, D. Hirshfeld Surface Analysis. *CrystEngComm* **2009**, *11*, 19–32. [CrossRef]
45. McKinnon, J.J.; Spackman, M.A.; Mitchell, A.S. Novel Tools for Visualizing and Exploring Intermolecular Interactions in Molecular Crystals. *Acta Crystallogr. B* **2004**, *60*, 627–668. [CrossRef] [PubMed]
46. Rybalova, T.V.; Bagryanskaya, I.Y. C-F ... π, F ... H, and F ... F Intermolecular Interactions and F-Aggregation: Role in Crystal Engineering of Fluoroorganic Compounds. *J. Struct. Chem.* **2009**, *50*, 741–753. [CrossRef]
47. Arkhipov, D.E.; Lyubeshkin, A.V.; Volodin, A.D.; Korlyukov, A.A. Molecular Structures Polymorphism the Role of F ... F Interactions in Crystal Packing of Fluorinated Tosylates. *Crystals* **2019**, *9*, 242. [CrossRef]
48. Bauzá, A.; Frontera, A. Electrostatically Enhanced F···F Interactions through Hydrogen Bonding, Halogen Bonding and Metal Coordination: An Ab Initio Study. *Phys. Chem. Chem. Phys.* **2016**, *18*, 20381–20388. [CrossRef]
49. Cavallo, G.; Metrangolo, P.; Milani, R.; Pilati, T.; Priimagi, A.; Resnati, G.; Terraneo, G. The Halogen Bond. *Chem. Rev.* **2016**, *116*, 2478–2601. [CrossRef]
50. Varadwaj, P.; Varadwaj, A.; Marques, H. Halogen Bonding: A Halogen-Centered Noncovalent Interaction Yet to Be Understood. *Inorganics* **2019**, *7*, 40. [CrossRef]
51. Scheiner, S. F-Halogen Bond: Conditions for Its Existence. *J. Phys. Chem. A* **2020**, *124*, 7290–7299. [CrossRef]

52. Bernstein, J.; Davis, R.E.; Shimoni, L.; Chang, N.-L. Patterns in Hydrogen Bonding: Functionality and Graph Set Analysis in Crystals. *Angew. Chem. Int. Ed. Engl.* **1995**, *34*, 1555–1573. [CrossRef]
53. Bissantz, C.; Kuhn, B.; Stahl, M. A Medicinal Chemist's Guide to Molecular Interactions. *J. Med. Chem.* **2010**, *53*, 5061–5084. [CrossRef]
54. Dunitz, J.D.; Gavezzotti, A. How Molecules Stick Together in Organic Crystals: Weak Intermolecular Interactions. *Chem. Soc. Rev.* **2009**, *38*, 2622. [CrossRef] [PubMed]
55. Scilabra, P.; Terraneo, G.; Resnati, G. The Chalcogen Bond in Crystalline Solids: A World Parallel to Halogen Bond. *Acc. Chem. Res.* **2019**, *52*, 1313–1324. [CrossRef] [PubMed]

crystals

MDPI

Article

Synthesis, Characterization, Hirshfeld Surface Analysis, Crystal Structure and Molecular Modeling Studies of 1-(4-(Methoxy(phenyl)methyl)-2-methylphenoxy)butan-2-one Derivative as a Novel α-Glucosidase Inhibitor

Chandra Shivanna [1,†], Shashank M. Patil [2,†], C. Mallikarjunaswamy [3], Ramith Ramu [2,*], Prabhuswamy Akhileshwari [4], Latha Rani Nagaraju [4], Mandayam A. Sridhar [4], Shaukath Ara Khanum [5], V. Lakshmi Ranganatha [6,*], Ekaterina Silina [7], Victor Stupin [8] and Raghu Ram Achar [9]

1 Department of Physics, The National Institute of Engineering, Manandavadi Road, Mysore 570008, India; chandru@nie.ac.in
2 Department of Biotechnology and Bioinformatics, JSS Academy of Higher Education and Research, Mysuru 570015, India; shashankmpatil@jssuni.edu.in
3 Postgraduate Department of Chemistry, JSS College of Arts, Commerce and Science, JSS Research Centre (A Recognized Research Centre of University of Mysore), Mysuru 570025, India; mallik.aanekere@gmail.com
4 Department of Studies in Physics, University of Mysore, Manasagangotri, Mysuru 570006, India; akhila@uomphysics.net (P.A.); rani04latha@gmail.com (L.R.N.); mas@physics.uni-mysore.ac.in (M.A.S.)
5 Department of Chemistry, Yuvaraja's College, University of Mysore, Mysore 570005, India; shaukathara@yahoo.co.in
6 Department of Chemistry, The National Institute of Engineering, Manandavadi Road, Mysore 570008, India
7 Department of Human Pathology, I.M. Sechenov First Moscow State Medical University (Sechenov University), 119991 Moscow, Russia; silinaekaterina@mail.ru
8 Department of Hospital Surgery 1, N.I. Pirogov Russian National Research Medical University (RNRMU), 117997 Moscow, Russia; stvictor@bk.ru
9 Division of Biochemistry, School of Life Sciences, JSS Academy of Higher Education and Research, Mysuru 570015, India; rracharya@jssuni.edu.in
* Correspondence: ramith.gowda@gmail.com (R.R.); lranganath.v@nie.ac.in (L.R.); Tel.: +91-09-98-6380-920 (R.R.); +91-09-88-0680-493 (L.R.)
† These authors contributed equally to this work.

Citation: Shivanna, C.; Patil, S.M.; Mallikarjunaswamy, C.; Ramu, R.; Akhileshwari, P.; Nagaraju, L.R.; Sridhar, M.A.; Khanum, S.A.; Ranganatha, V.L.; Silina, E.; et al. Synthesis, Characterization, Hirshfeld Surface Analysis, Crystal Structure and Molecular Modeling Studies of 1-(4-(Methoxy(phenyl)methyl)-2-methylphenoxy)butan-2-one Derivative as a Novel α-Glucosidase Inhibitor. *Crystals* **2022**, *12*, 960. https://doi.org/10.3390/cryst12070960

Academic Editor: Waldemar Maniukiewicz

Received: 15 May 2022
Accepted: 6 July 2022
Published: 9 July 2022

Abstract: The crystal compound was synthesized and characterized using conventional analytical techniques. The compound $C_{19}H_{21}O_3$ crystallizes in a monoclinic crystal system with the space group *P21/c*. The crystal structure is stabilized by C-H ... O interactions. The structure is further reinforced by π-π interactions. During in vitro inhibition of α-glucosidase, the crystal compound exhibited a significant inhibition of the enzyme (IC_{50}: 10.30 ± 0.25 μg/mL) in comparison with the control, acarbose (IC_{50}: 12.00 ± 0.10 μg/mL). Molecular docking studies were carried out for the crystal compound with the α-glucosidase protein model, which demonstrated that the crystal molecule has a good binding affinity (−10.8 kcal/mol) compared with that of acarbose (−8.2 kcal/mol). The molecular dynamics simulations and binding free energy calculations depicted the stability of the crystal molecule throughout the simulation period (100 ns). Further, a Hirshfeld analysis was carried out in order to understand the packing pattern and intermolecular interactions. The energy difference between the frontier molecular orbitals (FMO) was 4.95 eV.

Keywords: crystal structure; Hirshfeld surfaces; α-glucosidase inhibition; molecular docking simulation; molecular dynamics simulations; binding free energy calculations

1. Introduction

Diabetes mellitus is a common chronic metabolic condition that causes high blood sugar levels due to damage to the specialized cells (islets of Langerhans) that produce insulin in the human body. Diabetic individuals either do not produce enough insulin or cannot use it effectively [1,2]. Hyperglycemia is caused by a malfunction that causes high blood glucose levels in the body. Diabetes mellitus has two pathways, with type 2 (non-insulin dependent) diabetes mellitus being more common than type 1 (insulin dependent) [3]. The enzyme α-amylase is linked to diabetes type 2 in a direct way. α-Amylase is a pancreatic and salivary gland secretory substance that hydrolyzes complex carbohydrates into polysaccharides, most commonly starch to glucose and maltose in the intestine. By the action of α-glucosidase, they are further degraded to monosaccharides and released into the bloodstream, raising blood sugar levels [4,5].

Instant hydrolysis of carbohydrates can be slowed by limiting the actions of α-amylase, which controls the quick rise in blood sugar levels [6]. The current medication options for amylase and glucosidase have a number of adverse effects that limit their utility in diabetic treatment. As a result, alternative medicines with low side-effects are urgently needed to act as an option to the treatment of diabetes mellitus [7].

Phenoxyacetates are very robust moieties in the face of various harsh reaction conditions. Phenyl acetate is an aromatic fatty acid metabolite of phenylalanine with potential antineoplastic activity, and its stability is documented by numerous transformations on the aryl system without affecting the side chain [8]. Phenoxyacetic acids are very important chemicals due to their wide distribution and extensive use as plant growth regulators, and they are employed on a large scale for weed control on cereal crops and lawns. Phenoxyacetic acid induces hematopoietic cell proliferation, providing potential for oral therapeutics. Particularly, ethyl phenoxyacetate and its derivatives exhibit potential anti-inflammatory and plant growth regulation activity [9]. Modification of the oxyacetamideureido-phenyl moieties of compounds in the phenoxy acetic acid series is considered likely to lead to more potent antagonists. Therefore, phenoxyacetic acid analogues are interesting to study by various chemical and physical means, as these derivatives are very useful in hyperglycemia and insulin resistance treatment [10–12].

In view of their broad spectrum of important medicinal applications and as a part of our ongoing research on the synthesis and characterization of novel compounds, the title molecule 1-(4-(methoxy(phenyl)methyl)-2-methylphenoxy)butan-2-one was synthesized. Herein we report on the crystal structure, Hirshfeld surface analysis, and molecular modeling studies.

2. Materials and Methods

2.1. Chemicals and Instrumentation

All chemicals were purchased from Sigma-Aldrich Chemicals Pvt Ltd., St. Louis, Missouri, United States, and all are analytical grade 99.0% pure. Melting points were determined on an electrically heated VMP-III melting point apparatus. The FTIR spectra were documented using KBr discs on a FTIR Jasco 4100 infrared spectrometer. The 1H NMR spectra were recorded using a Bruker DRX 400 spectrometer at 400 MHz with TMS as an internal standard. Mass spectra were recorded on a LC-MS/MS (API-4000) mass spectrometer. An additional elemental analysis of the compound was performed on a Perkin Elmer 2400 elemental analyzer.

2.2. Synthesis of 1-(4-(Methoxy(phenyl)methyl)-2-methylphenoxy)butan-2-one (3)

1-(4-(Methoxy(phenyl)methyl)-2-methylphenoxy)butan-2-one (3) was attained by refluxing a mixture of 4-(methoxy(phenyl)methyl)-2-methylphenol (0.01 mol) and 1-chlorobutan-2-one (2) (0.02 mol) in dry distilled acetone (75 mL) and anhydrous potassium carbonate (0.02 mol) for 4 h. The reaction mixture was cooled to room temperature and the solvent was removed using a flash evaporator. The residual mass was triturated with ice-cold water to remove potassium carbonate and extracted with ether (3 × 50 mL). The ether layer was washed with a 10% sodium hydroxide solution (3 × 50 mL), followed by water (3 × 30 mL), and then dried over anhydrous sodium sulphate and evaporated to dryness to obtain a crude solid. Further recrystallization with ethanol afforded the title compound in a pure state (Figure 1).

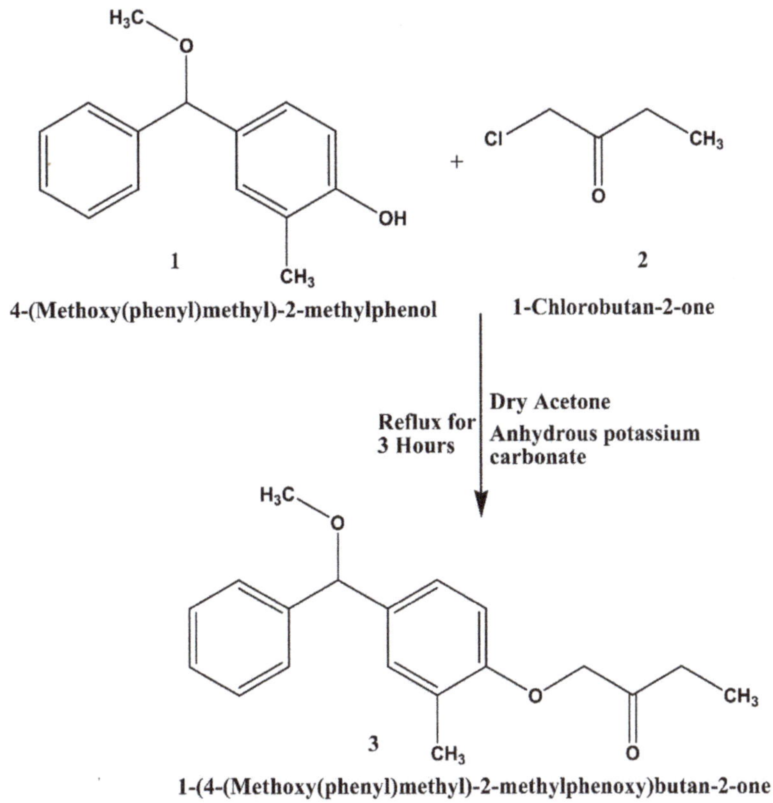

Figure 1. Synthesis of the crystal molecule.

1-(4-(Methoxy(phenyl)methyl)-2-methylphenoxy)butan-2-one (3): Yield 75%. M.p.80–82 °C; IR (KBr): 1745 (C=O), 1510–1620 (aromatic) 2860–2800 cm^{-1} (O-CH_3). 1H NMR (CDCl3): δ 1.3 (t, J = 6 Hz, 3H, CH_3 of ester), 2.3 (s, 3H, Ar-CH_3), 2.55 (q, J = 6 Hz, 2H, CH_2), 3.25 (s, 3H, OCH_3), 4.35 (s, 2H, OCH_2), 5.45 (s, 1H, Ar-CH-O), 7.2–7.8 (bm, 8H, Ar-H). Mol. Wt.: 298.38 (M + 1): 299.16 (100.0%), Anal. Cal. for C19H22O3 (298.38): C, 76.48; H, 7.43. Found: C, 76.45; H, 7.47%.

2.3. X-ray Crystallographic Analysis

A single crystal with of appropriate dimensions was selected for X-ray diffraction analysis. Data were collected using a Bruker Kappa Apex II Single Crystal X-ray Diffractometer equipped with Cu Kα radiation and a CCD detector [13]. The crystal structure was solved and refined by using SHELXS/L-18 software [14]. The obtained model was refined by isotropic thermal parameters, and later by anisotropic thermal parameters. The geometric calculations were carried out using the program PLATON [15]. The molecular and packing diagrams were generated using Mercury CSD 2.0 [16].

2.4. α-Glucosidase Inhibition Assay and Kinetics

The inhibition test for yeast α-glucosidase (EC 3.2.1.20, a type-1 α-glucosidase, was conducted as defined earlier [3]. The inhibitory activity of the test compound was represented by the least-squares regression line of logarithmic concentrations plotted against percentage inhibition, which yielded the IC_{50} values (μg/mL). When compared to the control, this number (IC_{50} values) shows the concentration of samples that can inhibit enzyme activity by 50%. The inhibition kinetics of the compound against α-glucosidase was determined using the method described by Maradesha et al. [17].

2.5. Molecular Docking Simulation

The protein sequence of *Saccharomyces cerevisiae* α-glucosidase MAL-32 obtained from UniProt (UniProt ID: P38158) was used to construct a homology model using SWISS-MODEL. The model was constructed using the X-ray crystal structure of *S. cerevisiae* isomaltase (PDB ID: 3AXH), which revealed a 72% identical and 84% similar sequence at a resolution of 1.8 Å. Construction of this protein model was essential, as the human α-glucosidase protein is yet to be characterized. Since the authors used the yeast α-glucosidase in the in vitro studies, homology model of *S. cerevisiae* α-glucosidase MAL-32 from UniProt was constructed. Protein and ligand preparation was performed according to the previous study conducted by Patil et al. [18]. Since the constructed model had already been evaluated in previous work by the authors [17], the same model was used in the present study. The binding site prediction and positioning of the binding pocket was established according to the previous work of the authors [18]. The binding residues were placed in a grid box measuring 30 Å × 30 Å × 30 Å positioned at the coordinates x = −17.489 Å, y = −8.621 Å and z = −19.658 Å using the software AutoDock Tools 1.5.6. For molecular docking simulation, the protein and ligand preparations were undertaken according to the previous work by the authors [18] using AutoDock Tools 1.5.6 software. The ligand molecule was docked into the protein target using the software AutoDock Vina 1.1.2. Acarbose was used as a control [19].

2.6. Molecular Dynamics Simulation

A command-line interface software package, GROMACS-2018.1, was used to perform molecular dynamics (MD) simulation. It is specifically designed for biochemical molecules such as proteins, lipids, and nucleic acids that possess a great many complex bonded interactions. For systems with hundreds to millions of particles, the program can simulate the Newtonian equations of motion, as well as calculate nonbonded interactions swiftly. Based on the previous study conducted by Patil et al. [20], docked complexes of α-glucosidase protein with the crystal compound, as well as acarbose with the most negative binding affinities, were submitted for simulation. The simulation boxes, consisting of a protein–crystal compound (9463 residues) complex and a protein–acarbose (9472 residues) complex, were simulated for 100 ns at a temperature of 310 K and 1 bar pressure [19]. A trajectory analysis of root-mean-square deviation (RMSD), root-mean-square fluctuation (RMSF), radius of gyration (Rg), ligand–hydrogen bonds, and solvent-accessible surface area (SASA) parameters was performed and the results plotted using XMGRACE, a GUI based software for plotting the results of MD simulation [19,20].

2.7. Binding Free Energy Calculations

Using the MD simulation results, both protein–ligand complexes were subjected to binding free energy calculations using the Molecular Mechanics/Poisson–Boltzmann Surface Area (MM–PBSA) technique. This is an efficient and reliable free energy simulation method used to model molecular recognition, such as for protein–ligand binding interactions. A GROMACS program, g_mmpbsa [21] with the MmPbSaStat.py [22] script was exploited to evaluate the binding free energy for each protein–ligand complex. The g_mmpbsa program calculates binding free energy using three components: molecular mechanical energy, polar and apolar solvation energies, and molecular mechanical energy. The binding free energy was computed using the molecular dynamics trajectories of the last 50 ns and dt 1000 frames. Equations (1) and (2) were used to calculate the free binding energy [23,24].

$$\Delta G_{Binding} = G_{Complex} - (G_{Protein} + G_{Ligand}) \quad (1)$$

$$\Delta G = \Delta E_{MM} + \Delta G_{Solvation} - T\Delta S = \Delta E_{(Bonded + non\text{-}bonded)} + \Delta G_{(Polar + non\text{-}polar)} - T\Delta S \quad (2)$$

$G_{Binding}$: binding free energy; $G_{Complex}$: total free energy of the protein–ligand complex; $G_{Protein}$ and G_{Ligand}: total free energies of the isolated protein and ligand in solvent, respectively; ΔG: standard free energy; ΔE_{MM}: average molecular mechanics potential energy in vacuum; $G_{Solvation}$: solvation energy; ΔE: total energy of bonded as well as non-bonded interactions; ΔS: change in entropy of the system upon ligand binding; T: temperature in Kelvin [25,26].

3. Results and Discussion

3.1. X-ray Crystallographic Details

The asymmetric crystal structure consists of two molecules (A and B). Visualization (ORTEP and packing) of the synthesized molecule has been given in Figure 2. Crystallization data and structure refinement of the crystal molecule have been detailed in Table 1.

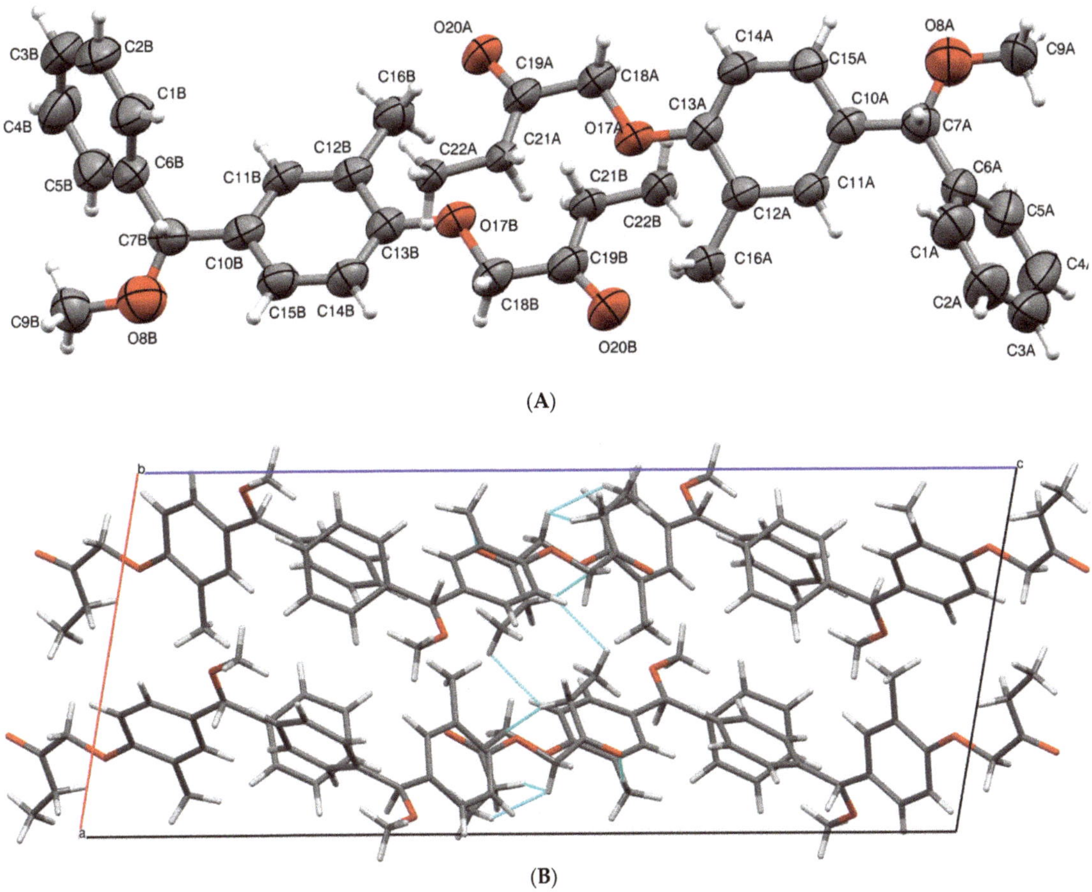

Figure 2. (**A**) ORTEP of the molecule (3) with thermal ellipsoids drawn at 50% probability; (**B**) Packing diagram of the molecule (3).

Table 1. Crystal data and structure refinement for 3.

Parameter	Data
Identification code	3
Empirical formula	$C_{19}H_{22}O_3$
Formula weight	298.36
Temperature	296 K
Wavelength	1.54178 Å
Crystal system, space group	Monoclinic, *P21/c*
Unit cell dimensions	a = 11.1757(2) Å, b = 10.2148 Å
	c = 27.0700(9) Å, β = 99.853°
Volume	3044.66(17) $Å^3$
Z, Calculated density	8, 1.302 Mg/m^3
F_{000}	1280
Crystal size	0.21 × 0.21 × 0.21 mm
Theta range for data collection	3.31 to 64.60°
Limiting indices	$-12 \leq h \leq 13, -11 \leq k \leq 8, -31 \leq l \leq 24$
Reflections collected/unique	12,552/4959
Refinement method	Full-matrix least-squares on F^2
Data/restraints/parameters	4959/27/404
Goodness-of-fit on F^2	1.716
Final R indices [I > 2sigma(I)]	R1 = 0.1196, wR2 = 0.2847
Largest diff. peak and hole	0.720 and −0.760 e.$Å^{-3}$

In molecule A, the rings are planar. The RMSD of the ring C1A-C6A from the mean plane is 0.009(7) Å (atom C6A deviate by 0.003 Å from the mean plane defined for the ring). The RMSD of the ring C10A-C15A from the mean plane is 0.012(4) Å (atom C10A deviates by 0.009(4) Å from the mean plane defined for the ring). The phenyl rings are sp2 hybridized. The atoms C9A-O8A-C7A-C6A show torsion angles of −6.6(7)°, and suggests that they adopt +*anti*-clinal (+ac) conformation. Similarly, in molecule B, the rings are planar. The RMSD of the ring C1B-C6B from the mean plane is 0.006(5) Å (atom C6B deviates by 0.002(5) Å from the mean plane defined for the ring). The RMSDof the ring C10B-C15B from the mean plane is 0.013(4) Å (atoms C12B and C13B deviate by 0.012(4) Å from the mean plane defined for the ring). The phenyl rings are sp2 hybridized. The atoms C9B-O8B-C7B-C6B show torsion angle of −3.9(7)°, and suggests that they adopt +*anti*-clinal (+ac) conformation. Further, the structure is stabilized by C-H . . . O intermolecular hydrogen bond interactions. The details of the hydrogen bond geometry are given in Table 2. The molecule is reinforced by various π–π interactions. The π–π interactions exist between Cg2 and Cg4. Cg2 is the center of gravity of the phenyl ring (C10A-C15A) and Cg4 is the center of gravity of the ring C10B-C15B. The molecule exhibits medium to weak π–π interactions as the Cg2-Cg4 distance is 4.768(2) Å. The molecular packing showing C-H . . . O interactions is depicted in Figure 3.

Table 2. Hydrogen bond geometry(Å).

D-H-A	D-H	H-A	D-A	D-H-A (°)	Symmetric Code
C22A-H22C-O20A	0.96	2.33	2.736(5)	104	Intramolecular interaction
C22B-H22D-O20B	0.96	2.33	2.721(6)	103	Intramolecular interaction
C9A-H23B-O20A	0.96	2.17	2.911(6)	133	1-x, −y, 1-z
C9B-H23D-O20B	0.96	2.28	2.924(6)	124	−x, 2-y, 1-z

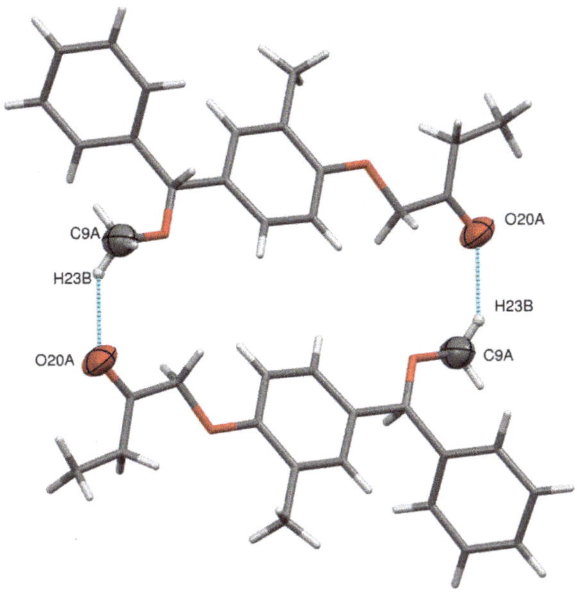

Figure 3. Molecular packing showing C-H . . . O interactions.

3.2. *Effect of Crystal Compound on α-Glucosidase Diabetic Enzyme Inhibition and Kinetics*

The crystal compound inhibited the α-glucosidase enzyme (IC_{50}: 10.30 ± 0.25 µg/mL). Acarbose (positive control) showed IC_{50} values of 12.00 ± 0.10 µg/mL under the same conditions, indicating that the crystal compound inhibition occurred at considerably higher ($p \leq 0.05$) IC_{50} values than with acarbose. To elucidate the manner of α-glucosidase inhibition, a kinetic analysis of the crystal compound was carried out by incubating it with diverse doses of *p*NPG (0.25–4 mmol L^{-1} in the absence (control) or presence of the crystal compound at IC_{20}, IC_{40}, and IC_{60} inhibitory concentrations (µgmL^{-1})). Lineweaver Burk (LB) plots in the reaction were used to define the type of inhibition, as well as the Vmax and Km values. Figure 4 presents the LB plots of the crystal compound against the inhibition of α-glucosidase. Other than the various slopes and x-intercepts, the LB plots demonstrated that the intersecting point for diverse concentrations of the crystal compound came from the same y-intercept as the uninhibited enzyme. The slope and vertical axis intercept rose as crystal compound concentrations increased, with a corresponding increase in the horizontal axis intercept (−1/Km). The kinetic data indicated that with increasing concentrations of the crystal compound, the maximum velocity (Vmax) catalyzed by α-glucosidase remained constant. These findings suggest that the mechanism of α-glucosidase inhibition was reversible, and that it followed the conventional pattern of competitive inhibition. Dixon plots revealed that the inhibitory constant (Ki) for α-glucosidase was 0.41.

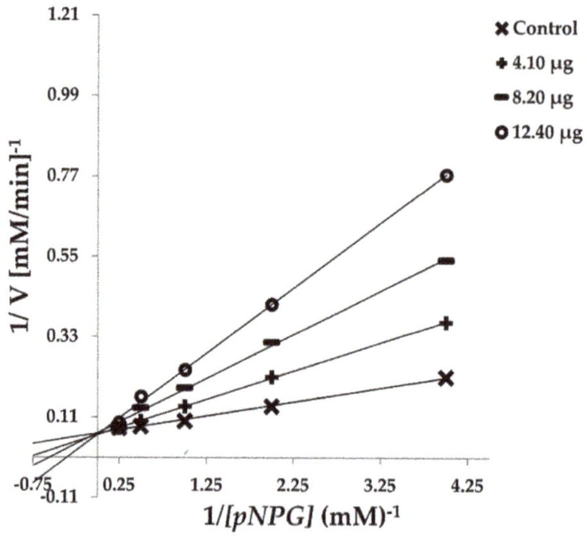

Figure 4. Lineweaver–Burk plot showing substrate-dependent enzyme kinetics when the crystal compound inhibits α-glucosidase enzyme.

3.3. Molecular Docking Simulation

Molecular docking simulation is used to determine how ligands interact with target proteins on a molecular level. It determines the degree of ligand binding, which indicates whether or not the protein is inhibited or activated. While binding with the inhibitor binding site of α-glucosidase, the molecule was found to be docked deep inside the binding pocket, occupying the cleft present in the active site. The compound was predicted to form a total of seven non-bonding interactions, including two hydrogen bonds with GLN 350 (2.82 Å) and ARG 312 (2.57 Å). The molecule also formed two hydrophobic π–π interactions with PHE 157 (3.98), HIS 239 (4.33) and a pi-alkyl bond with ARG 312 (4.93 Å). In addition, the molecule formed electrostatic π-anion and π-cation interactions with ASP 408 (4.98 Å) and HIS 239 (4.99 Å), respectively. With these interactions, the molecule had a binding affinity of −10.8 kcal/mol. However, acarbose was not able to bind with a higher binding affinity. It was also found that acarbose was not able to bind to the deep cleft of the active site, as the molecule did. Acarbose formed seven non-bonding interactions, all of them being hydrogen bonds. They included GLU 304 (2.36 Å), THR 307 (2.55 Å), SER 308 (1.92 Å), PRO 309 (1.77 Å), HIS 279 (2.49 Å), HIS 239 (3.02 Å), and PHE 157 (3.56 Å). With these interactions, acarbose had a binding affinity of −8.2 kcal/mol. Figure 5 depicts the binding interactions of the molecule and acarbose with the α-glucosidase. The outcomes from the docking simulation depict that the molecule can bind within the binding site of the enzyme and can induce biological activity, as was observed in the in vitro studies. According to Patil et al. [18] and Maradesha et al. [17], the docking was accurate, and the binding interactions were similar. The binding interactions validate the results obtained from XRD analysis. This shows that the C-H … O interaction is also observed by docking analysis. The intermolecular C-H … O hydrogen interaction which is observed in XRD analysis is also seen on the Hirshfeld surface map. The presence of C-H … O interaction in the molecule was confirmed the by the XRD analysis, Hirshfeld surface analysis and molecular electrostatic interactions.

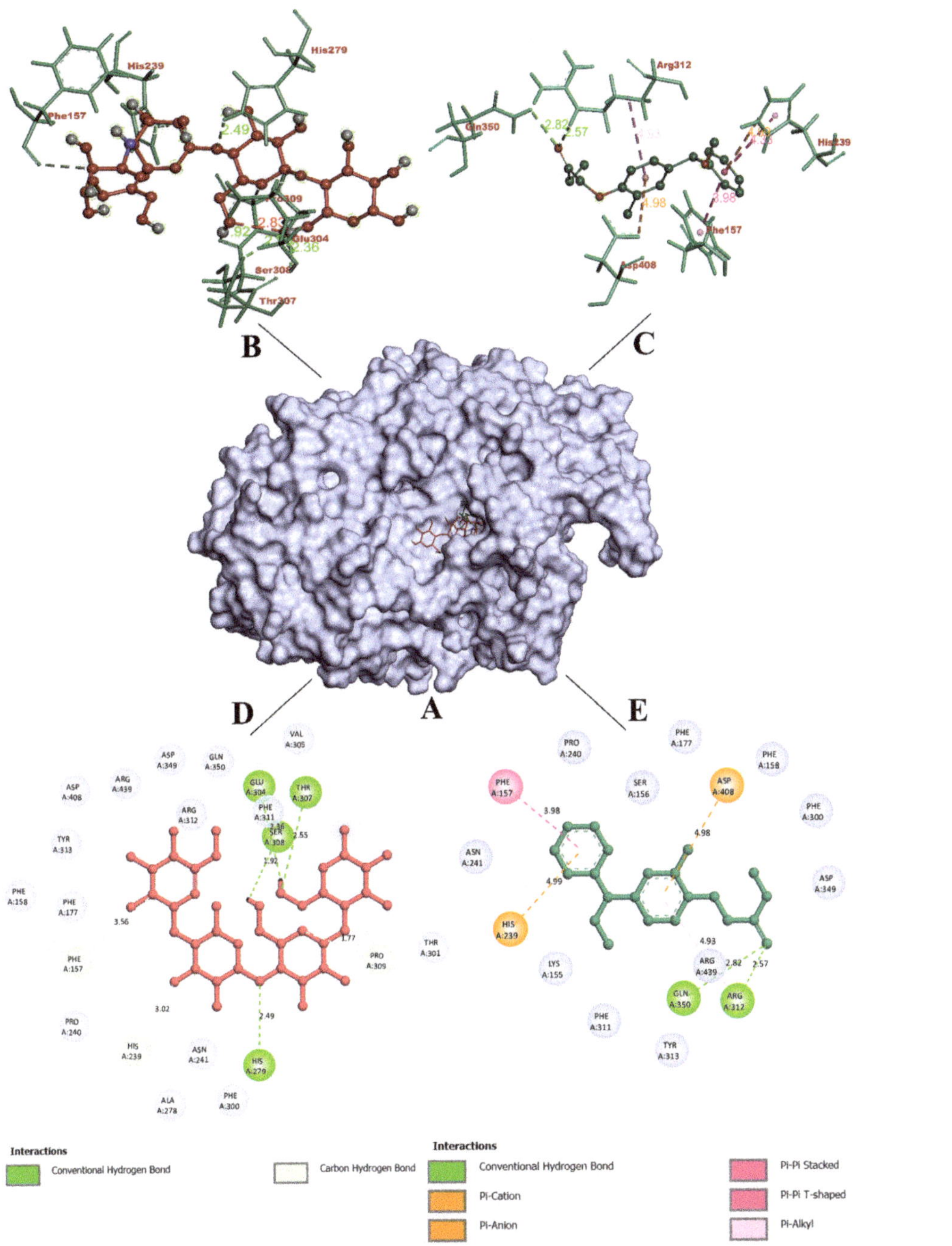

Figure 5. Visualization of binding interactions of the crystal compound and acarbose with the α-glucosidase; (**A**) surface diagram showing bound acarbose (red) and molecule (green); (**B**,**C**) 3D representation of acarbose (red) and molecule (green), respectively; (**D**,**E**) 2D representation of acarbose (red) and molecule (green), respectively.

3.4. Molecular Dynamics Simulation

The overall stability of the protein–ligand combination kept in a particular environment for a specific period of time was assessed using molecular dynamics simulation. Several metrics were evaluated during molecular dynamics simulations to determine the complex's overall stability, including the protein–ligand complex's RMSD, RMSF, Rg, SASA, ligand RMSD and ligand–hydrogen bonds. Over the course of a 100 ns simulation, the RMSD plot of the protein–ligand combination depicts the ligand's stability inside the binding pocket. On the other hand, the RMSF of a protein–ligand complex is used to calculate the average deviation of a particle (e.g., a protein residue) over time from a reference site. As a result, RMSF concentrates on the protein structural regions that differ the most from the mean. Furthermore, by calculating the root-mean-square distances with respect to the central axis of rotation, the radius of gyration (Rg) reflects the structural compactness of the molecules. For all protein–ligand complexes, SASA plots showed the area around the hydrophobic core generated between them. Only a few bonds were simultaneously broken and re-established during the simulation, with the majority of H-bonds remaining consistent with molecular docking. As a result, in dynamic trajectory analysis, ligand–hydrogen bonds are also important.

The RMSD plots depict that during molecular dynamics simulation, both the protein backbone atoms and the protein–crystal compound had a concurrent equilibration point, ranging from 0.30 to 0.35 nm, whereas the protein–acarbose plot was predicted with the RMSD value of 0.3 nm. In case of RMSF, the protein–acarbose plot was predicted with more fluctuations in comparison with the protein–molecule plot. All the plots were found with N-terminal, C-terminal, and loop fluctuations. In case of the Rg, the protein–acarbose and protein–molecule plots were both equilibrated within a range of 2.25–2.50 nm. In addition, the protein backbone atoms were found within a range of 3.0–3.25 nm. A similar pattern of results was obtained in the case of the SASA plots. The protein backbone atoms were predicted with a SASA value of 350 nm^2, whereas the protein–acarbose and the protein–molecule plots were both found within a SASA value range of 225–250 nm^2. During the ligand–hydrogen bond analysis, the molecule was found to have more hydrogen bonds (9), in comparison with the acarbose (3). It can therefore be asserted that the MD simulation results support the docking simulation outcomes. Visualizations of the MD trajectories are depicted in Figure 6. The outcomes of the MD simulation detail the overall stability of the molecule over the acarbose control during the 100 ns long simulation period. The concurrent plots of the protein–molecule complex with the protein backbone atoms indicate stronger binding affinity during the simulation study. The results obtained in this study are in accordance with previous studies in which MD simulation was performed for α-glucosidase [27,28].

3.5. Binding Free Energy Calculations

The binding free energy calculations revealed that the α-glucosidase–molecule complex has a better binding efficiency than the α-glucosidase–acarbose complex. Both the compounds used mainly Van der Waals energy to form the complex, followed by binding energy. The α-glucosidase–crystal compound complex had the highest amount of Van der Waals binding free energy (−108.593 kJ/mol). In all types of binding free energy, the α-glucosidase–crystal compound complex had more binding free energy than the α-glucosidase–acarbose. Results obtained from binding free energy calculations support the outcomes of both the docking and the MD simulation in terms of binding efficiency (Table 3). In addition, these outcomes were on par with those in previous studies that performed binding free energy calculations for α-glucosidase [17,18].

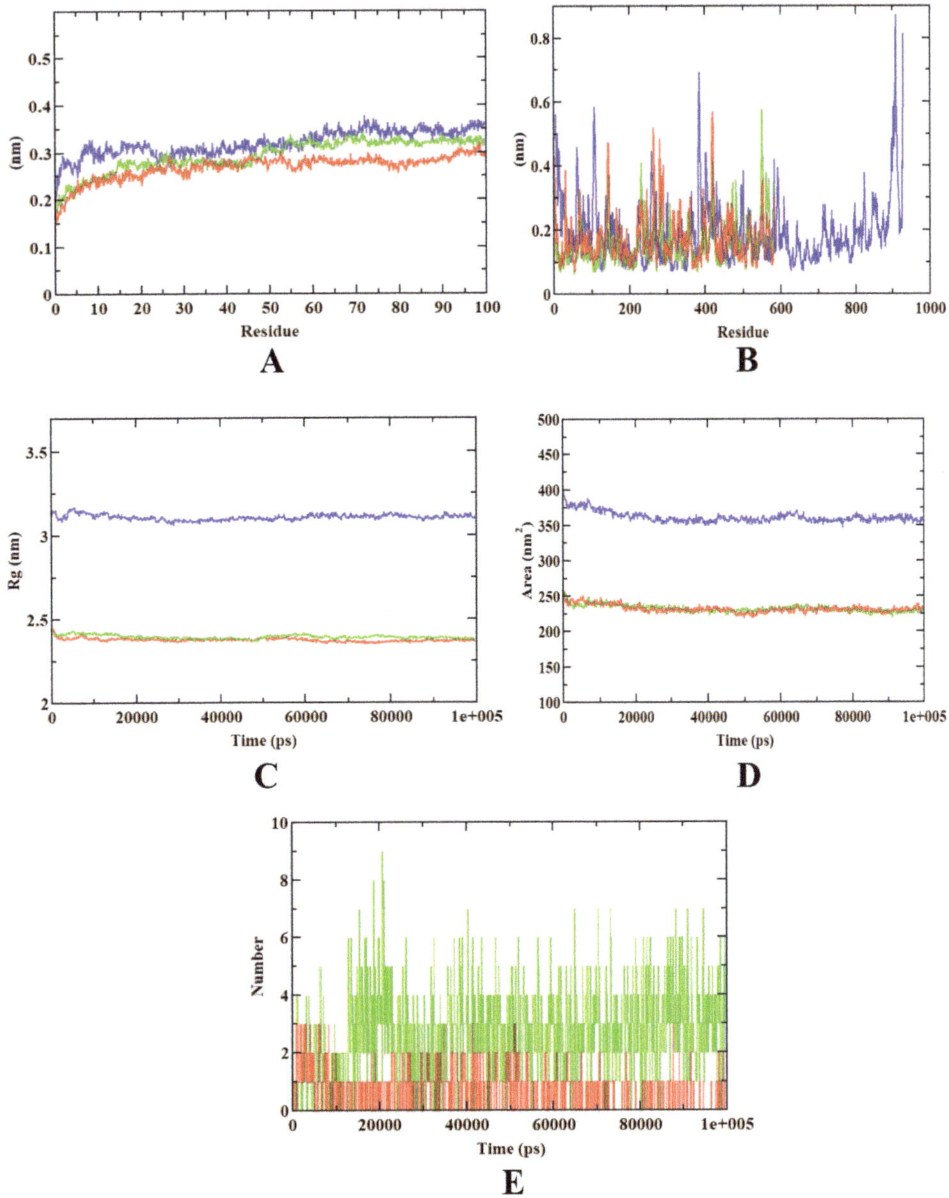

Figure 6. Visualization of the acarbose and crystal compound complexed with α-glucosidase MD simulation run for 100 ns; (**A**) RMSD; (**B**) RMSF; (**C**) Rg; (**D**) SASA; and (**E**) ligand hydrogen bonds. Blue: protein backbone atoms; red: protein–acarbose complex; green: protein–molecule complex.

Table 3. Binding free energy calculations for α-glucosidase–crystal compound complex and α-glucosidase–acarbose complexes.

Types of Binding Free Energies	Values and Standard Deviations	Van Der Waal's Energy	Electrostatic Energy	Polar Solvation Energy	SASA Energy	Binding Energy
α-glucosidase-crystal compound complex	Values (kJ/mol)	−108.593	−31.892	65.071	−9.127	−71.615
	Standard deviation (kJ/mol)	±12.178	±8.561	±9.298	±3.726	±10.028
α-glucosidase-acarbose	Values (kJ/mol)	−99.716	−27.716	49.918	−7.561	−61.239
	Standard deviation (kJ/mol)	±9.257	±7.145	±8.769	±2.539	±8.751

3.6. Hirshfeld Surface Analysis

Hirshfeld surface analysis is used to visualize the intermolecular interactions in a crystal. It is a powerful tool for analyzing intermolecular interactions, such as hydrogen bonds and C-H contacts [29]. These intermolecular interactions can be summarized in two-dimensional fingerprint plots. The distance from the nearest atoms inside and outside the Hirshfeld surface are characterized by the quantities d_i and d_e, respectively. The normalized contact distance (d_{norm}) is

$$d_{norm} = (d_i - r_i^{vdW})/r_i^{vdW} + (d_e - r_e^{vdW})/r_e^{vdw} \quad (3)$$

where r_i^{vdw} and r_e^{vdw} are the van der Waals radii internal and external to the surface. The close intermolecular contacts are represented by red-colored regions on the Hirshfeld surface. The Hirshfeld surface was generated using CrystalExplorer17 software [30]. Figure 7 shows the Hirshfeld surface mapped with d_{norm}. The d_{norm} surface is drawn in the range −0.40 to 1.70. The bright red regions on the Hirshfeld surface represent C-H . . . O intermolecular interaction. The fingerprint plot gives the atomic pair-wise interactions. The precise two-dimensional fingerprint plots are shown in Figure 8. The major contribution is from H-H (63.4%) contacts to the total Hirshfeld surface area and the least contribution is from O-H/H-O (14.5%) contacts. The remaining interactions observed are C-C and O-O and C-O/O-C, which contribute less to the Hirshfeld surface area. They do not show dominant contributions to the total surface.

3.7. Frontier Molecular Orbitals

The frontier molecular orbitals (FMOs) are the highest occupied molecular orbital (HOMO) and lowest unoccupied molecular orbital (LUMO). GAMESS software [31] was used to perform density functional theory (DFT) calculations. The energy levels of the FMOs were computed using the B3LYP/6-31 G (d, p) basis set and are displayed in Figure 9. The calculations show that the HOMO and LUMO are mainly localized on the benzene ring. The energy difference between the HOMO and the LUMO indicates the energy gap. The energy gap between the HOMO and the LUMO in this case was 4.952 eV. This energy difference predicts that the molecule is kinetically stable. As the value of the HOMO–LUMO gap increases, the molecule becomes less stable. A larger energy gap between the HOMO and the LUMO is associated with high kinetic stability [32]. The molecular descriptors calculated from the energy values of the frontier molecular orbitals are listed in Table 4.

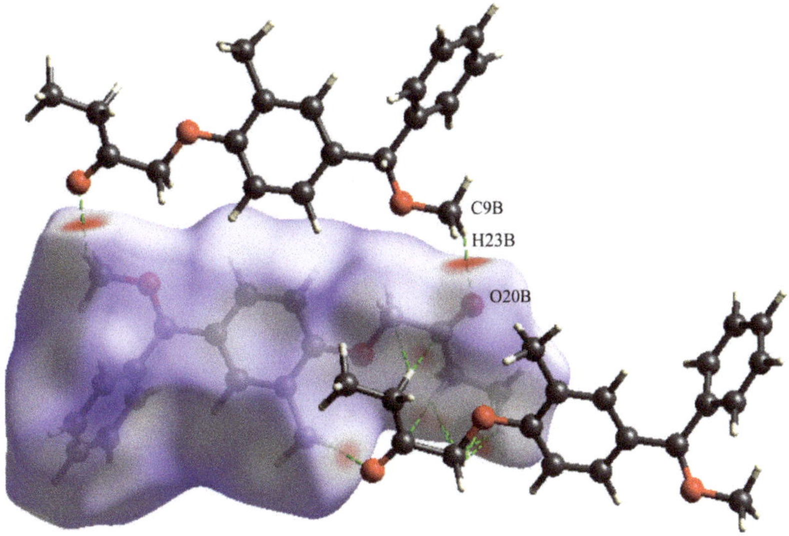

Figure 7. Hirshfeld surface mapped with d_{norm}.

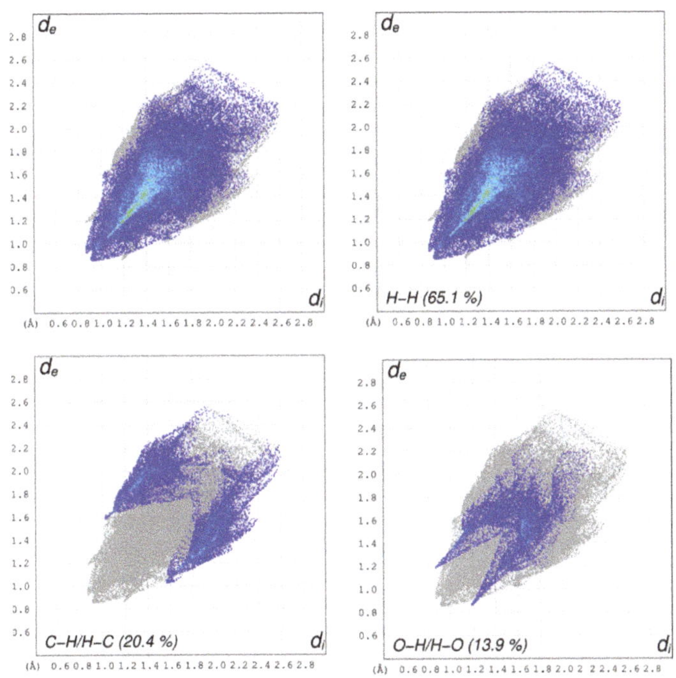

Figure 8. Fingerprint plots for the title molecule.

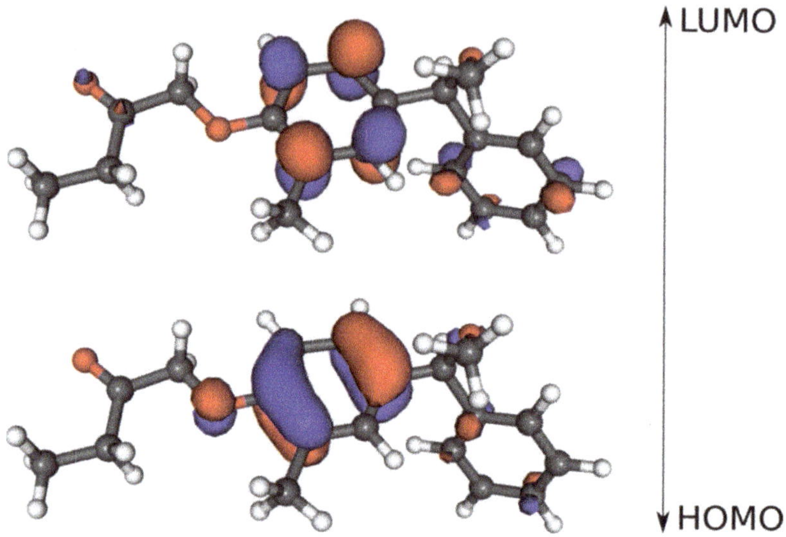

Figure 9. Energy levels of the frontier molecular orbital of the title molecule.

Table 4. Calculated energy values of molecular descriptors of the title compound.

Molecular Descriptor	Value
HOMO	−8.489 eV
LUMO	3.537 eV
Energy gap (ΔE)	4.952 eV
Ionization potential (I)	8.489 eV
Electron affinity (E)	−3.537 eV
Chemical potential (μ)	−6.013 eV
Electronegativity (χ)	6.013 eV
Global hardness (σ)	2.476 eV
Global softness (η)	0.2019 eV^{-1}
Electrophilicity (ω)	7.301 eV

3.8. Molecular Electrostatic Potential (MEP)

The molecular electrostatic potential (MEP) map is a tool to analyze the charge distribution in the molecule. The MEP surface was generated using the Gaussian09 program [33] and is shown in Figure 10. The molecular electrostatic potential is represented by different colors. The value of potential increases, with red < orange < yellow < blue. The deep red areas indicate regions with a negative electrostatic potential and the blue sites indicate regions with a positive electrostatic potential. Figure 10 shows that an electropositive region (blue) is observed around the hydrogen atom and the negative region (red) is concentrated over the oxygen atom, indicating an electrophilic area.

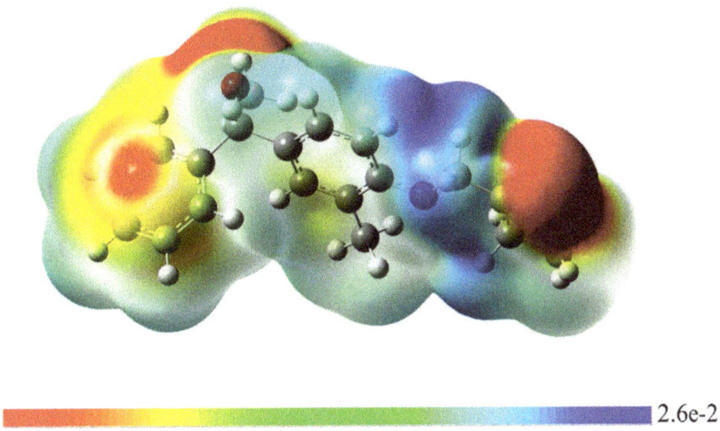

Figure 10. Molecular electrostatic map of the compound.

4. Conclusions

The crystal compound was synthesized and the single crystal developed by means of the slow solvent evaporation technique using ethanol as a solvent. The compound was characterized using NMR and mass spectroscopy, and, finally, the molecular structure of the compound was confirmed by single crystal X-ray diffraction. In vitro α-glucosidase inhibition for the crystal compound was better than with the acarbose. A molecular docking study revealed that the probable binding interaction of molecule (3) with the protein target showed a low binding free energy, which prompted us to look for further experimental studies. The outcomes from in silico studies depicted that the molecule was able to inhibit the enzyme by interacting with the binding pocket residues. The molecule was stable throughout the simulation run of 100 ns, indicating that it can efficiently carry out the biological activity of enzyme inhibition The energy difference between the frontier molecular was 4.95 eV, which predicts the title molecule is kinetically stable. The molecular electrostatic potential surface revealed the electronegative and electropositive sites present around the oxygen and hydrogen atoms, respectively.

Author Contributions: Planning and conceptualization of the manuscript: L.R. and R.R.; data analysis and method development: S.A.K., E.S. and C.M.; writing: preparation of original draft and writing—V.L.R., R.R., L.R.N., P.A., V.S., R.R.A. and M.A.S.; supervision and editing—C.S. and S.M.P. All authors have read and agreed to the published version of the manuscript.

Funding: This research received no external funding.

Institutional Review Board Statement: Not applicable.

Informed Consent Statement: Not applicable.

Data Availability Statement: Not applicable.

Acknowledgments: Chandra and Lakshmi Ranganatha V gratefully acknowledge the principal and NIE-Management for the support and encouragement to carry out this research. Additionally, Akhileshwari P. thanks DST-KSTePS, Government of Karnataka, Bengaluru for providing the fellowship. All the authors thank JSS Academy of Higher Education and Research (Mysore, Karnataka, India) for their kind support, encouragement, and provision of the necessary facilities.

Conflicts of Interest: All authors declare no conflict of interest.

References

1. Patil, S.M.; Shirahatti, P.S.; Ramu, R. *Azadirachta indica* A. Juss (neem) against diabetes mellitus: A critical review on its phytochemistry, pharmacology, and toxicology. *J. Pharm. Pharmacol.* **2022**, *74*, 681–710. [CrossRef] [PubMed]
2. Ramu, R.; Patil, S.M. A perspective on the effective conduction of functional-based coaching program on diabetic Indonesian communities. *Oman Med. J.* **2021**, *36*, e281. [CrossRef] [PubMed]
3. Ramu, R.; Shirahatti, P.S.; Deepika, T.H.; Bajpe, S.N.; Sreepathi, N.; Patil, S.M.; Prasad, N. Investigating *Musa paradisiaca* (Var. Nanjangud rasa bale) pseudostem in preventing hyperglycemia along with improvement of diabetic complications. *J. Appl. Biol. Biotechnol.* **2022**, *10*, 56–65. [CrossRef]
4. Shaw, J.E.; Sicree, R.A.; Zimmet, P.Z. Global estimates of the prevalence of diabetes for 2010 and 2030. *Diabetes Res. Clin. Pract.* **2010**, *87*, 4–14. [CrossRef]
5. Storr, S.J.; Royle, L.; Chapman, C.J.; Hamid, U.M.; Robertson, J.F.; Murray, A.; Dwek, R.A.; Rudd, P.M. The O-linked glycosylation of secretory/shed MUC1 from an advanced breast cancer patient's serum. *Glycobiology* **2008**, *18*, 456–462. [CrossRef]
6. Bingol, Z.; Kızıltaş, H.; Gören, A.C.; Kose, L.P.; Topal, M.; Durmaz, L.; Alwasel, S.H.; Gulcin, İ. Antidiabetic, anticholinergic and antioxidant activities of aerial parts of shaggy bindweed (*Convulvulus betonicifolia* Miller subsp.)–profiling of phenolic compounds by LC-HRMS. *Heliyon* **2021**, *7*, e06986. [CrossRef]
7. Simsek, E.; Lu, X.; Ouzounov, S.; Block, T.M.; Mehta, A.S. α-Glucosidase inhibitors have a prolonged antiviral effect against hepatitis B virus through the sustained inhibition of the large and middle envelope glycoproteins. *Antivir. Chem. Chemother.* **2006**, *17*, 259–267. [CrossRef]
8. Al-Ghorbani, M.; Vigneshwaran, V.; Ranganatha, V.L.; Prabhakar, B.T.; Khanum, S.A. Synthesis of oxadiazole–morpholine derivatives and manifestation of the repressed CD31 Microvessel Density (MVD) as tumoral angiogenic parameters in Dalton's Lymphoma. *Bioorg. Chem.* **2015**, *60*, 136–146. [CrossRef]
9. Khanum, S.A.; Shashikanth, S.; Deepak, A.V. Synthesis, and anti-inflammatory activity of benzophenone analogues. *Bioorg. Chem.* **2004**, *32*, 211–222. [CrossRef]
10. Mohammed, Y.H.; Naveen, S.; Issa, H.H.; Manjunath, H.R.; Lokanath, N.K.; Khanum, S.A. Synthesis, Characterization, Hirshfeld Analysis, Crystal and, Molecular Structure Studies of 2,6-difluoro phenoxy Acetic Acid. *Der. Pharma. Chem.* **2016**, *8*, 286–291.
11. Srivastav, A.; Saxena, S.R.; Mishra, S.; Singh, Y. Molecular structure, experimental and theoretical spectroscopic studies and quantum chemical calculation of phenoxyacetic acid and its p-chloro derivative. *Orient J. Chem.* **2012**, *28*, 1517–1529. [CrossRef]
12. Timchalk, C. Comparative inter-species pharmacokinetics of phenoxyacetic acid herbicides and related organic acids. evidence that the dog is not a relevant species for evaluation of human health risk. *Toxicology* **2004**, *200*, 1–19. [CrossRef]
13. Tahir, M.N.; Anwar-Ul-Haq, M.; Choudhary, M.A. Crystal structure of 4-[€-(2-carbamo-thio-ylhydrazinyl-idene)meth-yl]benzoic acid. *Acta. Crystallogr. E. Crystallogr. Commun.* **2015**, *71*, 771–772. [CrossRef]
14. Sheldrick, G.M. *A Program for Crystal Structure Determination*; SHELX 18; University of Cambridge: Cambridge, UK, 2018.
15. Spek, A.L. PLATON SQUEEZE: A tool for the calculation of the disordered solvent contribution to the calculated structure factors. *Acta Crystallogr. C Struct. Chem.* **2015**, *71*, 9–18. [CrossRef]
16. Macrae, C.F.; Bruno, I.J.; Chisholm, J.A.; Edgington, P.R.; McCabe, P.; Pidcock, E.; Rodriguez-Monge, L.; Taylor, R.; van de Streek, J.; Wood, P.A. Mercury CSD 2.0—New Features for the Visualization and Investigation of Crystal Structures. *J. Appl. Cryst.* **2008**, *41*, 466–470. [CrossRef]
17. Maradesha, T.; Patil, S.M.; Al-Mutairi, K.A.; Ramu, R.; Madhunapantula, S.V.; Alqadi, T. Inhibitory effect of polyphenols from the whole green jackfruit flour against α-glucosidase, α-amylase, aldose reductase and glycation at multiple stages and their interaction: Inhibition kinetics and molecular simulations. *Molecules* **2022**, *27*, 1888. [CrossRef]
18. Patil, S.M.; Martiz, R.M.; Ramu, R.; Shirahatti, P.S.; Prakash, A.; Kumar, B.P.; Kumar, N. Evaluation of flavonoids from banana pseudostem and flower (quercetin and catechin) as potent inhibitors of α-glucosidase: An in silico perspective. *J. Biomol. Struct. Dyn.* **2021**, *7*, 1–5. [CrossRef]
19. Kumar, V.; Ramu, R.; Shirahatti, P.S.; Kumari, V.C.; Sushma, P.; Mandal, S.P.; Patil, S.M. α-Glucosidase; α-Amylase Inhibition; Kinetics and Docking Studies of Novel (2-Chloro-6-(trifluoromethyl) benzyloxy) arylidene) Based Rhodanine and Rhodanine Acetic Acid Derivatives. *Chem. Select.* **2021**, *6*, 9637–9644. [CrossRef]
20. Patil, S.M.; Martiz, R.M.; Ramu, R.; Shirahatti, P.S.; Prakash, A.; Chandra, J.S.; Ranganatha, L.V. In silico identification of novel benzophenone-coumarin derivatives as SARS-CoV-2 RNAdependent RNA polymerase (RdRp) inhibitors. *J. Biomol. Struct. Dyn.* **2021**, *11*, 1–17.
21. Martiz, R.M.; Patil, S.M.; Ramu, R.; Jayanthi, M.K.; Ashwini, P.; Ranganatha, L.V.; Khanum, S.A.; Silina, E.; Stupin, V.; Achar, R.R. Discovery of novel benzophenone integrated derivatives as anti-Alzheimer's agents targeting presenilin-1 and presenilin-2 inhibition: A computational approach. *PLoS ONE* **2022**, *17*, e0265022. [CrossRef] [PubMed]
22. Martiz, R.M.; Patil, S.M.; Abdulaziz, M.; Babalghith, A.; Al-Areefi, M.; Al-Ghorbani, M.; Kumar, J.M.; Prasad, A.; Mysore Nagalingaswamy, N.P.; Ramu, R.; et al. Defining the Role of Isoeugenol from *Ocimum tenuiflorum* against Diabetes Mellitus-Linked Alzheimer's Disease through Network Pharmacology and Computational Methods. *Molecules* **2022**, *27*, 2398. [CrossRef]
23. Gurupadaswamy, H.D.; Ranganatha, V.L.; Ramu, R.; Patil, S.M.; Khanum, S.A. Competent synthesis of biaryl analogs via asymmetric Suzuki–Miyaura cross-coupling for the development of anti-inflammatory and analgesic agents. *J. Iran. Chem. Soc.* **2022**, *1*, 2421–2436. [CrossRef]

24. Patil, S.M.; Maruthi, K.R.; Bajpe, N.S.; Vyshali, V.M.; Sushmitha, S.; Chagalamari, A.; Ramith, R. Comparative molecular docking and simulation analysis of molnupiravir and remdesivir with SARS-CoV-2 RNA dependent RNA polymerase (RdRp). *Bioinformation* **2021**, *7*, 932–939.
25. Ganavi, D.; Ramu, R.; Kumar, V.; Patil, S.M.; Martiz, R.M.; Shirahatti, P.S.; Sathyanarayana, R.; Poojary, B.; Holla, B.S.; Poojary, V.; et al. In vitro and in silico studies of fluorinated 2,3-disubstituted thiazolidinone-pyrazoles as potential α-amylase inhibitors and antioxidant agents. *Arch. Pharm.* **2021**, *12*, e2100342. [CrossRef]
26. Poojary, B.; Kumar, V.; Arunodaya, H.S.; Chandra, S.; Ramu, R.; Patil, S.M.; Baliga, A.; Rai, V.M.; Vishwanatha, U.; Vishwanatha, P.; et al. Potential fluorinated anti-MRSA thiazolidinone derivatives with antibacterial, antitubercular activity and molecular docking studies. *Chem. Biodivers.* **2021**, *19*, e202100532.
27. Nipun, T.S.; Khatib, A.; Ibrahim, Z.; Ahmed, Q.U.; Redzwan, I.E.; Saiman, M.Z.; Supandi, F.; Primaharinastiti, R.; El-Seedi, H.R. Characterization of α-glucosidase inhibitors from *Psychotria malayana* jack leaves extract using LC-MS-based multivariate data analysis and in-silico molecular docking. *Molecules* **2020**, *25*, 5885. [CrossRef]
28. Murugesu, S.; Ibrahim, Z.; Ahmed, Q.U.; Uzir, B.F.; Yusoff, N.I.; Perumal, V.; Abas, F.; Shaari, K.; Khatib, A. Identification of α-glucosidase inhibitors from *Clinacanthus nutans* leaf extract using liquid chromatography-mass spectrometry-based metabolomics and protein-ligand interaction with molecular docking. *J. Pharm. Anal.* **2019**, *9*, 91–99. [CrossRef]
29. Hökelek, T.; Yavuz, V.; Dal, H.; Necefoğlu, H. Crystal structure and Hirshfeld surface analysis of aqua-bis-(nicotinamide-κN)bis-(4-sulfamoylbenzoato-κO1)copper(II). *Acta. Crystallogr. E. Crystallogr. Commun.* **2018**, *74*, 45–50. [CrossRef]
30. Spackman, M.A.; McKinnon, J.J.; Jayatilaka, D. Electrostatic potentials mapped on Hirshfeld surfaces provide direct insight into intermolecular interactions in crystals. *Cryst. Eng. Comm.* **2008**, *10*, 377–388. [CrossRef]
31. Schmidt, M.W.; Baldridge, K.K.; Boatz, J.A.; Elbert, S.T.; Gordon, M.S.; Jensen, J.H.; Koseki, S.; Matsunaga, N.; Nguyen, K.A.; Su, S.; et al. General atomic and molecular electronic structure system. *Comput. Chem.* **1993**, *14*, 1347–1363. [CrossRef]
32. Aihara, J.I. Weighted HOMO-LUMO energy separation as an index of kinetic stability for fullerenes. *Theor. Chem. Acc.* **1999**, *102*, 134–138. [CrossRef]
33. Frisch, M.J.; Trucks, G.W.; Schlegel, H.B.; Scuseria, G.E.; Robb, M.A.; Cheeseman, J.R.; Scalmani, G.; Barone, V.; Mennucci, B.; Petersson, G.A. *Gaussian 09*; Gaussian, Inc.: Wallingford, CT, USA, 2009.

crystals MDPI

Article

Synthesis of Two Novel Copper (II) Complexes as Potential Inhibitors of HIV-1 Protease Enzyme: Experimental and Theoretical Investigations

Meriem Hamlaoui [1,*], Ikram Hamlaoui [2], Maamar Damous [1,3], Youghourta Belhocine [3,*], Najoua Sbei [4], Fatima Adam Mohamed Ali [5], Mashael A. Alghamdi [5], Sarra Talab [6], Seyfeddine Rahali [7] and Hocine Merazig [1,*]

1 Unité de Recherche de Chimie de l'Environnement et Moléculaire Structurale (URCHEMS), Département de Chimie, Université Frères Mentouri de Constantine, Constantine 25000, Algeria; damousmmaamar@yahoo.fr
2 Département de Pharmacie, Faculté de Médecine, Université Saleh Boubnider, Constantine 3, Constantine 25000, Algeria; ikram.hamlaoui@univ-constantine3.dz
3 Department of Process Engineering, Faculty of Technology, 20 August 1955 University of Skikda, El Hadaik Road, Skikda 21000, Algeria
4 Institute of Nanotechnology, Karlsruhe Institute of Technology, Eggenstein Leopoldshafen, 76344 Karlsruhe, Germany; najwasbei89@hotmail.fr
5 Department of Chemistry, College of Science, Imam Mohammad Ibn Saud Islamic University (IMSIU), Riyadh 11432, Saudi Arabia; famohamedali@imamu.edu.sa (F.A.M.A.); mabalghamdi@imamu.edu.sa (M.A.A.)
6 Department of Chemistry, College of Applied and Industrial Science, University of Bahri, P.O. Box 11111, Khartoum 1660, Sudan; stalab9@gmail.com
7 Department of Chemistry, College of Science and Arts, Qassim University, Ar Rass 82571, Saudi Arabia; s.rahali@qu.edu.sa
* Correspondence: hamlaoui.meriem@yahoo.fr (M.H.); y.belhocine@univ-skikda.dz (Y.B.); hmerazig@gmail.com (H.M.)

Citation: Hamlaoui, M.; Hamlaoui, I.; Damous, M.; Belhocine, Y.; Sbei, N.; Ali, F.A.M.; Alghamdi, M.A.; Talab, S.; Rahali, S.; Merazig, H. Synthesis of Two Novel Copper (II) Complexes as Potential Inhibitors of HIV-1 Protease Enzyme: Experimental and Theoretical Investigations. *Crystals* **2022**, *12*, 1066. https://doi.org/10.3390/cryst12081066

Academic Editor: Waldemar Maniukiewicz

Received: 29 June 2022
Accepted: 22 July 2022
Published: 30 July 2022

Abstract: In this study, we report the synthesis of two new copper complexes: $[Cu(C_{11}H_7O_2)(SCN)(C_{10}H_8N_2)]$, denoted as (C-1), and $[Cu(C_{11}H_7O_2)(C_{12}H_8N_2)Cl]\cdot H_2O$, denoted as (C-2). They are based on 2,2′-bipyridine or 1,10-phenanthroline and 2-hydroxy-1-naphtaldehyde ligands. The obtained complexes were characterized by FT-IR, UV-visible spectroscopy, and single-crystal X-ray diffraction analysis. Molecular docking was employed to predict the binding mode involved in the interaction between the two synthetic copper (II) complexes and HIV-1 protease enzyme. The X-ray structural analysis revealed that the crystal structures of both complexes are mainly stabilized by several intra- and intermolecular hydrogen bonds. The fingerprint plots associated with the Hirshfeld surfaces of both complexes clearly show that H⋯H interactions provide the largest contributions. According to the docking results, the synthesized complexes exhibit promising features which enable them to be bound to the HIV-protease enzyme.

Keywords: copper (II) complex; synthesis; HIV-1 protease enzyme inhibitor; crystal structure; non-covalent interactions; Hirshfeld surface; molecular docking

1. Introduction

In recent years, there have been several reports dedicated to the development of transition metal complexes that are based on the 2-hydroxy-1-naphthaldehyde $C_{11}H_8O_2$ ligand [1]. Indeed, 2-hydroxy-1-naphthaldehyde ligands can be used as a starting block for the synthesis of new ligands [2,3], or as a potential chelating agent for ions; however, only a few studies were focused on the synthesis of copper (II) complexes of bidentate ligands [4,5].

Moreover, the 2,2′-bipyridine $C_{10}H_8N_2$ and 1,10-phenanthroline $C_{12}H_8N_2$ bidentate chelating ligands are widely used as complexing moieties to form stable coordination complexes with miscellaneous transition metal ions [6]. Transition metal complexes that have these ligands are recognized in the literature as potent antibacterial and antimicrobial agents [7–9]. It was also reported that copper complexes are good candidates for the design of novel types of protease enzyme inhibitors; here, the main goal is to treat the acquired immunodeficiency syndrome (AIDS) infection induced by the human immunodeficiency virus (HIV). Since the protease of the human immunodeficiency virus is key to the replication of the virus, the enzyme is therefore considered an attractive target for antiviral therapy; thus, the treatment strategy for AIDS involves the inhibition of HIV-1 protease [10]. To date, real progress in terms of therapy for AIDS has been achieved by using peptide-based therapeutics; however, the toxicity of peptides reduces the effectiveness of this treatment [11,12]. The copper complexes, which contain non-peptide ligands that have therapeutic potential, could be an alternative methodology for the treatment of AIDS-induced diseases.

In this contribution, we describe the synthesis of two new complexes, namely, the [(2,2′-bipyridine-K^{2}N,N′) (1-formyl-2-naphtholato-K^{2}O,O′) (Thiocyanate-KN) copper(II)], denoted as C-1, and [(1-formyl-2-naphtholato-K^{2}O,O′) (1,10-phenanthroline-K^{2}N,N′) chloride copper], denoted as C-2. FT-IR, UV–visible spectroscopy, and single-crystal X-ray diffraction analysis were used to characterize and analyze the structure of the two synthesized complexes. DFT calculations using the B3LYP [13–16] functional were conducted on the molecular structures of C-1 and C-2, and the findings were compared with the X-ray crystal data. The docking of C-1 and C-2 into the HIV-1 protease active site was assessed through the calculation of binding energy with the Autodock vina program [17], and the obtained results were compared with those of the (dimethanol)bis[N-(4-methyl-2-pyridyl)-2,3,6-trimethoxybenzamide] copper (II) dimethanol diperchlorate complex as a reference (complex-ref), which was found to be a non-peptide HIV-1 protease inhibitor with an IC50 of 1.5 μM [18]. It is well known that the most effective inhibitor is the one that forms a bond deep in the hydrophobic cavity of the HIV-1 protease active site, and is positioned in the axes of an octahedron with the four subsites S1/S2 and S′1/S′2 forming the equatorial plane. The binding pocket of this enzyme includes the Asp25 (25′)-Thr26 (26′)-Gly27(27) catalytic triad and flap regions, which are most likely involved in the binding process of substrates or inhibitors [19,20].

2. Experimental

2.1. Materials and Instruments

All reagents and solvents were purchased from commercial sources (Sigma-Aldrich, St. Louis, MI, USA), and they were used in the form in which they were received.

Infrared spectra were recorded in the range of 4000–400 cm^{-1} on a FT-IR Bruker ATR Vertex 70 Spectrometer. UV-visible spectra were recorded on a UV–visible Spectrophotometer Optizen 1220. The absorbance of ligands and their metals complexes were measured in DMSO in a concentration of 10^{-5} mol·L^{-1}, in a wavelength range of 280–440 nm.

Crystals suitable for single crystal X-ray diffraction were selected, and the lattice parameters were determined using an APEXII Bruker diffractometer.

2.2. General Procedure for the Synthesis of C-1 and C-2

2.2.1. Synthesis of [(2,2′-Bipyridine-K^{2}N,N′) (1-formyl-2-Naphtholato-K^{2}O,O′) (Thiocyanate-KN)-Copper (II)] (C-1)

For the synthesis of C-1, $Cu(CH_3COO)_2$. H_2O (0.25 mmol) and 2-hydroxy-1-naphthaldehyde (0.125 mmol) were added and stirred into a DMSO solution containing 0.25 mmol of 2,2′-bipyridine and 0.125 mmol of KSCN. The mixture was stirred for 1 h at T = 25 °C (Scheme 1a). After several days of slow evaporation, light green crystals appeared. This complex is stable in the air and soluble only in DMSO. Yield: 61%. IR (cm^{-1}): 2063.83(s), 1604.77(s), 1577.77(s), 1531.48(s), 1423.47(s), 1392.61(s), 1365.60(s), 1311.59(m), 1184.29(m), 1141.86(w), 1014.56(s), 972.12(w), 948.98(m), 829.39(m), 721.38(s), 636.51(w), 563.21(w), 497.63(w). UV–Vis (DMSO) (λmax, nm): 290, 300 (π-π*) 305, 315.

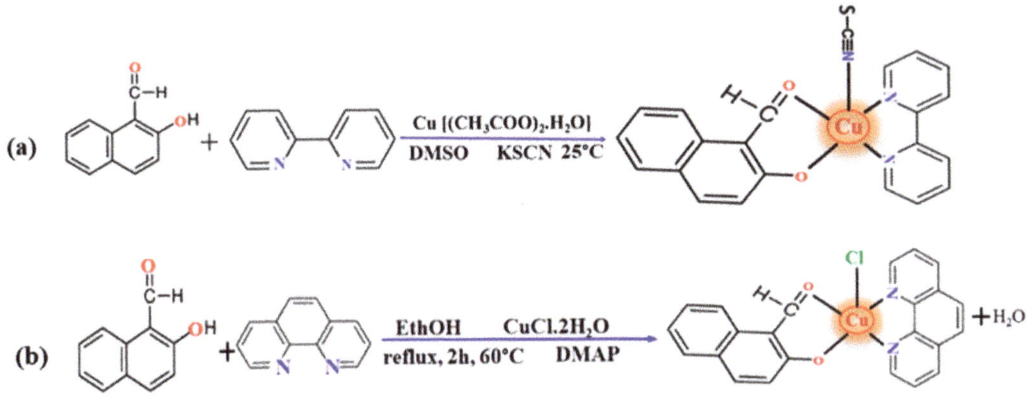

Scheme 1. Synthesis of (**a**) C-1, (**b**) C-2.

2.2.2. Synthesis of [(1-Formyl-2-naphtholato-K^2O,O′) (1,10-Phenanthroline-K^2N,N′) Chloride Copper] (C-2)

A mixture of $CuCl_2.2H_2O$ (0.167 mmol), 2-hydroxy-1-naphthaldehyde (0.167 mmol), 1,10-phenanthroline (0.167 mmol), and 4-dimethyl aminopyridine (0.167 mmol) was dissolved in ethanol. The solution was refluxed for 2 h at 60 °C (Scheme 1b). After the slow evaporation of the solution, green single crystals were obtained and characterized by X-ray diffraction. Yield: 68%. IR (cm^{-1}): 1608.63(s), 1581.63(s), 1539.20(m), 1423.47(s), 1392.61(s); 1365.60(m), 852.54(s), 825.53(s), 752.24(m), 717.52(s), 543.93(s), 520.78(s), 493.78(s). UV–Vis (DMSO) (λ_{max}, nm): 290, 300 (π-π^*) 310, 325.

2.3. X-ray Diffraction

Data collection was performed using Mo$K\alpha$ radiation with a BRUKER APEX2 diffractometer [21]. Cell refinement and data reduction were performed through the SAINT program. SHELXS and SHELXL [22] were used for structure solution and refinement. Structure refinement and crystal data details are reported in Table 1. All hydrogen atoms were refined in geometrically idealized positions, with C-H = 0.93 Å for aromatic rings (2,2′-bipyridine, 1,10-phenanthroline), and they were allowed to ride on their parent atoms, with Uiso(H) = 1.2 Ueq(C). The molecular structures of the two complexes are illustrated in Figure 1.

2.4. Computational Methodology

The structural optimization of C-1 and C-2 was performed using the framework of the density functional theory (DFT). The B3LYP hybrid functional was used with a 6–31 + G(d) basis set for C, H, N, Cl, O, S, and SDD for Cu. The DFT computations were conducted using the Gaussian 09W software package [23]. The PDB file of the ligand was generated using OpenBabel, version 2.4.1 [24,25]. The HIV-1 protease–ligand interaction properties were assessed by performing docking studies using AutoDock vina software. The three-dimensional structure of HIV-1 protease (PDB ID: 3TLH) was retrieved from the Protein Data Bank (http://www.rcsb.org/pdb (accessed on 8 March 2022)), and it was used to prepare the target site for docking calculation. The water molecules and ligand were removed from the protein, and the missing hydrogen atoms and charge were corrected. The search grid of the HIV-1 protease was identified as center_x: −6.118, center_y: 23.382, center_z: 36.63 with dimensions size_x: 62, size_y: 82, and size_z: 76. The best-scoring conformation was visualized and analyzed using Autodock vina tools.

Table 1. Crystallographic data and refinement parameters of C-1 and C-2.

Complex	C-1	C-2
Empirical formula	$CuC_{22}H_{15}N_3O_2S$	$CuC_{23}H_{17}N_2O_3Cl$
Formula weight	448.97	468.39
Crystal size (mm)	0.2 × 0.1 × 0.01	0.2 × 0.1 × 0.01
Temperature (K)	273	296
Crystal system	Monoclinic	Triclinic
Space group	$P2_1/n$	*P*-1
a (Å)	9.5832(4)	8.5321(8)
b (Å)	12.4567(5)	9.5749(9)
c (Å)	16.2129(6)	13.6821(14)
α (°)	90	93.512(5)
β (°)	102.715(2)	106.802(4)
γ (°)	90	113.281(4)
V ($Å^3$)	1887.95(13)	963.13(17)
Z	4	2
Dx (g cm^{-3})	1.580	1.615
Abs. coefficient (mm^{-1})	1.29	1.30
Transmission factors (min, max)	0.851, 0.987	0.851, 0.987
θ Range (°)	4.4 to 27.1	4.5 to 39.4
Reflections measured	19,345	27,567
Independent reflections; R_{int}	0.068	0.0147
Reflections with I > 2σ(I)	3341	5844
Number of parameters	262	271
R(F) [I > 2σ(I)]	0.040	0.025
wR(F^2) (all data)	0.1048	0.0681
Goodness-of-fit on F^2	1.070	1.044
$\Delta\rho_{max}$, $\Delta\rho_{min}$ (e $Å^{-3}$)	1.32, −0.77	0.53, −0.29

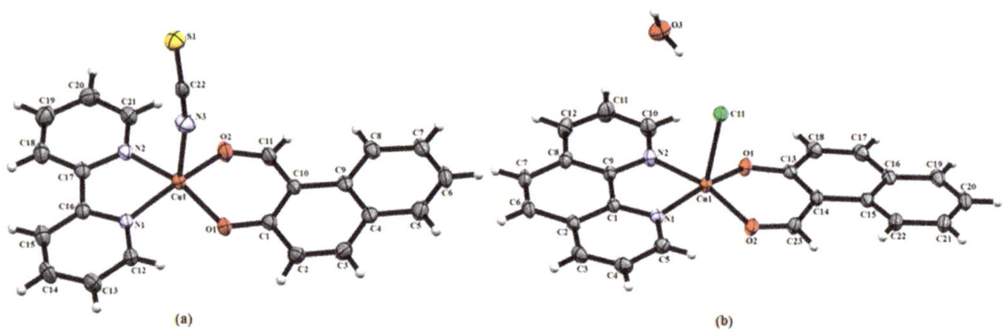

Figure 1. An ORTEP view of the asymmetric unit of the two complexes. (**a**) [Cu ($C_{11}H_7O_2$) (SCN) ($C_{10}H_8N_2$)] (C-1), (**b**) [Cu($C_{11}H_7O_2$) ($C_{12}H_8N_2$) Cl]·H_2O (C-2).

3. Results and Discussions

3.1. Crystal Structure of C-1 [Cu ($C_{11}H_7O_2$) (SCN) ($C_{10}H_8N_2$)]

The single-crystal X-ray diffraction shows that the [Cu ($C_{11}H_7O_2$) (SCN) ($C_{10}H_8N_2$)] complex crystallizes in the monoclinic space group $P2_1/n$ with Z = 4. The asymmetric unit of this structure is formed by a Cu(II) that coordinates with a thiocyanate ligand in an apical position, together with a nitrogen atom (N3). The coordination sphere of Cu(II) is completed by one 2,2′-bipyridine molecule acting as a bidentate ligand (through the two nitrogen atoms (N1, N2)), and one 2-hydroxy-1-naphthaldehyde (through the two oxygen atoms (O1, O2)). The Cu(II) atom has five coordinates, and it displays a square-pyramidal coordination geometry, as quantified by the value of the structural index [26] τ_5 = 0.019 ($\tau_5 = (\beta - \alpha)/60$, where β = N1-Cu1-O2 = 164.11° and α = N2-Cu1-O1 = 162.97°. (Perfect coordination geometries of square pyramidal and trigonal bipyramidal arrangements are associated with the τ_5 values of zero and unity, respectively). The basal plane is formed by (O1, O2, N1, N2) and N3 atoms in the axial site (Figure 2a).

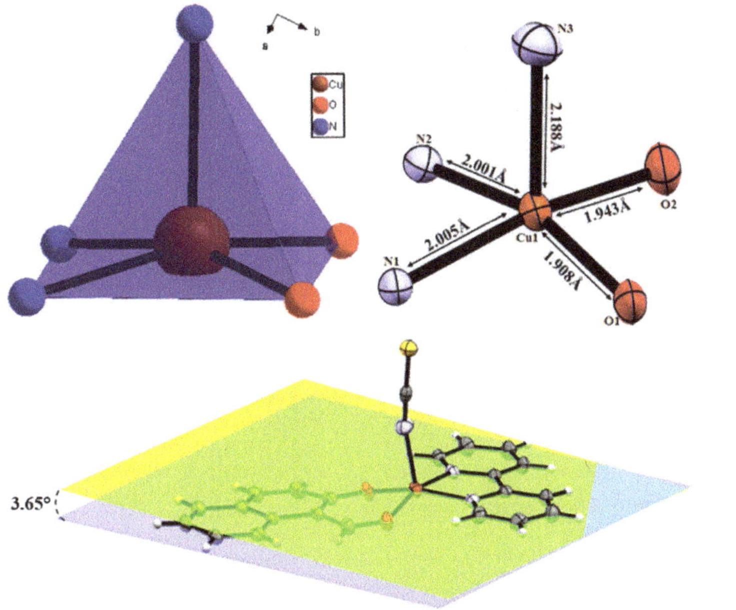

Figure 2. The copper square-pyramidal environment (**a**), and the dihedral angle between the two ligands (**b**).

The dihedral angle between the two planes that contain ligands is 3.65°. The Cu-N and Cu-O distances in the basal plane are [(Cu1-N1 = 2.005 Å, Cu1-N2 = 2.001 Å, Cu1-O1 = 1.908 Å, Cu1-O2 = 1.943 Å], and the apical Cu1-N3 bond 2.188 Å is longer than those of the basal plane (Table S1); this is in accordance with the data in the literature [27]. These parameters were also calculated in the gas phase using the B3LYP functional, and the results indicate that the calculated bond lengths in the copper coordination sphere agree with the corresponding experimental values within a maximum deviation of 2.3% (Table S1). The bond angles exhibit a maximum deviation of 10%. This discrepancy between the experimental and theoretical values may be due to environmental effects; indeed, the theoretical values were obtained from gas-phase DFT calculations, whereas the experimental values were taken from crystallographic data.

The crystalline structure of C-1 consists of four hydrogen bond networks (Table 2). Two intermolecular hydrogen bonds are observed between H7 of C11 with S1 ($\frac{1}{2} - x$, $\frac{1}{2} + y$, $\frac{1}{2} - z$), and H14 of C20 with S1($x + 1, y, z$). Moreover, two intramolecular hydrogen bonds of the C-H···O type are established between 2,2′-bipyridine and 2-hydroxy-1-naphthaldehyde, subsequently generating S(5) ring motifs [28,29]; thus, the presence of hydrogen-bond networks points to the significant role of intra and intermolecular hydrogen bonding in terms of the stabilization of complex structures. Indeed, the networks facilitate intramolecular contact and they hold the chains in a zig-zag arrangement along the a-axis (Figure 3a).

Table 2. Hydrogen-bond geometry (Å, °) in C-1 and C-2 complexes.

D-H···A	D-H	H···A	D···A	D-H···A
C-1 complex				
C11- H7···S1 [i]	0.93	2.84	3.729 (3)	161
C20- H14···S1 [ii]	0.93	2.97	3.621 (3)	128
C12- H8···O1	0.93	2.46	2.951(3)	113
C21-H15···O2	0.93	2.50	3.002 (3)	114
C-2 complex				
D-H···A	D-H	H···A	D···A	D-H···A
O3-H (3A)···Cl1	0.85	2.73	3.206 (14)	169
O3-H (3B)···Cl1 [III]	0.85	2.40	3.246 (13)	178
C3-H3···O3 [II]	0.93	2.54	3.436 (19)	162
C4-H4···O3 [V]	0.93	2.32	3.243 (2)	174
C5-H5···O2	0.93	2.54	3.003 (16)	111
C6-H6···Cl1 [II]	0.93	2.82	3.657 (15)	151
C12-H12···Cl1 [IV]	0.93	2.77	3.475 (18)	133
C19-H19···Cl1 [I]	0.93	2.82	3.597 (15)	142

Symmetry codes: (i) $-x + 1/2, y + 1/2, -z + 1/2$; (ii) $x + 1, y, z$; (I): $1 - x, -y, 1 - z$; (II): $-x, 1 - y, -z$; (III): $-x, -y, -z$; (IV): $-1 + x, y, z$; (V): $1 + x, 1 + y, z$.

(a)

(b)

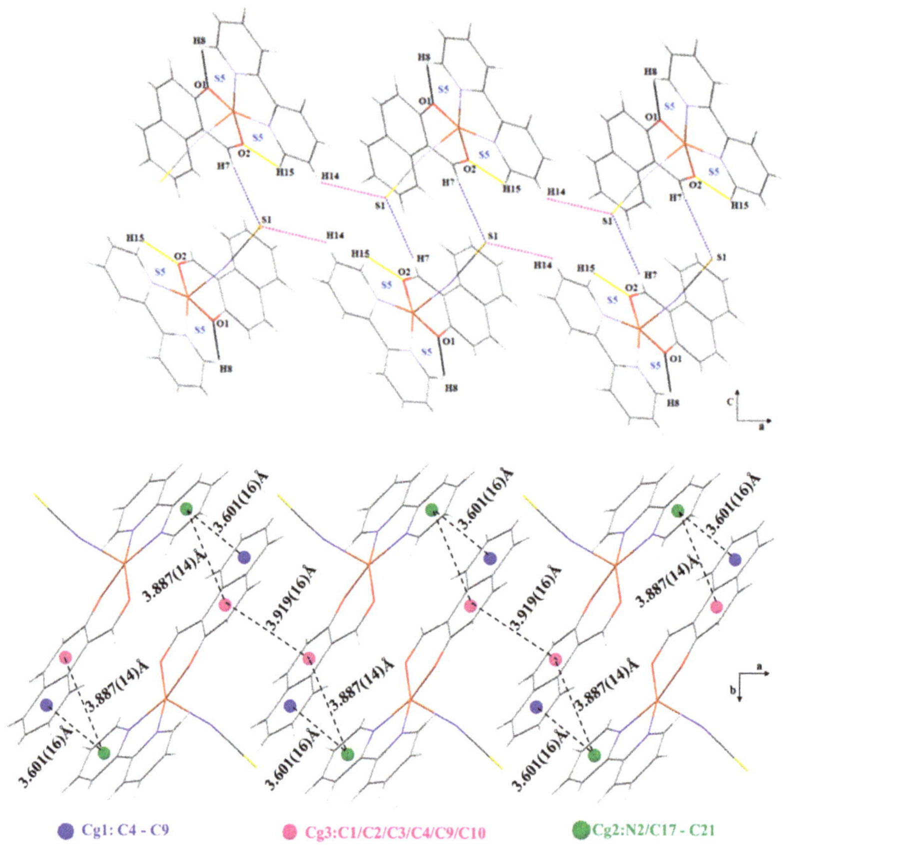

Figure 3. (**a**) Zig-zag chains of C-1 formed by hydrogen bonds (dashed lines), (**b**) the aromatic ring organization with centroid–centroid distances (Å).

In addition, the C-1 complex exhibits three π···π stacking interactions; the first is between the centroid of the 2-hydroxy-1-naphthaldehyde ring Cg1 (C4 to C9) and the centroid of the 2,2′-bipyridine ring Cg2 (N2 and C17 to C21), with a Cg1 ... Cg2 distance of 3.601 Å, the second is between the Cg3 (C1/C2/C3/C4/C9/C10) rings of the 2-hydroxy-1-naphthaldehyde, with a Cg3···Cg3 distance of 3.919 Å, and the third is between Cg2 and Cg3, with a distance of 3.887 Å (Figure 3b).

3.2. *Crystal Structure of C-2 [Cu($C_{11}H_7O_2$) ($C_{12}H_8N_2$) Cl]·H_2O*

This complex crystallizes in the triclinic system, space group *P*-1, with Z = 2. The asymmetric unit in this structure is composed of a copper (II) cation that is coordinated with the nitrogen atoms of the 1,10-phenanthroline, two oxygen atoms of the 2-hydroxy-1-naphtaldehyde in the equatorial plane, and a chlorine anion in the axial position. We also note the presence of a water molecule.

The Cu(II) atom adopts a square-pyramidal geometry, with a tau value of 0.02 ($\tau_5 = (\beta - \alpha)/60$, where β = N1-Cu1-O2 = 164.59° and α = N2-Cu1-O1 = 163.34°). In the CuN_2O_2Cl coordination sphere, the bond lengths of Cu_1-N_1 and Cu_2-N_2 are 2.012 (10)Å and 2.007(10)Å, respectively. The bond lengths of Cu_1-O1, Cu_1-O_2, and Cu_1-Cl_1 are 1.917 Å, 1.931 Å, and 2.537 Å, respectively (Table S2). The apical bond is longer than those of the basal plane, which is in accordance with the literature [30]. These experimental parameters were compared with those calculated in the gas phase; the calculated bond lengths in the copper coordination sphere agree with the corresponding experimental values within a maximum deviation of 3.4%, whereas the bond angles exhibit a maximum deviation of 7.4% (Figure S1a). The dihedral angle between the two planes containing the ligands is 7.33° (Figure S1b).

The structure of C-2 is stabilized as a result of the occurrence of an extensive network of C-H···O, C-H···Cl and O-H···Cl hydrogen bonding interactions. One intramolecular hydrogen bond is observed between H5 of C5 with O2, between 1,10-phenanthroline and 2-hydroxy-1-naphthaldehyde; this generated S(5) ring motifs and seven intermolecular hydrogen bonds, as shown in Table 2 and Figure 4a. Moreover, it is worth noting that the cohesion in the crystal is also ensured by the three weak interactions between the π···π stacking type. The first interaction is observed between the centroid of the 1,10-phenanthroline ring (Cu1/N1/C1/C9/N2) (Cg1) and the centroid of the 2-hydroxy-1-naphthaldehyde ring (C13-C18) (Cg2), with a Cg1 ... Cg2 distance of 3.915 Å. The second interaction occurs at a distance of 3.594 Å, between Cg1 and the centroid Cg3 (C15/C16/C19/C20/C21/C22). The third interaction is between Cg4 (C1/C2/C6/C7/C8/C9) and Cg3, with the ring centroids separated by 3.653 Å (Figure 4b).

3.3. *FT–IR and UV-Visible Absorption Spectra*

Figure 5a shows the FT-IR spectra of the two complexes. Overall, both complexes show similar IR bands. The strong absorptions at 1604 cm^{-1} and 1119 cm^{-1}, correspond to the C=O and C-O bonds of the aldehyde, respectively. The C-1 complex exhibits a broad band in the region of 3100–2800 cm^{-1}, which is attributed to the υ(C-H) of the aromatic ring. The bands appearing at 1562 cm^{-1} and 1014 cm^{-1} are due to the υ (C=N) and υ(C-N) of the 2-2′-bipyridine, respectively. The υ (C=N) band of the thieno ligand is observed at 2063 cm^{-1}. In the FT-IR spectrum of C-2, the bands appearing at 3100 cm^{-1} and 2855 cm^{-1} correspond to the aromatic υ(C-H). The spectrum also shows bands at 1434, 1274, and 730 cm^{-1}, which are assigned to the coordinated phenanthroline molecule [31,32].

The two complexes exhibit a d–d transition band at 624–640 cm^{-1}, and a band at 821–833 cm^{-1}, which is typical for square pyramidal geometry [30].

Figure 5b shows the UV-visible spectra in the range of 280–440 nm for C-1 and C-2. The absorption peaks of these two complexes are very similar. Nevertheless, those of C-2 are very weak. In the UV region, the two complexes show peaks near 290–300 nm due to the $\pi \rightarrow \pi^*$ aromatic ring. Moreover, we noticed two intense peaks, or a shoulder, in the (305 → 315 nm) (I)—(310 → 325 nm) (II) region, which can be assigned to the charge transfer between the ligand and the metal.

(a)

(b)

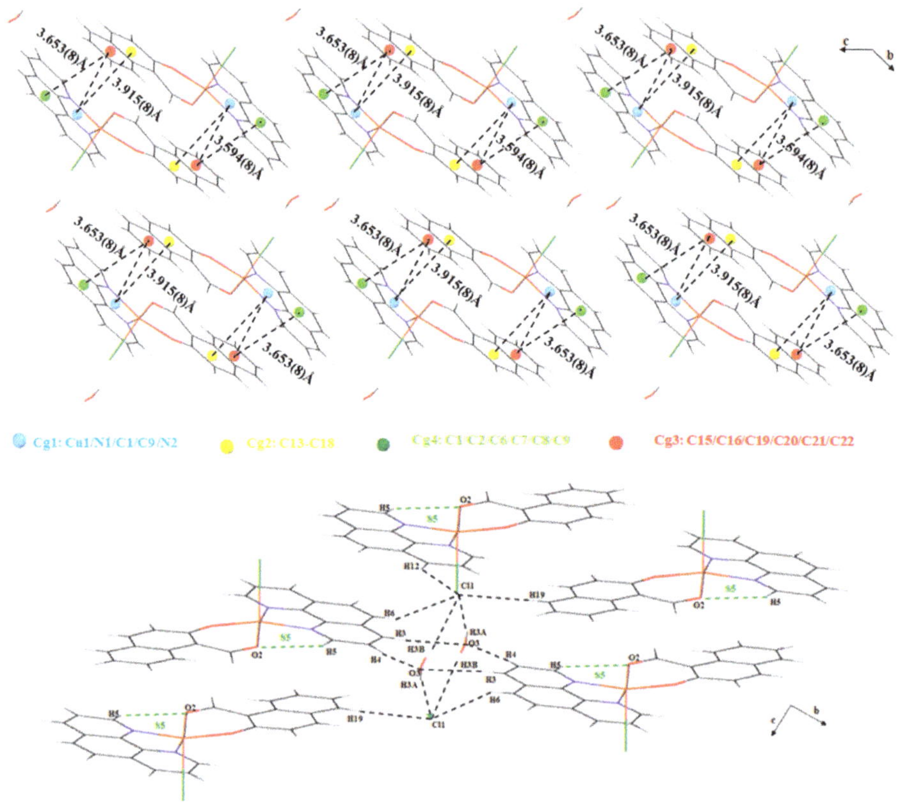

Figure 4. (**a**) Diagram packing of hydrogen bonding, parallel to (**b**,**c**). Dashed lines (green, black) indicate intra- and intermolecular H-bonds, respectively. (**b**) A view of the aromatic ring organization, with centroid–centroid distances (Å) for C-2.

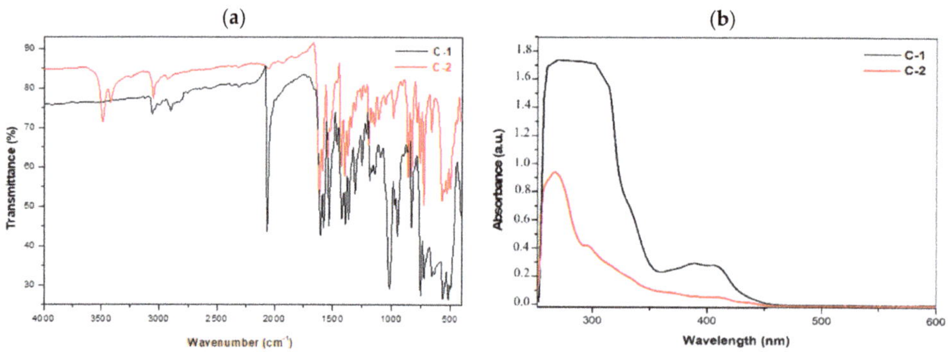

Figure 5. The FT–IR Spectrum (**a**) and the overlapping UV-visible spectrum in the DMSO (**b**) of complexes C-1 and C-2.

No d–d bands were observed in the spectra of the two complexes. These bands should be observed in the region of 500–600 nm. Their absence is due to the low concentration (10^{-5} mol·L^{-1}) of the solutions of the complexes.

3.4. Hirshfeld Surface Analysis

The intermolecular interactions and the packing modes that occur in crystalline structures can be studied quantitatively using Hirshfeld surface analysis [33]. Hirshfeld surfaces (HS) and fingerprint plots [34] were carried out using Crystal Explorer software [35].

The HS of the C-1 complex was mapped over d_{norm} in the range of −0.1832 to +1.3833 a.u., as shown in Figure 6a. Color coding is used for each specific region to identify the nature of the occurring intermolecular interactions, wherein contacts that are shorter, equal to, and longer than the sum of the van der Waals radii, correspond to the red, white and blue areas, respectively. The major interactions are caused by H···H, H···S/S···H, H···C/C···H, C···C and H···N/N···H contact. (Figure 6a).

The shape–index surface (Figure 6b) indicates the presence of blue and red triangles that are characteristic of $\pi\cdots\pi$ interactions between neighboring molecules, whereas the flat surface patches in the curvedness plots (Figure 6c) reveal planar stacking.

The 2-D fingerprint plots for the C-1 complex show the relative contributions of H···H, H···S/S···H, H···C/C···H, C···C and H···N/N···H contacts to the Hirshfeld surface; these are illustrated in Figure 6d. Contact between H···H comprise the predominant form of interactions (33.7%).

The HS of the C-2 complex is depicted in Figure 7a. The dominant interactions between (water) O-H, C-H, and Cl atoms are represented by the bright-red areas, which are marked as 1, 2, and 3. The light-red spots denoted as 4, 5, 6 and 7 are due to C-H···O and C-H···Cl interactions.

The $\pi\cdots\pi$ interactions in the C-2 complex are revealed by the characteristic blue and red triangles on the shape–index surface (Figure 7b), whereas planar stacking is identified by the curvedness plots (Figure 7c) that exhibit characteristic flat surface patches.

The relative contributions of the H···H, H···C/C···H, H···O/O···H, H···Cl/Cl···H and H···N/N···H contacts to the Hirshfeld surface in the C-2 complex are determined using the 2-D fingerprint plots, as illustrated in Figure 7d. As in the C-1complex, the fingerprint plots of the C-2 complex show that the H···H contacts provide the most significant contribution to the Hirshfeld surface (45.1%).

3.5. Inhibitory Activity of the C-1 and C-2 Complexes against the HIV-1 Protease Enzyme

To provide insight into the binding mode of C-1 and C-2, a molecular docking investigation was carried out. As shown in Figure 8 (upper panel), the tested complexes occupied a common cavity in the receptor, suggesting that C-1 and C-2 block access to the active site of the HIV-1 protease. They are located in the hydrophobic pocket of HIV-1 protease, and they are surrounded by the Val32, Pro81, and Pro79 residues of the active site, and Gly48, Gly49, Ile50, Ile84, Ile47, and Leu84 of the subsites S1/S1′, S2/S2′, and S3/S3′, wherein a strong hydrophobic interaction was observed, as illustrated in Figure 8 (lower panel). The ligand–receptor complex indicates that most of the ligand is positioned in the flap region of the enzyme, as represented by the Ile50, Gly51, Gly52, and Gly49 residues. No hydrogen bonds were found in the docked structures.

The estimated binding energies were −7.6 kcal/mol for C-1 and −7.1 kcal/mol for complex-ref, respectively. This result suggests that the tested compounds that were produced with HIV-1 protease formed stable complexes. The best ligand/receptor complex was formed by using C-2 with a binding energy equal to −8.3 kcal/mol.

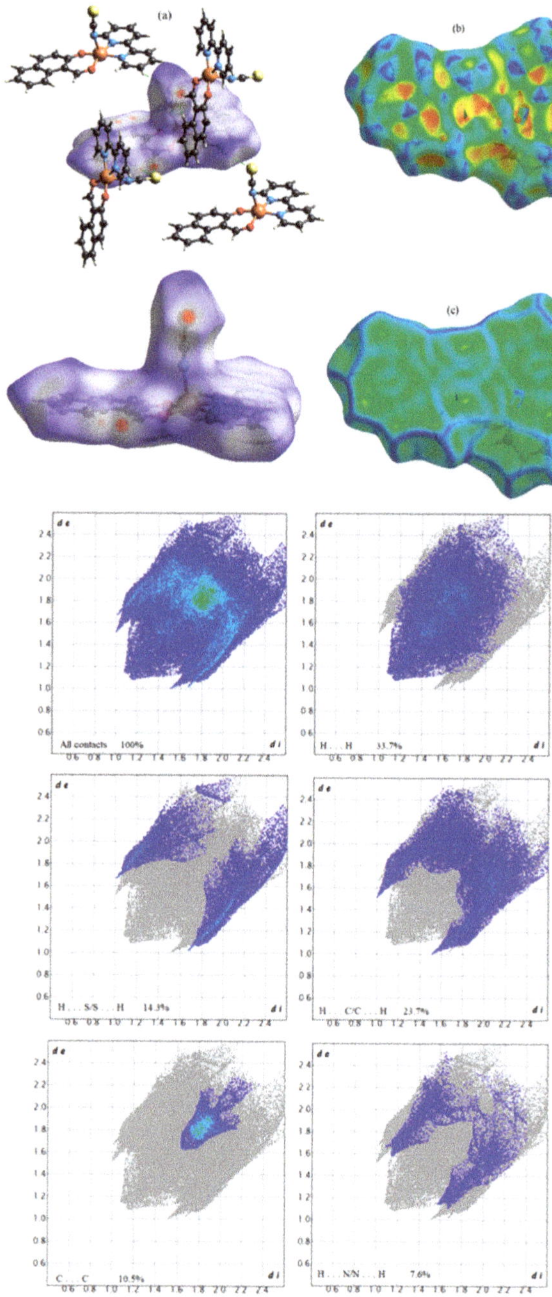

(d)

Figure 6. A view of the Hirshfeld surface for (I), which is mapped over (**a**) d_{norm} in the range of −0.1832 to +1.3833 arbitrary units. (**b**) Shape–index, (**c**) curvedness, and (**d**) 2-D fingerprint plots for the C-1 complex, which reveal the contributions of all contacts.

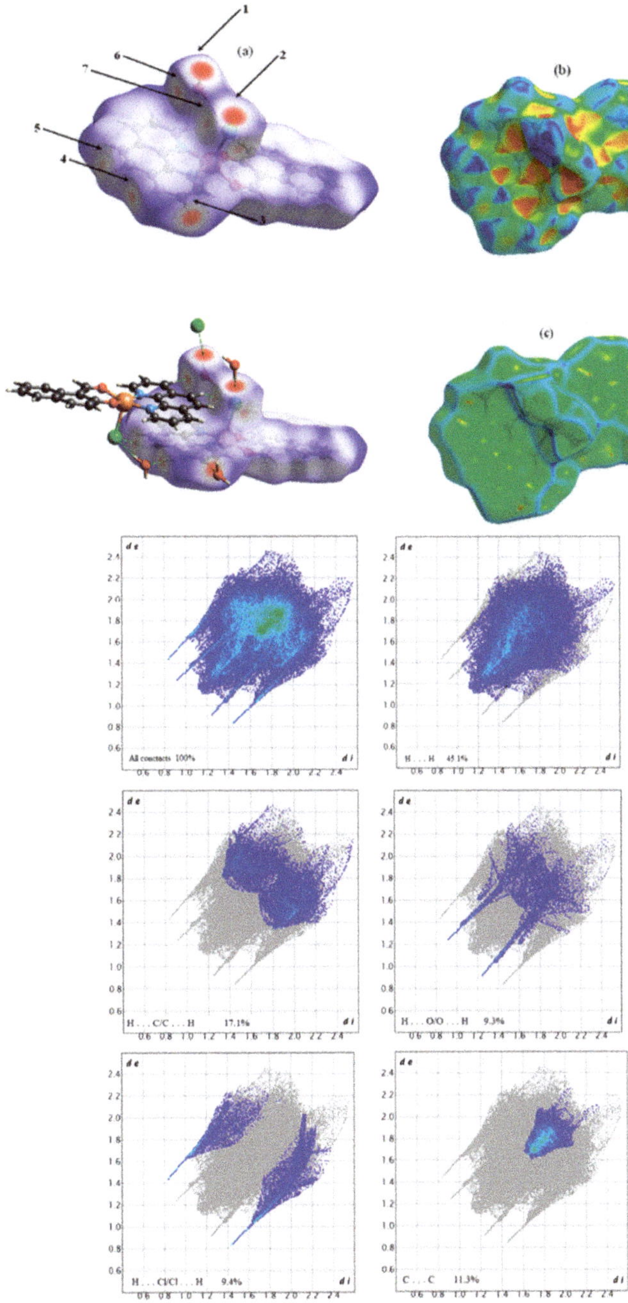

(d)

Figure 7. A view of the Hirshfeld surface for (II), which is mapped over (**a**) d_{norm} in the range of −0.4175 to +1.3253 arbitrary units. (**b**) Shape–index, (**c**) curvedness, and (**d**) 2-D fingerprint plots for the C-2 complex, which reveal the contributions of all contacts.

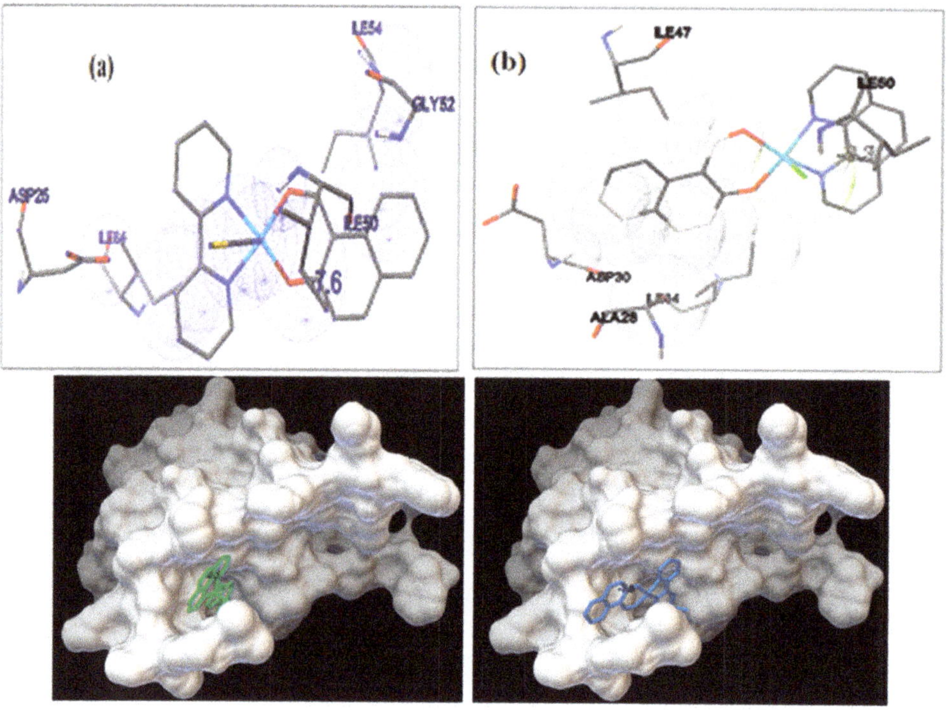

Figure 8. The upper panel shows the molecular docked model of C-1 (blue color) and C-2 (green color) in the HIV-1 protease active pocket. The enzyme is shown as a surface; the lower panel shows the best-docked conformation of C-1 (**a**) and C-2 (**b**) in the binding site of HIV-1.

4. Conclusions

Two new copper (II) complexes, based on 2,2′-bipyridine and 1,10-phenanthroline ligands, were synthesized and characterized using X-ray crystal diffraction, FT–IR, and UV-visible spectroscopies, as well as DFT calculations.

The crystalline structure of synthesized complexes consists of five coordinated Cu (II) ions in a square-pyramidal geometric structure, wherein the packing arrangement is mainly stabilized using hydrogen bonding networks and π–π stacking interactions. Hirshfeld surface analysis indicates that H···H interactions account for 33.7% and 45.1% of the total Hirshfeld surface of the C-1 and C-2 complexes, respectively.

Based on computational results, the docking analysis revealed that both C-1 and C-2 Cu (II) complexes were more potent than the inhibitor reference when binding to the HIV-1 protease; the interactions that occurred were hydrophobic, resulting in their stabilization in the enzyme cavity.

Supplementary Materials: The following supporting information can be downloaded at: https://www.mdpi.com/article/10.3390/cryst12081066/s1, Figure S1: The copper square-pyramidal environment (a), and the dihedral angle between the two ligands (b) of C-2; Table S1: Selected experimental and calculated bond lengths (Å) and angles (°) for C-1; Table S2: Selected experimental and calculated bond lengths (Å) and angles (°) for C-2.

Author Contributions: Conceptualization, M.H. and H.M.; Data curation, M.H., I.H., M.D. and S.R.; Formal analysis, M.H., I.H., M.D. and S.R.; Funding acquisition, F.A.M.A., M.A.A. and S.T.; Investigation, M.H. and I.H.; Methodology, M.D., Y.B. and S.R.; Project administration, H.M.; Resources, I.H., M.D. and H.M.; Software, M.H. and M.D.; Supervision, H.M.; Validation, F.A.M.A., M.A.A., S.T. and H.M.; Writing—original draft, M.H., I.H., M.D. and Y.B.; Writing—review & editing,

Y.B., N.S., F.A.M.A., M.A.A. and S.T. All authors have read and agreed to the published version of the manuscript.

Funding: This research received no external funding.

Institutional Review Board Statement: Not applicable.

Informed Consent Statement: Not applicable.

Data Availability Statement: The data presented in this study are available on request from the corresponding author.

Conflicts of Interest: The authors declare no conflict of interest.

References

1. Janzen, D.E.; Wang, X.; Carr, P.W.; Mann, K.R. Synthesis and characterization of three geometric isomers of ruthe-niumII(2,2′-bipyridyl) (salicylaldehyde) 2. *Inorg. Chim. Acta* **2004**, *357*, 3317–3324. [CrossRef]
2. Vigato, P.A.; Tamburini, S. The challenge of cyclic and acyclic schiff bases and related derivatives. *Coord. Chem. Rev.* **2004**, *248*, 1717–2128. [CrossRef]
3. Yamada, S. Advancement in stereochemical aspects of Schiff base metal complexes. *Coord. Chem. Rev.* **1999**, *190*, 537–555. [CrossRef]
4. Jayamani, A.; Bellam, R.; Gopu, G.; Ojwach, S.O.; Sengottuvelan, N. Copper(II) complexes of bidentate mixed ligands as artificial nucleases: Synthesis, crystal structure, characterization and evaluation of biological properties. *Polyhedron* **2018**, *156*, 138–149. [CrossRef]
5. Elmali, A.; Elerman, Y. Crystal structure of (2,2′-Dipyridyl)-(2-hydroxynaphthaldehydato) copper (II) perchlorate. *Anal. Sci.* **2002**, *18*, 855–856. [CrossRef] [PubMed]
6. Seco, J.M.; Gonzalez Garmendia, M.J.; Quiros, M. Synthesis characterization of tris (2,2′-bipyridine) and tris (1,10-phenanthroline) copper (II) hexafluorophosphate. Crystal structure of the phenanthroline complex. *J. Coord. Chem.* **2002**, *55*, 345–351. [CrossRef]
7. Adelaide, O.M.; Abidemi, O.O.; Olubunmi, A.D. Synthesis, characterization and antibacterial studies of some copper (II) complexes of 2,2′-bipyridine and 1.10-phenanthroline. *J. Chem. Pharm. Res.* **2013**, *5*, 69–73.
8. Agwaral, M.O.; Ndifon, P.T.; Ndosiril, N.B.; Paboudam, A.G.; Yufanyi, D.M.; Mohamadou, A. Synthesis, Characterisa-tion and Antimicrobial activities of Cobalt (II), Copper (II) and Zinc (II) mixed-ligand complexes containing 1,10 Phe-nanthroline and 2,2-Bipyridine. *Bull. Chem. Soc. Ethiop.* **2010**, *24*, 383–389.
9. Lobana, T.S.; Indoria, S.; Jassal, A.K.; Kaur, H.; Arora, D.S.; Jasinski, J.P. Synthesis, structures, spectroscopy and anti-microbial properties of complexes of Copper(II) with salicylaldehyde N-substituted thiosemicarbazones and 2,20-bipyridine or 1,10-phenanthroline. *Eur. J. Med. Chem.* **2014**, *76*, 145–154. [CrossRef]
10. Sierra-Aragón, S.; Walter, H. Targets for Inhibition of HIV Replication: Entry, Enzyme Action, Release and Maturation. *Intervirology* **2012**, *55*, 84–97. [CrossRef]
11. Ghosh, A.K.; Osswald, H.L.; Prato, G. Recent Progress in the Development of HIV-1 Protease Inhibitors for the Treatment of HIV/AIDS. *J. Med. Chem.* **2016**, *59*, 5172–5208. [CrossRef]
12. Wang, Y.; Lv, Z.; Chu, Y. HIV protease inhibitors: A review of molecular selectivity and toxicity. *HIV/AIDS-Res. Palliat. Care* **2015**, *7*, 95–104. [CrossRef]
13. Lee, C.; Yang, W.; Parr, R.G. Development of the Colic-Salvetti correlation-energy formula into a functional of the electron density. *Phys. Rev. B* **1988**, *37*, 785. [CrossRef]
14. Becke, A.D. Density-functional thermochemistry. I. The effect of the exchange-only gradient correction. *J. Chem. Phys.* **1992**, *96*, 2155–2160. [CrossRef]
15. Becke, A.D. Density-functional thermochemistry. III. The role of exact exchange. *J. Chem. Phys.* **1993**, *98*, 5648–5652. [CrossRef]
16. Becke, A.D. A new mixing of Hartree-Fock and local density-functional theories. *J. Chem. Phys.* **1993**, *98*, 1372–1377. [CrossRef]
17. Trott, O.; Olson, A.J. AutoDock Vina: Improving the speed and accuracy of docking with a new scoring function, efficient optimization and multithreading. *J. Comp. Chem.* **2010**, *31*, 455–461. [CrossRef]
18. Lebon, F.; Ledecq, M.; Benatallah, Z.; Sicsic, S.; Lapouyade, R.; Kahn, O.; Garçon, A.; Reboud-Ravaux, M.; Durant, F. Synthesis and structural analysis of copper(II) pyridine amide complexes as HIV-1 protease inhibitors. *J. Chem. Soc. Perkin Trans.* **1999**, *2*, 795–800. [CrossRef]
19. Mager, P.P. The active site of HIV-1 protease. *Med. Res. Rev.* **2001**, *21*, 348–353. [CrossRef]
20. Mudgal, M.M.; Birudukota, N.; Doke, M.A. Applications of Click Chemistry in the Development of HIV Protease Inhib-itors. *Inter. J. Med. Chem.* **2018**, *2018*, 2946730.
21. Bruker. *APEX2 and SAINT*; Bruker AXS Inc.: Madison, WI, USA, 2011.
22. Sheldrick, G.M. A short history of SHELX. *Acta Crystallogr. Sect. A* **2008**, *A64*, 112–122. [CrossRef]
23. Frisch, M.J.; Trucks, G.W.; Schlegel, H.B.; Scuseria, G.E.; Robb, M.A.; Cheeseman, J.R.; Scalmani, G.; Barone, V.; Men-nucci, B.; Petersson, G.A.; et al. *Gaussian 09, Revision B.01*; Gaussian, Inc.: Wallingford, CT, USA, 2010.

24. O'Boyle, N.M.; Banck, M.; James, C.A.; Morley, C.; Vandermeersch, T.; Hutchison, G.R. Open babel: An open chemical toolbox. *J. Cheminform.* **2011**, *3*, 33. [CrossRef]
25. The Open Babel Package, Version 2.3.1. Available online: http://openbabel.org (accessed on 30 March 2022).
26. Addison, A.W.; Rao, T.N.; Reedijk, J.; van Rijn, J.; Verschoor, G.C. Synthesis, structure, and spectroscopic properties of copper(II) compounds containing nitrogen–sulphur donor ligands; the crystal and molecular structure of aq-ua[1,7-bis(N-methylbenzimidazol-2′-yl)-2,6-dithiaheptane]copper(II) pe. *J. Chem. Soc. Dalt. Trans.* **1984**, *7*, 1349–1356. [CrossRef]
27. Asgharpour, Z.; Farzaneh, F.; Abbasi, A. Synthesis, characterization and immobilization of a new cobalt(ii) complex on modified magnetic nanoparticles as catalyst for epoxidation of alkenes and oxidation of activated alkanes. *RSC Adv.* **2016**, *6*, 95729–95739. [CrossRef]
28. Bernstein, J.; Davis, R.E.; Shimoni, L.; Chang, N.-L. Patterns in Hydrogen Bonding: Functionality and Graph Set Analysis in Crystals. *Angew. Chem. Int. Ed. Engl.* **1995**, *34*, 1555–1573. [CrossRef]
29. Etter, M.C.; Macdonald, J.C.; Bernstein, J. Graph-set analysis of hydrogen-bond patterns in organic crystals. *Acta Crystallogr. Sect. B Struct. Sci.* **1990**, *46*, 256–262. [CrossRef] [PubMed]
30. Hangan, A.; Bodoki, A.; Oprean, L.; Alzuet, G.; Liu-Gonzalez, M.; Borras, J. Synthesis, crystallographic, spectroscopic characterization and magnetic properties of dimer and monomer ternary copper(II) complexes with sulfonamide deriv-atives and 1,10-phenantroline. Nuclease activity by the oxidative mechanism. *Polyhedron* **2010**, *29*, 1305–1313. [CrossRef]
31. Lara, J.G.; Sanchez, A.A. The infra-red spectra of 2-hydroxy-l-naphthaldehyde complexes. *Spectrochim. Acta Part A* **1967**, *23*, 1299–1306. [CrossRef]
32. Mohapatra, R.K.; Saikishore, V.P.; Azam, M.; Biswal, S.K. Synthesis and physicochemical studies of a series of mixed-ligand transition metal complexes and their molecular docking investigations against Coronavirus main protease. *Open Chem.* **2020**, *18*, 1495–1506. [CrossRef]
33. Spackman, M.A.; Jayatilaka, D. Hirshfeld surface analysis. *CrystEngComm* **2009**, *11*, 19–32. [CrossRef]
34. Spackman, M.A.; McKinnon, J.J. Fingerprinting intermolecular interactions in molecular crystals. *CrystEngComm* **2002**, *4*, 378–392. [CrossRef]
35. Turner, M.J.; McKinnon, J.J.; Wolff, S.K.; Grimwood, D.J.; Spackman, P.R.; Jayatilaka, D.; Spackman, M.A. *Crystal Ex-Plorer17.5*; The University of Western Australia: Perth, Australia, 2017.

crystals

MDPI

Article

Bis-Citrullinato Copper(II) Complex: Synthesis, Crystal Structure, and Non-Covalent Interactions

Diego Ramírez-Contreras [1], Amalia García-García [1,2], Brenda L. Sánchez-Gaytán [1], Laura E. Serrano-de la Rosa [3], Francisco J. Melendez [4], Duane Choquesillo-Lazarte [5], Antonio Rodríguez-Diéguez [2], María Eugenia Castro [1,*] and Enrique González-Vergara [1,*]

1 Centro de Química del Instituto de Ciencias, Benemérita Universidad Autónoma de Puebla, 18 sur y Av. San Claudio, Col. San Manuel, Puebla 72570, Mexico
2 Departamento de Química Inorgánica, Facultad de Ciencias, Universidad de Granada, Av. Fuentenueva S/N, 18071 Granada, Spain
3 Laboratorio Central del Instituto de Física "Luis Rivera Terrazas" (IFUAP), Benemérita Universidad Autónoma de Puebla, 18 sur y Av. San Claudio, Col. San Manuel, Puebla 72570, Mexico
4 Lab. de Química Teórica, Centro de Investigación, Depto. de Fisicoquímica, Facultad de Ciencias Químicas, Benemérita Universidad Autónoma de Puebla, 22 sur y Av. San Claudio, Col. San Manuel, Puebla 72570, Mexico
5 Laboratorio de Estudios Cristalográficos, IACT, CSIC-UGR, Av. Las Palmeras n°4, 18100 Granada, Spain
* Correspondence: mareug.castro@correo.buap.mx (M.E.C.); enrique.gonzalez@correo.buap.mx (E.G.-V.); Tel.: +52-222-363-0623 (E.G.-V.)

Abstract: Citrulline ($C_6H_{13}N_3O_3$) is a non-protein amino acid found in watermelon. In physiological conditions, it is almost entirely present as a zwitterion, so its carboxylic and amine groups can act as Lewis donors, chelating metallic cations. In addition, Citrulline possesses a terminal ureide group of the aliphatic chain, which appears to be non-innocent. Although Citrulline is similar to other classical amino acids, only one coordination complex has been reported in the Cambridge Crystallographic Database. As part of our search for Casiopeina® analogs, we synthesized and characterized the copper bis-citrullinato complex, $[Cu(Citr)_2]_n$. The compound was described using UV-Vis, Infrared, and Raman spectroscopy, together with single-crystal X-ray diffraction. Computational tools were also used. The optimized structure, MEP map, IR and Raman spectra, and ^1H and ^{13}C chemical shifts were obtained with functional mPW1PW91 using 6-31G(d) basis set for N, O, C, and H atoms, and LANL2DZ basis set and ECP=LANL2DZ for the Cu atom. TD-mPW1PW91 calculations generated the UV-Vis spectrum. Finally, AIM and Hirshfeld surface analysis were used to examine non-covalent interactions. Previous investigations suggest Casiopeina®-like complexes can interact with DNA/RNA, creating potential anticancer chemicals. The $[Cu(Citr)_2]_n$ complex's polymeric nature and insolubility make it difficult for such purposes. However, the facile synthesis of D-Citrulline could be a novel way to find new applications for this interesting amino acid.

Keywords: Citrulline; copper; DFT calculations; Hirshfeld surface analysis; non-covalent interactions

Citation: Ramírez-Contreras, D.; García-García, A.; Sánchez-Gaytán, B.L.; Serrano-de la Rosa, L.E.; Melendez, F.J.; Choquesillo-Lazarte, D.; Rodríguez-Diéguez, A.; Castro, M.E.; González-Vergara, E. Bis-Citrullinato Copper(II) Complex: Synthesis, Crystal Structure, and Non-Covalent Interactions. *Crystals* **2022**, *12*, 1386. https://doi.org/10.3390/cryst12101386

Received: 6 September 2022
Accepted: 26 September 2022
Published: 29 September 2022

Publisher's Note: MDPI stays neutral with regard to jurisdictional claims in published maps and institutional affiliations.

1. Introduction

The discovery of Citrulline that we know today has been traditionally attributed to Mitsunori Wada in 1930, who separated and determined its chemical formula from the watermelon juice *Citrullus vulgaris* [1]. However, it was first isolated in 1914 by Yotaro Koga and Ryo Ohtake, although they did not describe its structure, nor did they name it [2]. Today, we know that L-Citrulline is also present in other foods, such as cucumbers, pumpkins, melons, squashes, and gourds [3].

L-Citrulline ($C_6H_{13}N_3O_3$) is not involved in protein formation since it is a non-essential amino acid. The body produces it endogenously, predominantly in the intestine [4,5], by biosynthesis from principally L-glutamine, L-arginine, and proline. During the urea cycle, glutamine is transformed into ornithine, which is turned into L-Citrulline in the mitochondria. The second pathway uses arginine as a precursor to making Citrulline and nitric oxide; lastly, proline is metabolized into ornithine, which is converted into Citrulline [6]. Although this biomolecule has received little attention because it is non-protein, its complex metabolism has made it more relevant in recent decades [7]. It may be utilized as a biomarker or treatment for many diseases. If Citrulline metabolism is insufficient, the urea cycle and NO regulation do not work properly [8,9]. Citrulline supplementation may be used to treat hypertension since the NO cycle affects blood pressure, arginine shortage, liver damage, or cerebral malaria [10–13]. In addition, the consumption of Citrulline could also be beneficial in the treatment and prevention of other important disorders such as erectile dysfunction, sickle cell anemia, cancer chemotherapy, and cardiovascular disease; to enhance protein content in malnourished and elderly patients; or to treat Alzheimer's disease and multi-infarct dementia, among others [14,15]. Another research points out the role of Citrulline as an antibody to citrullinated peptides. Protein citrullination could produce, as the final step, autoantibodies which can be used as biomarkers in various disorders such as rheumatoid arthritis, intestinal pathology, and short bowel syndrome [16]. It is clear that Citrulline plays a critical role in human health. Still, much more investigation is needed before Citrulline is used in therapy with confidence against all mentioned disorders [17].

Coordination complexes based on biological ligands and d block metal ions, such as Cu(II), Co(II), and Zn(II), have several applications in environmental, biological, pharmaceutical, and chemical and electrochemical industries [18]. In 1938, Fox S. reported the preparation of Citrulline from hydrolysis of arginine. In this procedure, copper oxide was introduced to form the copper-Citrulline complex, which serves as an intermediate for the isolation of Citrulline after the copper was removed; thus, the copper complex was mentioned for the first time [19]. In the same year, Kurtz described the synthesis of DL-Citrulline, proposed the compound's structure, and mentioned its insolubility [20]. Trikha et al., (1968) and Clarke and Martell (1970) reported the formation constants for 1:1 and 1:2 molar ratio Cu(II)-Citrulline complexes in aqueous solutions [21,22]. Later, Yamauchi et al., crystallized [Cu(L-His)(L-Citr)] complex in 1979 [23], and Ganadu et al., examined Pd(II)-Citrulline interaction using ^{1}H and ^{13}C NMR spectroscopy and potentiometry in 1991 [24]. The $[Pd(Citr)_2]$ complex was subsequently synthesized and characterized by Mascaliovas et al., (2015). Its crystal structure was solved by powder X-ray diffraction data and simulated annealing in real space [25]. The structure showed Pd(II) ions in a roughly square planar environment and deprotonated Citrulline. Finally, equilibrium studies of Hg(II)—L-Citrulline(A)—Uracil (B) with d-block metal ions M = Cu(II), Zn(II), Co(II), and Ni(II) were reported in 2017 [26]. The same group recently explored the equilibrium constants of Citrulline/tryptophan and thymine with Cu(II), Zn(II), Co(II), and Ni(II) [27].

As part of our attempts to generate possible metallodrugs based on Casiopeina® analogs, we have previously reported copper-mixed complexes incorporating lysine, ornithine, and glutamine, and docking studies reveal their interaction with DNA/RNA [28–30]. Here we report the synthesis and experimental-theoretical characterization of the $[Cu(Citr)_2]$ complex and its crystal structure for the first time. In addition to the complete structural characterization by spectroscopic techniques, DFT studies and Hirschfeld analysis have been carried out to describe important electronic and structural characteristics such as relevant non-covalent interactions.

2. Materials and Methods

2.1. Synthesis of $[Cu(Citr)_2]_n$

L-Citrulline ($C_6H_{13}N_3O_3$; M.W. = 175.188 g/mol) was purchased from Merck Mexico, while copper chloride (II) dihydrate ($CuCl_2{\cdot}2H_2O$; M.W. = 170.480 g/mol), copper nitrate (II) trihydrate $Cu(NO_3)_2{\cdot}3H_2O$; M.W. = 241.600 g/mol), and copper perchlorate (II) hexahydrate $Cu(ClO_4)_2{\cdot}6H_2O$; M.W. = 370.540 g/mol) were purchased from Sigma-Aldrich. Methanol (CH_3OH) and deuterated water (D_2O) (99.9%) were purchased from Fisher Scientific (J.K. Baker). Purification or pretreatment of reagents was not necessary.

The bis-citrullinato copper (II) complex was synthesized using a similar methodology described by Su et al. [31]. In a flask, 2 mmol (0.340 g) of Citrulline was mixed with 1 mmol of $CuCl_2{\cdot}2H_2O$ (0.170 g) in 10 mL of methanol. The mixture was stirred for two hours (due to the low solubility of Citrulline in methanol) under slight heating (the temperature shall not exceed 50 °C). Then, 2 mmol of sodium hydroxide (0.080 g) dissolved in methanol (10 mL) was added, and the mix continued to be stirred in the same conditions for 2 h until a deep blue powder was formed. After this process, the suspension was filtered using a Whatman 40 filter paper, and the solid was washed with methanol and air-dried at room temperature. Yield of $[Cu(Citr)_2]$: (0.358 g) 68.89%. The same procedure was repeated separately using $Cu(NO_3)_2{\cdot}3H_2O$ and $Cu(ClO_4)_2{\cdot}6H_2O$; a blue solid with the same IR spectrum was obtained regardless of the copper salt used.

The same compound was obtained by electrosynthesis to give good quality crystals [32,33]. A piece of standard copper wire was used as a sacrificial anode, a pencil carbon electrode as a cathode, and an AC eliminator was used as a power supply. In this case, 5.0 mmol of Citrulline was dissolved in 50 mL of water, and this solution was used as an electrolyte. The conditions were fixed to 2 A and 9 V. The electrosynthesis was carried out for 6–10 h or until the formation of crystals in the copper anode. The resulting crystals were separated from the anode, washed with methanol, and air-dried at room temperature. The new compound was fully characterized by single-crystal X-ray diffraction and spectroscopic techniques.

2.2. Characterization Methods

The UV-Vis experiment was performed in a Cary 5000 UV-Vis-NIR Spectrophotometer-Agilent. A powder holder was used to compensate for the very small amount, and a Teflon tablet with a cavity was used to deposit the material. The scanning was performed from 200 to 800 nm, at a scanning speed of 600 nm per minute, with 1 nm steps in the reflectance mode. The accessory used is the DRA-2500 with an internal PbS detector. The Infrared spectrum was recorded from 4000 to 650 cm^{-1} using a Nicolet 6700 FTIR spectrophotometer coupled with the iTR accessory with a diamond tip. The Raman spectrum was obtained at room temperature in a backscattering configuration using the 633 nm line of a He-Ne laser as an excitation source by using a LabRAM HR-Olympus Micro Raman. The ^1H- and ^{13}C-NMR spectra were recorded by adding 10 mg of the blue powder in 500 uL of D_2O and 50 uL of HCl 1M to enhance the compound's solubility using a Bruker AVANCE III 500 MHz spectrometer at 25 °C. Quality crystals of the copper-Citrulline complex were subject to X-ray diffraction studies. Data were collected with a Bruker D8 Venture diffractometer with a Photon detector equipped with graphite monochromated CuK radiation (λ = 1.54178 Å). The data were processed with the APEX3 suite, and SADABS-2016/2 was used for absorption correction [34]. The structure was solved by direct methods using the ShelXT program and refined by full-matrix least squares on F^2 with SHELXL-2019 [35]. The positional and anisotropic atomic displacement parameters were refined for all non-hydrogen atoms. Hydrogen atoms were located in different Fourier maps and included as fixed contributions riding their parent atoms with isotropic thermal factors chosen as 1.2 times their carrier atoms. The OLEX2 software [36] was used as a graphical interface. Details of the structure determinations and refinement are summarized in Table 1 and Tables S1–S3 in the Supplementary Materials Section. Crystallographic data (excluding structure factors) for the structure reported in this paper have been deposited with the Cambridge Crystallographic Data Center as supplementary publication No. 2172421. Copies of the data can be obtained free of charge on application to the Director, CCDC, 12 Union Road, Cambridge, CB2 1EZ, UK (Fax: +44-1223-335033; e-mail: deposit@ccdc.cam.ac.uk or http://www.ccdc.cam.ac.uk (accessed on 12 July 2022).

Table 1. Experimental details of the X-ray collection of data.

	$[Cu(Citr)_2]_n$
Chemical formula	$C_{12}H_{24}CuN_6O_6$
Formula weight (g mol^{-1})	411.91
CCDC	2,172,421
Crystal system	Monoclinic
Space group	$P2_1$
Temperature (K)	300.7
a, b, c (Å)	5.1270 (2), 9.2086 (4), 17.1099 (7)
β (°)	93.972 (2)
V ($Å^3$)	805.86 (6)
Z	2
Radiation type	Cu Kα (1.54178 λ)
ρ_{calcd} (g cm^{-3})	1.698
μ (mm^{-1})	2.339
Crystal size (mm)	0.12 × 0.1 × 0.1
GoF on F^2	1.147
R_1 [I > 2σ(I)]	0.0497
R_1 [all data]	0.0512
wR_2 [I > 2σ(I)]	0.1469
wR_2 [all data]	0.1520

2.3. *Computational Details*

Density functional theory (DFT) [37] and time-dependent DFT (TD-DFT) [38] calculations were used to obtain the optimized structure and the electronic properties of the compound $[Cu(Citr)_2]_n$ in both its electronic ground and excited states. The functional mPW1PW91 [39] with the split-valence 6-31G(d) basis set [40] was used for the C, H, O, and N atoms. The LANL2DZ basis set [41] and an effective core potential (ECP) were used for the copper atom (mPW1PW91/6-31G(d)-LANL2DZ and ECP=LANL2DZ). Calculations were performed without symmetry imposition. The vibrational frequencies were determined to establish the stationary points on the potential energy surface, using the second derivatives of the energy analytically. The solvent effect was implicitly included with the universal solvation model based on solute electron density (SMD) [42], using water as solvent. This theory level was used to calculate geometry optimization, the molecular electrostatic potential (MEP) map, IR and Raman, and ^1H and ^{13}C NMR spectra calculations. For IR and Raman, calculated spectra were obtained with a scaling factor of 0.961. The ^1H and ^{13}C NMR spectra were calculated using the Gauge-Independent Atomic Orbital (GIAO) method [43]. For the UV-Vis spectrum, TD-mPW1PW91 time-dependent density functional and the same basis set was applied to the vertical absorption calculation. It is a reliable tool used to investigate excited states of medium and large molecular systems with reasonable computational cost [37]. The calculations were carried out with the Gaussian16 program [44], and the results were visualized with the GaussianView 6.0.16 program [45]. The main non-covalent interactions were characterized by mPW1PW9/6-31G(d)—LANL2DZ wavefunction using the atoms in molecules (AIM) approach with AIMAll software [46]. Finally, the Hirshfeld surface analysis was performed from the crystallographic data using CrystalExplorer 17.5 software [47].

3. Results

3.1. *Structural Description*

$[Cu(Citr)_2]_n$ crystallizes in the monoclinic $P2_1$ space group and consists of a bi-dimensional coordination polymer. The asymmetric unit comprises one Cu(II) ion and two coordinated Citrulline ligand molecules, being the X-ray formula $[Cu(Citr)_2]_n$ (Figure 1a). Each metal center exhibits an octahedral geometry with an elongated deformation and coordinates to four Citrulline molecules. The CuO_4N_2 sphere is formed by two O1 atoms from the carboxylate groups of ligands A and B, two N1 atoms belonging to amino groups of the same ligands, and two O3 atoms of amide groups of two different ligands (Figure 1b). The coordination sphere's deformation results from the high Jahn–Teller effect around de d-orbitals of Cu(II) in octahedral geometry, which tend to split degenerate energy levels [48]; the basal positions are occupied by two Cu-N and two Cu-O bonds with distances in the range of 1.952(8)–1.978(8) Å and 1.951(8)–1.961(7) Å, respectively, similar to other reported structures [30,49–51]. According to the literature, the Cu-O bond distances at the apical positions range from 2.4 to 2.8 Å, corresponding to electrostatic interactions [52]. In our case, Cu-O3A ^{i} (i = 1 − x, −1/2 + y, −z) and Cu-O3B ^{ii} (ii = −x, 1/2 + y, 1 − z) 2.724(9) and 2.622(9) Å, respectively.

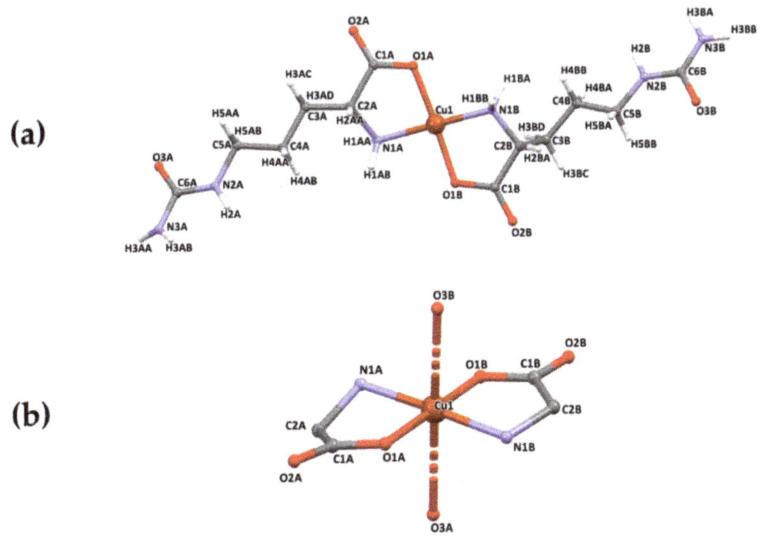

Figure 1. (**a**) Asymmetric unit of $[Cu(Citr)_2]_n$. (**b**) Detail of the coordination sphere. Color code: copper (brown), oxygen (red), nitrogen (blue), carbon (grey), and hydrogen atom (white).

The Citrulline molecule acts as a ditopic-bringing linker since carboxylate, amino, and amide groups interact with two metallic ions. The O1 atom of carboxylate and N1 from primary amine coordinate to one Cu(II) ion by a bidentate mode, as proposed for similar structures with Co(II), Cu(II), Pd(II), and Zn(II) already published [25,53,54]. However, for the first time, in this structure, the oxygen O3 of the amide group is also coordinated to one metal ion and forms the octahedron's axial bond. Moreover, it is essential to note that the reagent used in the synthesis was the L isomer of Citrulline. In contrast, the complex obtained gives rise to the D-Citrulline enantiomers due to the coordination with Cu(II) (Figures S1–S4). This arrangement of the ligands gives rise to layers parallel to the [202] plane (Figure 2).

In addition, the supramolecular structure is stabilized by intra- and intermolecular hydrogen bonds of type N-H··· O (Figure 3 and Table 2). Within the layers, two main hydrogen bonds that were analyzed by computational tools are formed from the primary and secondary amine to the oxygen of the carboxylate group of a different ligand, giving rise to a $R^1_2(6)$ and $R^2_2(8)$ graph-set motives, according to Etter's nomenclature [55] (Figure 3).

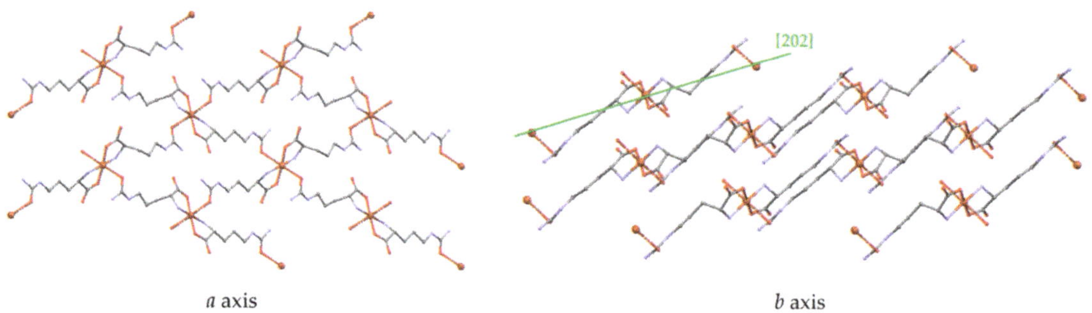

Figure 2. Crystal structure of the compound $[Cu(Citr)_2]_n$ along *a* and *b* crystallographic axis. The plane [202] is shown in green color. Hydrogen atoms have been omitted for clarity.

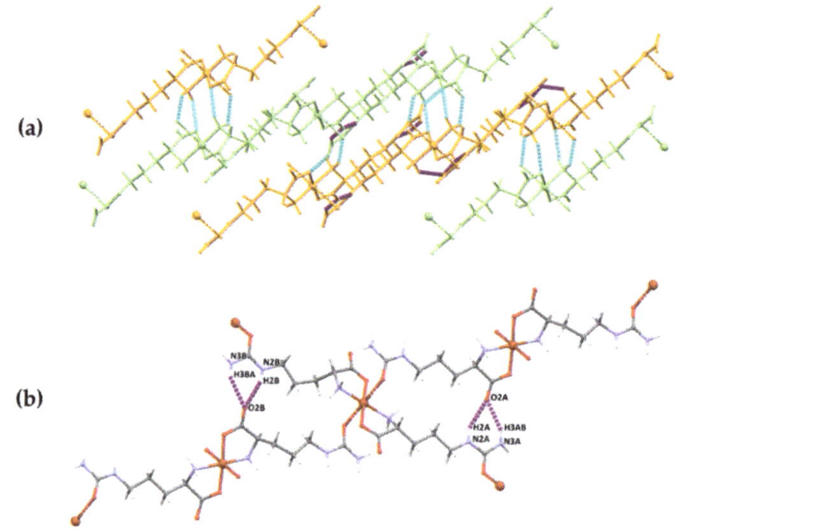

Figure 3. (**a**) Perspective view of all the hydrogen bonds in the structure. The different layers are indicated in orange and light green color, while intra- and intermolecular hydrogen bonds are in light blue and purple dotted lines, respectively. (**b**) Detail of the intramolecular H-bonds, which gives rise to a $R_2^1(6)$ and $R_2^2(8)$ ring motives.

Table 2. Hydrogen bonds for compound $[Cu(Citr)_2]_n$.

D-H···A	D-H (Å)	H···A (Å)	D-H···A (Å)	Angle (°)
N2A-H2A···O2A ^{i}	0.86	2.17	2.955(14)	150.9
N3A-H3AB···O2A ^{i}	0.87	2.14	2.935(16)	151.6
N2B-H2B···O2B ^{ii}	0.86	2.02	2.840(14)	159.1
N3B-H3BA···O2B ^{ii}	0.86	2.26	2.997	144.1
N1A ^{iii}-H1AA···O1A	0.89	2.34	3.153	151.7
N1A ^{iv}-H1AB···O3A	0.89	2.22	2.996	145.6
N3B ^{v}-H3BB···O1A	0.86	2.07	2.928(12)	171.4
N3A-H3AA···O1B ^{iv}	0.88	2.26	3.047	148.7

Symmetry operations: $^{i} = 1 - x, \frac{1}{2} + y, -z$; $^{ii} = -x, \frac{1}{2} + y, 1 - z$; $^{iii} = -1 + x, y, z$; $^{iv} = 2 - x, \frac{1}{2} + y, -z$; $^{v} = -1 - x, \frac{1}{2} + y, 1 - z$.

The selected parameters for the complex [Cu(Citr)$_2$] are shown in Table 3 for comparison between X-ray data and the calculated structure obtained at the level of theory mPW1PW91/6-31G(d)-LANL2DZ and ECP=LANL2DZ. The numbering convention is in agreement with Figure 1a. Similar values are observed in the bond lengths and angles. Some dihedral angles, such as C1B–C2B–N1B–Cu1 and C2B–C3B–C4B–C5B, present the major differences between the X-ray and the calculated data with 9 and 16°. Both in the X-ray data and the calculated data, the non-symmetrical structure of each Citrulline can be observed in the complex. In addition, according to the parameters around the equatorial positions of the atoms coordinated to the Cu ion, a distorted octahedral geometry is observed. The agreement between the optimized structure and that obtained with X-ray diffraction can be clearly observed in Figure 4, where both complexes were fitted using VMD molecular graphics viewer [56]. Only, small differences can be observed with an RMSD calculated value of 0.24 between them.

Table 3. Selected X-ray and optimized parameters for the compound [Cu(Citr)$_2$] obtained at the level of theory mPW1PW91/6-31G(d)-LANL2DZ and ECP=LANL2DZ. Bond lengths in (Å) and angles in (°).

Parameter	X-ray Data	Calculated
Cu1–O1A/Cu1–O1B	1.974/1.940	1.978/1.965
Cu1–N1A/Cu1–N1B	1.993/1.958	2.018/2.027
O1A–C1A/O1B–C1B	1.271/1.286	1.290/1.290
C1A–O2A/C1B–O2B	1.231/1.217	1.237/1.237
C1A–C2A/C1B–C2B	1.529/1.531	1.535/1.531
N1A–C2A/N1B–C2B	1.502/1.461	1.474/1.474
C2A–C3A/C2B–C3B	1.528/1.541	1.523/1.533
C3A–C4A/C3B–C4B	1.501/1.516	1.525/1.527
C4A–C5A/C4B–C5B	1.525/1.477	1.520/1.522
C5A–N2A/C5B–N2B	1.476/1.429	1.450/1.447
O1A–Cu1–N1B/O1B–Cu1–N1A	96.91/94.43	98.00/96.07
C1A–O1A–Cu1/C1B–O1B–Cu1	113.46/115.04	114.95/115.43
C2A–N1A–Cu1/C2B–N1B–Cu1	104.76/110.23	106.74/108.31
O1A–C1A–C2A/O1B–C1B–C2B	116.38/116.54	115.56/116.86
O2A–C1A–O1A/O2B–C1B–O1B	123.16/123.33	123.90/124.08
C2A–C3A–C4A/C2B–C3B–C4B	112.72/116.79	112.26/115.66
C3A–C4A–C5A/C3B–C4B–C5B	110.52/116.08	112.72/114.08
C4A–C5A–N2A/C4B–C5B–N2B	111.02/112.14	109.92/109.49
C1A–O1A–Cu1–N1A/C1B–O1B–Cu1–N1B	12.41/−10.77	13.22/−10.06
O2A–C1A–O1A–Cu1/O2B–C1B–O1B–Cu1	−173.90/−177.11	−177.30/179.08
C1A–C2A–N1A–Cu1/C1B–C2B–N1B–Cu1	40.41/−17.16	36.87/−26.97
C1A–C2A–C3A–H3AC/C1B–C2B–C3B–H3BC	−53.01/−51.68	−48.62/−50.53
C2A–C3A–C4A–C5A/C2B–C3B–C4B–C5B	176.50/81.02	177.12/65.95
C3A–C4A–C5A–N2A/C3B–C4B–C5B–N2B	176.51/174.14	−178.41/178.61

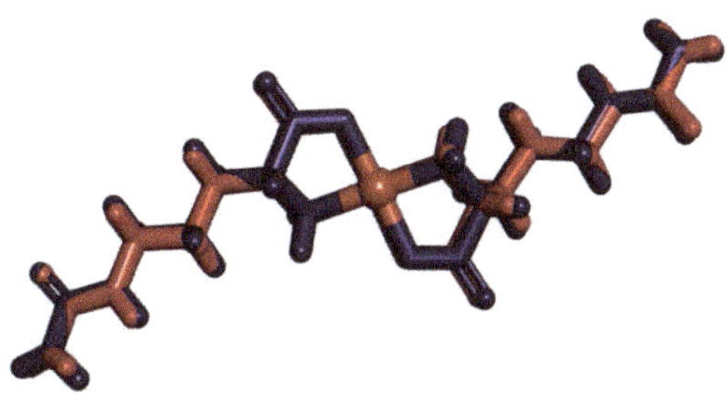

Figure 4. View of the X-ray complex and the optimized complex of [Cu(Citr)$_2$] fitted. X-ray complex is shown in red and calculated complex in blue.

3.2. Characterization of the Complex

3.2.1. UV-Visible Spectroscopy

Visible spectra of copper complexes with aminoacidato reveal a *d–d* transition as a broad band with a maximum of 650–550 nm and two lower-intensity bands at 540 and 720 nm—such an attribute for a square planar CuO_2N_2 chromophore. For the compound obtained, a typical unsymmetrical band at 600 nm was obtained by diffuse reflectance due to the insolubility of the complex. It corresponds to a *d–d* transition due to the coordination of two aminoacidato ligands and two carbonyl oxygens of the ureide groups comprising a tetragonally distorted octahedral environment (Figure 5). The *d–d* band consists of three overlapping electronic dipole transitions which are related to the three spin-allowed transitions, $^2A_{1g}(dz^2) \rightarrow {}^2B_{1g}(dx^2{-}y^2)$, $^2E_g(dyz{\approx}dxz) \rightarrow {}^2B_{1g}(dx^2{-}y^2)$, and $^2B_{2g}(dxy) \rightarrow {}^2B_{1g}(dx^2{-}y^2)$, respectively, for a distorted octahedral geometry of the copper(II) ion. The $n{\rightarrow}\pi^*$ characteristic band in the UV spectra at 267 nm is assigned to the intra-ligand transition of the amide C=O bond [57–62].

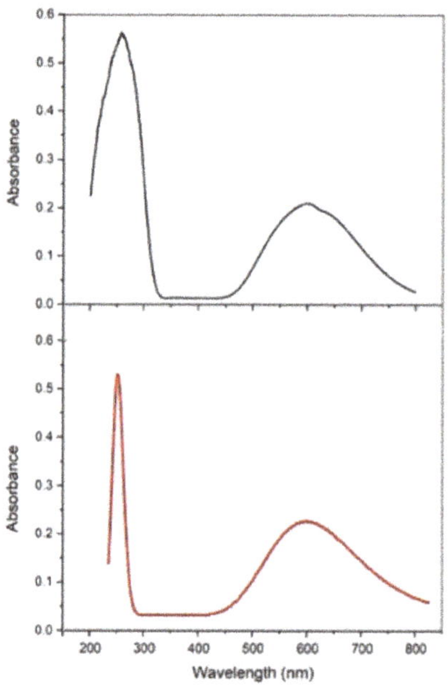

Figure 5. UV-Vis experimental spectrum of $[Cu(Citr)_2]_n$ obtained by diffuse reflectance (black line) and calculated spectrum at the level of theory mPW1PW91/6-31G(d)-LANL2DZ and ECP=LANL2DZ (red line).

The theoretical UV-Vis absorption spectrum of $[Cu(Citr)_2]_n$ was calculated using the TD-mPW1PW91 functional. The 6-31G(d)-LANL2DZ basis sets and the ECP=LANL2DZ for the Cu atom were used. The solvent effect was included using the SMD model with water as solvent. This UV-Vis spectrum was compared with the experimental one measured from 200 to 800 nm (Figure 5). Values such as absorption wavelengths, λ_{exp} and λ_{theo}, excitation energies, E_{xec}, oscillator strengths, and significant contributions of the electronic transitions were calculated using twenty states (n_{states} = 20).

Table 4 shows a good agreement between the experimental measurements and calculated values using the TD-DFT methodology. The experimental absorption band of $[Cu(Citr)_2]_n$ in the solid state is observed at 267 nm, corresponding to the $n{\rightarrow}\pi^*$ transition, which is calculated at 253 nm with the fundamental contribution of the HOMO and LUMO orbitals. The broadest band, which appears at 420–800 nm, is assigned to $d{\rightarrow}d$ transitions. This band shows two shoulders, the first between 523 (511 nm) and the second between 665

(642 nm). Both were found with low intensity and low contribution of the frontier molecular orbitals. Finally, the maximum of this broadband was calculated between 600 and 592 nm, with maximum intensity at 600 nm, in good agreement with the reported values. The TD-mPW1PW91 functional provided reliable results for the electronic transitions and excitation energies of compound $[Cu(Citr)_2]_n$ (see Figure 5 and Table 4). Due to that, the complexes of Cu(II) have one unpaired electron, located at the molecular orbitals β, and it is responsible for the electronic transitions. Only its spin was considered to analyze the UV-Vis spectrum. The mPW1PW91/6-31G(d)-LANL2DZ and ECP=LANL2DZ level of theory was used to calculate the gap energy, which corresponds to the difference between the Highest Occupied Molecular Orbital (HOMO) and the Lowest Unoccupied Molecular Orbital (LUMO). Therefore, the calculated value of the energy gap is 4.54 eV, using the UmPW1PW91 functional. On the other hand, a small energy gap value indicates good electronic conduction. Thus the electronic transitions between HOMO and LUMO are easy to carry out.

Table 4. Experimental absorption wavelengths (λ_{exp}), calculated absorption wavelengths (λ_{theo}), excitation energies ($E_{exc.}$), oscillator strengths, and significant contributions of the electronic transitions of $[Cu(Citr)_2]_n$ calculated at the level of theory mPW1PW91/6-31G(d)-LANL2DZ and ECP=LANL2DZ.

λ_{exp} (nm)	λ_{theo} (nm)	$E_{exc.}$ (eV)	Osc. Strengths	Major Contributions
720	665	1.87	0.0001	H-6→L+1 (2%)
	651	1.90	0.0003	H-8→L (4%)
	642	1.93	0.0003	H-4→L (3%)
600	600	2.06	0.0223	H→L (95%)
	598	2.07	0.0105	H-4→L (55%)
	592	2.09	0.0009	H-1→L (75%)
540	523	2.37	0.0003	H-8→L+1 (3%)
	516	2.40	0.0005	H-6→L+1 (2%)
	511	2.43	0.0001	H-2→L+1 (5%)
267	253	4.90	0.0436	H→L (91%)

3.2.2. Infrared Spectroscopy

The Infrared spectrum of the synthesized complex is presented in Figure 6, and it has three important band assignments; the first one refers to the peaks in the region of 3450–3150 cm^{-1} that are mainly attributed to amide (NH_2) asymmetric and symmetric stretching, amide (NH) stretching, and asymmetric and symmetric stretching vibrations of the amino group; the second one refers to two nearby weak peaks in the 2900–2800 cm^{-1} region, which correspond to asymmetric and symmetric vibrations of the methylene groups of the hydrocarbon chain. The third set of assignments corresponds to the region of 1700–1350 cm^{-1}. The peaks at 1652 and 1636 cm^{-1} correspond to C=O stretching vibrations of the carbonyl and carboxyl groups, respectively. The intense bands at 1575 and 1549 cm^{-1} correspond to scissoring vibrations of the (NH_2) amino and amide groups, respectively, and the peak at 1385 cm^{-1} is attributed to carboxylate symmetric stretching. The three assignments are concordant to a coordinated aminoacidato structure [25,63,64].

The theoretical IR vibrational frequencies of $[Cu(Citr)_2]_n$ are shown and compared with the experimental measurements obtained in this work presented in Table 5. In the high-frequency region at 3500–3400 cm^{-1}, the peaks are assigned to the stretching vibrational modes of the amide and amino groups (see Figure 6). In Table 5, it is possible to observe that the experimental values for other complexes of Pd(II) and Cu(II) with different ligands found in this region are in good agreement with those obtained experimentally and theoretically in this work. In addition, the double band assigned to the asymmetric and symmetric stretching modes, $\nu_{asym}(CH_2)$ and $\nu_{sym}(CH_2)$, of the methylene groups are found in the region 2951–2904 cm^{-1}. A low-intensity signal experimentally observed at 1652 cm^{-1}, corresponding to the stretching mode of the carbonylic group ν(C=O), is theoretically assigned by DFT calculations at 1708 cm^{-1}. The region 1700–1350 cm^{-1} at the experimental spectrum is relatively challenging to assign. However, with the help of theoretical calculations, three different vibrational modes can be observed. The first corresponds to the stretching modes of the carboxylic group, ν_{asym}(O=C–O$^-$), at 1695 cm^{-1}. The second is associated with the scissoring modes of amide $\delta(NH_2)$ and amino $\delta(NH_2)$ groups at 1583 and 1559 cm^{-1}, respectively. The third vibrational mode is assigned to the symmetric stretching of the carboxylic group, ν_{sym}(O=C–O$^-$), at 1390 cm^{-1}. A large displacement between the bands of asymmetric and symmetric modes of the carboxylic group is observed ($\Delta\nu_{complex}$ = 305 cm^{-1}), which suggests monodentate coordination of the carboxylate group through an oxygen atom. On the other hand, the metal-ligand vibrational modes are difficult to assign, however using DFT calculations and comparing with previous studies, it was possible to perform an adequate assignation of the stretching modes (Cu–N) in the higher frequency region than that region for stretching modes (Cu–O) calculated at 391 cm^{-1} in this work. The stretching mode ν(Cu–N) was calculated at 576 cm^{-1}, which is in good agreement with the values for some complexes of Cu(II) and Pd(II) with different amino acids reported in previous studies [24,25,62].

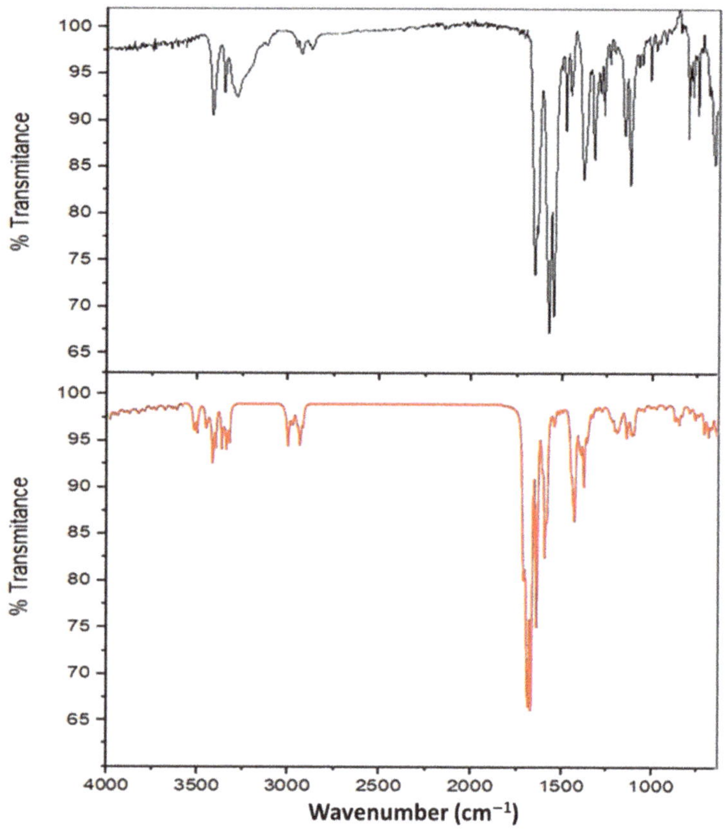

Figure 6. FTIR experimental spectrum of $[Cu(Citr)_2]_n$ in the region 4000–650 cm^{-1} (black line) and FTIR calculated spectrum at the level of theory mPW1PW91/6-31G(d)-LANL2DZ and ECP=LANL2DZ (red line).

Table 5. FTIR experimental vibrational frequencies of $[Cu(Citr)_2]_n$ and FTIR calculated vibrational frequencies obtained at the level of theory mPW1PW91/6-31G(d)-LANL2DZ and ECP=LANL2DZ.

Experimental (cm^{-1})	mPW1PW91 (cm^{-1})	Assignments
3450	3501	$\nu_{asym}(NH_2)$ amide
	3438	$\nu_{sym}(NH_2)$ amide + $\nu_{sym}(NH)$ amide
	3416	$\nu_{sym}(NH_2)$ amide
	3415	$\nu_{asym}(NH_2)$ amino
3150	3330	$\nu_{sym}(NH_2)$ amino
	2951	
	2947	$\nu_{asym}(CH_2)$
2900	2909	
	2910	$\nu_{sym}(CH_2)$
2800	2904	
1652	1708	ν(C=O) carbonylic group
1636	1695	$\nu_{asym}(O{=}C{-}O^-)$ carboxylic group
1575	1583	$\delta(NH_2)$ amide
1549	1559	$\delta(NH_2)$ amino
1385	1390	$\nu_{sym}(O{=}C{-}O^-)$ carboxylic group
	576	ν(Cu–N)

3.2.3. Raman Spectroscopy

The Raman spectrum is presented in Figure 7. The band at 3276 cm^{-1} is broadband, comprising all NH stretching frequencies of the amide and amino nitrogens. The most intense peak corresponds to the symmetric vibration of the terminal NH_2 of the amide group. The bands at 2924, 2910, 2895, 2881, and 2874 cm^{-1} are congruent with the bonds in the amino acid chain of carbons (CH_{str}, β, γ, δ). Between them, the peaks at 2924, 2910, and 2881 cm^{-1} present a slight shift with respect to the spectrum of pure Citrulline, which could be associated with the presence of Cu(II). The band at 1684 cm^{-1} appears minimized and corresponds to the C=O of the carboxylate. The peaks at 1463, 1460, 1327, and 1297 cm^{-1} coincide with those reported for L-Citrulline for CH_2 scissor, and at 1127, 990 cm^{-1} for CH twist. The band at 1453 cm^{-1} is associated with the CH_2 bend, 1312 cm^{-1} is associated with the wiggle of CH_2, 1279 cm^{-1} can be identified with the twisting of CH_2, 682 cm^{-1} belongs to the COO bend, 567 cm^{-1} associates with the C-O bond rocking in the carboxyl plane and 289 cm^{-1} is associated with a rocking C-C bond [65–70].

The experimental and the theoretical Raman vibrational frequencies of $[Cu(Citr)_2]_n$ are collected in Table 6. In the same way as the IR analysis, the theoretical Raman spectrum is in good agreement with experimental values obtained in this work and with values previously reported by Baran et al., for complexes of Cu(II) with different amino acids [64]. The stretching mode of the hydrogen linked to Cα, $\nu(C_\alpha H)$, is theoretically assigned at about 2932 cm^{-1}. In addition, the stretching mode ν(C–N) of the amino group is calculated at the 1189–1163 cm^{-1} region, which is in agreement with other values reported (see Table 6). The Raman vibrational bands observed in Figure 7 in the region at 3330–2894 cm^{-1} correspond to pure symmetric modes of the amino group $\nu_{sym}(-NH_2)$ and $\nu_{sym}(CH_2)$ and $\nu_{sym}(CH)$ modes of the hydrocarbon chain. In addition, the symmetric modes for the carbonylic and carboxylic groups are pure vibrational modes found in the 1642–1436 cm^{-1} range. A coupled mode corresponding to the scissoring mode, $\delta(CH_2)$, of the hydrocarbon chain coupled with the stretching mode of the carboxylic group, $\nu(O{=}C{-}O^-)$, was calculated at the 1354–1307 cm^{-1} range.

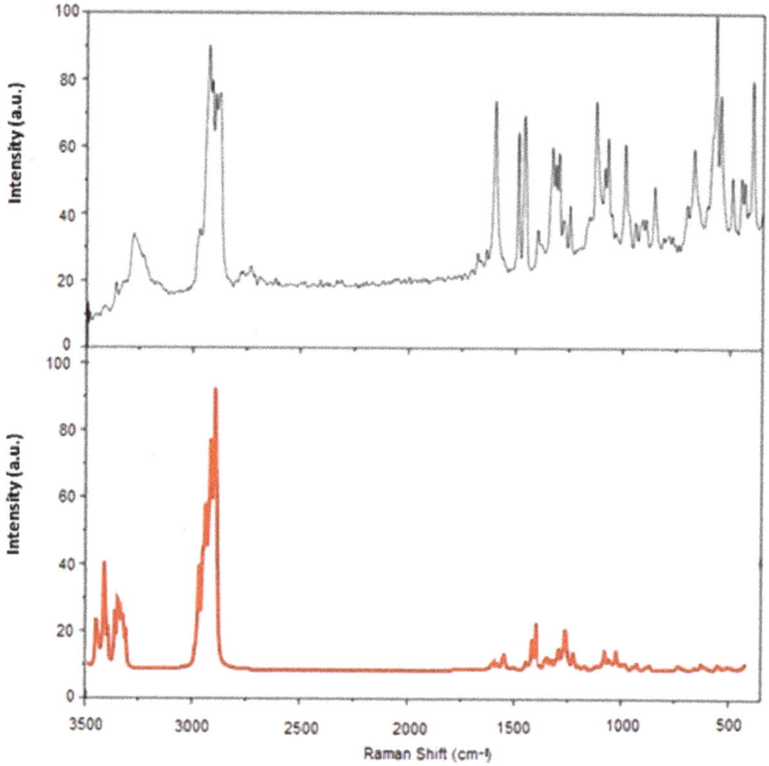

Figure 7. FT-Raman experimental spectrum of $[Cu(Citr)_2]_n$ in the region 3500–450 cm^{-1} (black line) and FT-Raman calculated spectrum at the level of theory mPW1PW91/6-31G(d)-LANL2DZ and ECP=LANL2DZ (red line).

Table 6. FT-Raman experimental and calculated vibrational frequencies of $[Cu(Citr)_2]_n$ obtained at the level of theory mPW1PW91/6-31G(d)-LANL2DZ and ECP=LANL2DZ.

Experimental	**mPW1PW91**	**Assignments**
3276	3330	$\nu_{sym}(-NH_2)$ amino
	2935–2929	$\nu(C_\alpha H)$
2924, 2910	2927, 2913	$\nu(CH_2)$
2895, 2881, 2874	2902–2894	$\nu(CH)$
1684	1642–1619	$\nu(C{=}O)$ amide
1463–1460	1465–1436	$\nu(C{=}O)$ carboxylic group
1327–1297	1354–1307	$\delta(CH_2) + \nu(O{=}C{-}O^-)$
1127	1189–1163	$\nu(C{-}N)$ amino

3.2.4. NMR Studies

In Figure 8, the NMR spectra of ^1H and ^{13}C of the paramagnetic copper(II) complex and free Citrulline are presented. Tetramethylsilane (TMS) was the reference to measure both spectra. According to the literature, the unpaired electron in Cu(II) complexes provokes a strong influence on the chemical shifts of ^1H and ^{13}C, as well as in several complexes with paramagnetic metals, through hyperfine interactions, causing difficult assignation of the conventional NMR spectra [70,71]. In this work, we have focussed our attention on one dimension (1D) assignation of ^1H and ^{13}C spectra for the paramagnetic Cu(II)-Citrulline complex using the Bruker AVANCE III 500 MHz spectrometer and calculations based on the density functional theory (DFT). In the experimental ^{13}C NMR spectrum in Figure 8a, the first peak is located downfield, assigned as C6A at 161.52 ppm. This signal is assigned to the carbon of the carbonyl moiety of the ureide group. The signals for the C5A/C5B at 57.40/48.83 ppm, C3A/C3B at 38.94/26.99 ppm, and C4A/C4B at 24.78/16.74 ppm were assigned. On the other hand, the ^{1}H NMR spectrum shows four signals in the upfield region. The first is observed at 1.51 ppm, corresponding to the H4AA/H4BA and H4AB/H4BB pairs of protons. The second signal at 1.87 ppm is assigned for H3AD/H3BD and H3AC/H3BC protons. The third signal at 3.05 ppm corresponds to H5AA/H5BA and H5AB/H5BB protons. The last signal at 3.96 ppm corresponds to a unique H2AA (H_{α}) proton linked to the Cu(II) atom, suggesting an interaction between the carboxylic and amino groups with the copper ion. A small single peak at 3.25 ppm corresponds to a residual methanol impurity [71,72].

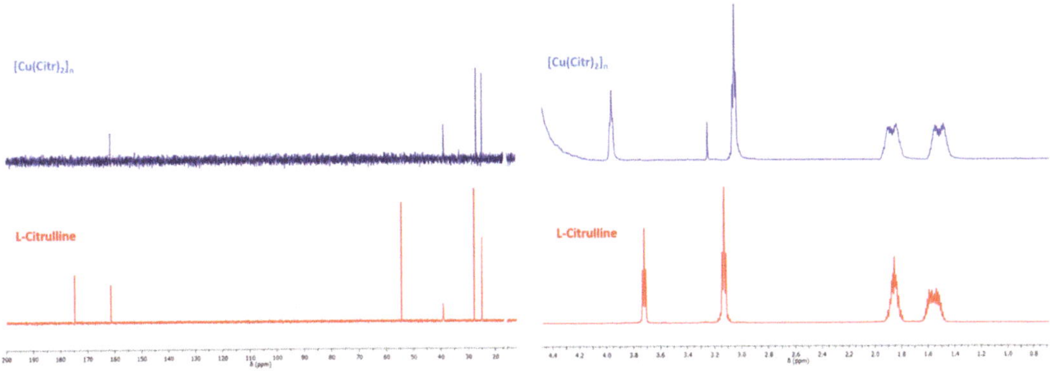

(a) **(b)**

Figure 8. (**a**) ^{13}C NMR spectrum of $[Cu(Citr)_2]_n$ in D_2O, and (**b**) ^{1}H NMR spectrum of $[Cu(Citr)_2]_n$ in D_2O.

The theoretical NMR characterization through hyperfine chemical shifts and the effect of the unpaired electron on the Cu(II) on the spin density in the $[Cu(Citr)_2]_n$ complex was analyzed. The adequate assignation of the ^{1}H and ^{13}C NMR hyperfine chemical shifts of the complex depends on how the spin density is distributed on the H and C nuclei in the vicinity of Cu(II) with an unpaired electron. The hyperfine shifts are calculated using the UmPW1PW91 functional according to Equations (S1a) and (S1b) shown in the Supplementary Materials Section. Figure 9a shows the labeling of the H and C nuclei for the assignation of the ^{13}C and ^{1}H NMR chemical shifts, while Figure 9b shows the spin density distribution for one unit of $[Cu(Citr)_2]_n$. The spin density distribution is mapped in the range of -1.4×10^{-4} u.a. (red color) to 1.4×10^{-4} u.a. (blue color). The Fermi contact shifts (δ_{FC}) are obtained when the spin density values are substituted in Equation (S1a). The total calculated shifts ($\delta_{Total}^{Calc.}$) are obtained from the Equation (S1b), when the diamagnetic shifts (δ_0) are added to the Fermi contact shifts, which are compared with the experimental chemical shifts (δ_{Exp}). Table 7 shows the assignment of the ^{13}C, and ^{1}H NMR experimental shifts, δ_{Exp}, the spin density, $\rho_{\alpha\beta}$, the Fermi contact shifts, δ_{FC}, the diamagnetic shifts, δ_0, the total calculated shifts, $\delta_{Total}^{Calc.}$ and the ^{13}C and ^{1}H nuclei assignment in the $[Cu(Citr)_2]_n$ complex; a good agreement is observed between the experimental and calculated results.

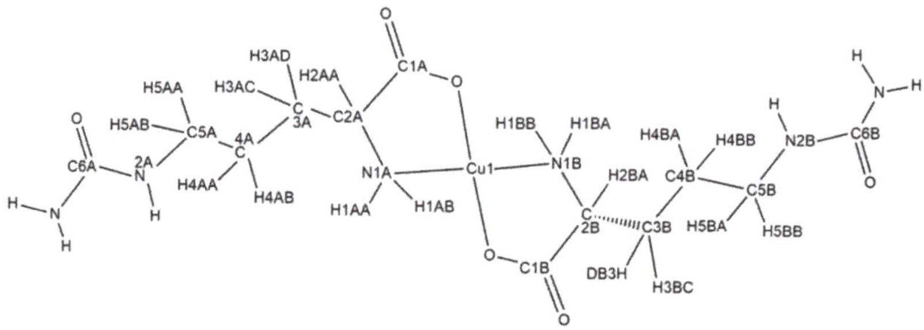

(a)

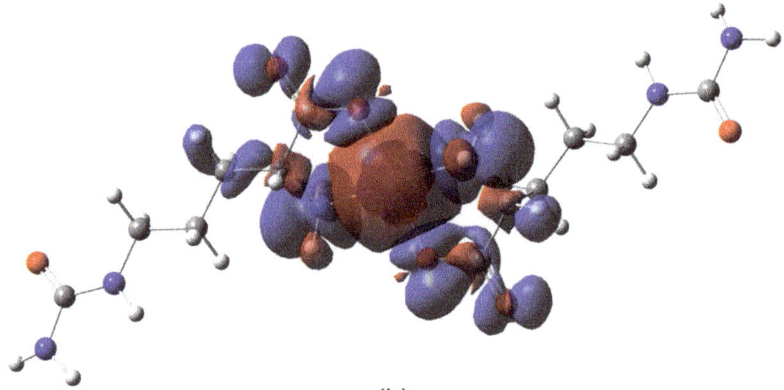

(b)

Figure 9. (**a**) Nomenclature for ^{1}H and ^{13}C chemical shifts assignation of compound $[Cu(Citr)_2]_n$, and (**b**) Spin density of $[Cu(Citr)_2]_n$ calculated at the level of theory mPW1PW91/6-31G(d)-LANL2DZ and ECP=LANL2DZ.

Table 7. Assignment of the ^{13}C and ^{1}H experimental and calculated hyperfine chemical shifts of the compound $[Cu(Citr)_2]_n$ obtained at the level of theory UmPW1PW91/6-31G(d)-LANL2DZ and ECP=LANL2DZ.

Nuclei	δ_{Exp} (ppm)		δ_{FC} (ppm)	δ_0 (ppm)	$\delta^{Calc.}_{Total}$ (ppm)	<Signal Assignment>
^{13}C	161.52	0.00000	0.00	151.32	151.32	<C6A>, <C6B>
	57.40, 48.83	0.00014, 0.00005	16.45, 5.88	41.93, 39.54	58.38, 45.42	<C5A>, <C5B>
	38.94, 26.99	0.00004, -0.00005	4.70, −5.88	32.61, 32.39	37.31, 26.52	<C3A>, <C3B>
	24.78, 16.74	−0.00003, −0.00007	3.53, −8.23	27.76, 24.85	24.24, 16.63	<C4A>, <C4B>
	Not observed	−0.00367, −0.00363	−431.23, −426.53	173.05, 176.55	−258.17, −249.97	<C1A>, <C1B>
	Not observed	−0.00313, −0.00319	−367.78, −374.83	56.75, 53.63	−311.02, −321.19	<C2A>, <C2B>
^{1}H	3.96	0.00001	1.07	3.32	4.38	<H2AA>
	Not observed	0.00101	107.56	3.19	110.75	<H2BA>
	3.05	0.00000	0.00	2.91, 3.03	2.91, 3.03	<H5AA>, <H5BA>
	3.05	0.00000	0.00	2.89, 2.89	2.89, 2.89	<H5AB>, <H5BB>
	1.87, 1.87	0.00057, 0.00000	60.70, 0.00	1.85, 1.77	62.55, 1.77	<H3AD>, <H3BD>
	1.87, 1.87	0.00001, 0.00000	1.07, 0.00	1.21, 2.01	2.27, 2.01	<H3AC>, <H3BC>
	1.51	0.00000	0.00	1.33, 1.68	1.33, 1.68	<H4AA>, <H4BA>
	1.51	0.00000	0.00	1.56, 1.57	1.56, 1.57	<H4AB>, <H4BB>
	Not observed	−0.00118, −0.00108	−125.66, −115.02	1.23, 1.68	−124.44, −113.34	<H1AA>, <H1BA>
	Not observed	−0.00126, −0.00136	−134.18, −144.83	1.98, 1.87	−132.20, −142.96	<H1AB>, <H1BB>

For the ^{13}C NMR, the signal downfield at 151.32 ppm corresponding to C6A is assigned to the carbonyl moiety of the ureide group. The assignation for the C5A/C5B, C3A/C3B, and C4A/C4B are in good agreement with the experimental chemical shifts. The average difference between values is 1.62 ppm (see Table 7). Two signals are calculated upfield at −311.02 and −321.19 ppm, corresponding to C2A (C_α) and C2B ($C_{\alpha'}$) in Figure 8a, respectively, which are not experimentally observed in the range of 10–200 ppm (see Figure 8b. The signals for the C1A and C1B of the carboxylate groups are calculated upfield at −249.97 and −258.17 ppm, respectively. Figure 9b shows the effect of the unpaired electron of the Cu(II) in the distribution of the isosurface of the spin density. It can be observed through the spin density (in blue color) that the complexes with square-plane or octahedral geometries involving the Cu(II) with an unpaired electron, which is in the $d_{x^2-y^2}$ orbital, excluding the Jahn-Teller effect, the atoms close to Cu(II) ion can be affected in their chemical shifts. In addition, the ^{13}C shifts must obey the following relation $\delta_{C_i/C_i'} \propto r^{-6}_{Cu-C_i}$ with the distance Cu–C for the pair of atoms C2A/C2B and C1A/C1B linked to the Cu(II) atom [66]. The values smaller than r^{-6} indicate the displacement of both signals at a very upfield region in the ^{13}C NMR spectrum. From X-ray data in this work, both pairs of carbons are very close to the Cu(II) atom at a distance of 2.74 Å for C2A/C2B and 2.81 Å for C1A/C1B.

Concerning ^{1}H NMR, as can be observed in Table 6, the chemical shifts obtained from DFT calculations show four signals upfield, which are assigned as follows: the signals at 1.56/1.57 and 1.33/168 ppm are assigned to the pairs of protons H4AB/H4BB, and H4AA/H4BA, respectively. The signal at 2.27/2.01 ppm is assigned to the protons H3AC/H3BC. The signal at 1.77 ppm is assigned to H3BD proton, while at 62.55 ppm is assigned to H3AD, which corresponds to the proton affected by the spin density of the unpaired electron of Cu(II) (see Figure 9b). The chemical shifts of 2.89/2.89 and 2.91/3.03 ppm are assigned to the pairs of protons H5AA/H5BA and H5AB/H5BB, respectively. In addition, the theoretical calculations corroborate that a unique signal is observed for the H2AA (H_α) proton at 4.83 ppm. In comparison, the signal for the H2BA ($H_{\alpha'}$) is calculated downfield at 110.75 ppm, which is not possible to be experimentally observed. Finally, the DFT calculations suggest that the upfield signals at (−113.34 to −142.96 ppm) can be assigned to the H1AA/H1BA and H1AB/H1BB, and neither are observed at the experimental NMR spectrum.

3.3. *Molecular Structure and Non-Covalent Interactions*

Figure 10a shows the optimized structure of a triad of copper of the compound $[Cu(Citr)_2]_n$, and Figure 10b shows the molecular electrostatic potential (MEP). It is also observed that the atom of Cu1 is coordinated with two Citrulline molecules through the two O3A and O3B atoms of carboxylate with distances of Cu–O 1.964–1.980 Å, and two N1A and N1B of amine groups in the equatorial sites with distances of Cu–N 2.018–2.027 Å (Citr1 and Citr2). The two axial positions of the distorted octahedral center are occupied by the O3A and O3B atoms of amide groups of one Citrulline molecule, each one (Citr3 and Citr4) with distances of O–Cu 3.051–3.263 Å. The hydrogen bonds are formed between O2A and O2B atoms of carboxylate with hydrogen atoms H2A/H2B and H3AB/H3BA of amide groups of the adjacent Citrulline. These hydrogens bonds are NH···O 1.964–2.101 Å, and N–H··· O valence angle of 146.1–153.4°. Figure 10b presents the MEP surface. The electrostatic potential was mapped on the total electronic density with isovalue = 0.004 a.u. in a range of -6.0×10^{-2} (red zones) to 6.0×10^{-2} (blue zones) of electronic density. In addition, the nucleophilic zones (negative charge density) are located on the carboxylate and amide groups of Citrulline. In contrast, the electrophilic zones (deficient density charge) are located on the amine groups and the rest of Citrulline molecules. In addition, yellow and green regions correspond to intermediate electron density zones where the hydrogen bonds are located. These electrophilic and nucleophilic regions are susceptible to interaction with adjacent chains of Citrulline.

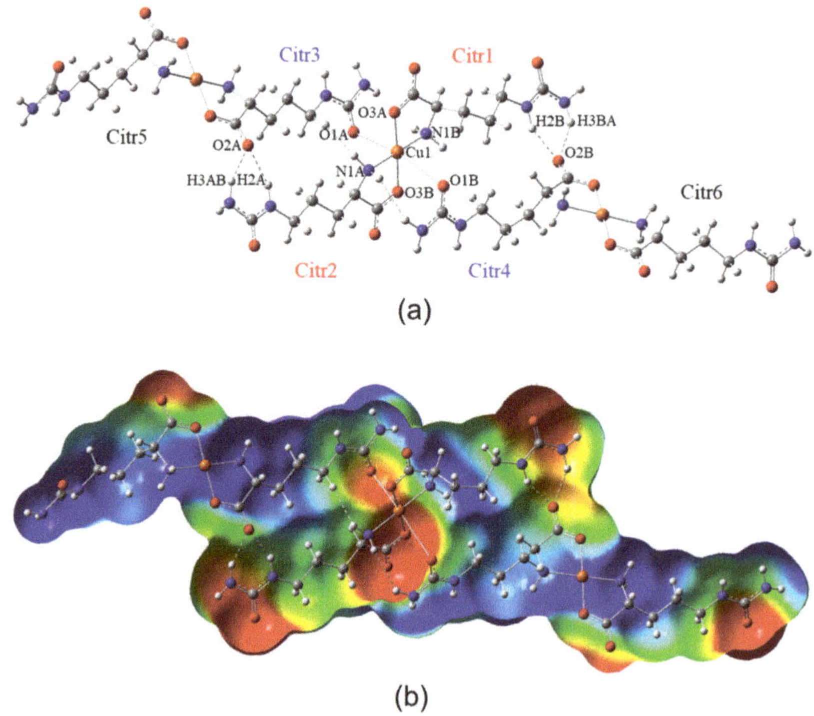

Figure 10. (**a**) Molecular structure of a representative triad of the compound. (**b**) Molecular electrostatic potential of compound $[Cu(Citr)_2]_n$ calculated at the level of theory mPW1PW91/6-31G(d)-LANL2DZ and ECP=LANL2DZ.

The hydrogen bonds were analyzed by electron density, ρ(r), the Laplacian of density, ∇^2 ρ(r), and the energy of interaction, $E_{H\cdots Y}$. The results are summarized in Table 8. Figure 11 shows the molecular graphs of compound $[Cu(Citr)_2]_n$. In this figure, green dots represent bond critical points (BCPs), purple dots represent ring critical points (RCPs), and cyan dots represent cage critical points (CCPs). From the results, it can be seen that the ρ(r) on the BCPs is at the range 0.0172–0.0246 a.u. for the hydrogen bonds N-H···O2A and N-H···O2B. The calculated interaction energy ($E_{H\cdots Y}$) is 4.49–6.78 kcal mol^{-1} for these hydrogen bonds connecting Citr1 with Citr4 and Citr2 with Citr3. RCPs with ρ(r) of 0.0080–0.0084 a.u. are observed between the interactions N-H3AB···O2A and N-H2A···O2A, and N-H2B···O2B and N-H3BA···O2B, forming stable ring structures of six-atoms.

Table 8. Topological parameters (in a.u.) and interaction energies $E_{H\cdots Y}$ (in kcal mol^{-1}) of compound $[Cu(Citr)_2]_n$. Atom labels correspond to those shown in Figure 10a.

BCP	ρ(r)	∇^2ρ(r)	$E_{H\cdots Y}$
N-H3AB/H3BA···O2A/O2B	0.0225–0.0246	0.0760–0.0898	4.52–5.93
N-H2A/H2B···O2A/O2B	0.0172–0.0178	0.0770–0.0994	4.49–6.78
Cu1···N1A/N1B	0.0790–0.0807	0.4261–0.4349	39.28–40.47
Cu1···O1A/O1B	0.0749–0.0775	0.5133–0.5334	44.43–57.73
Cu1···O3A/O3B	0.0057–0.0082	0.0156–0.0208	1.26–1.98

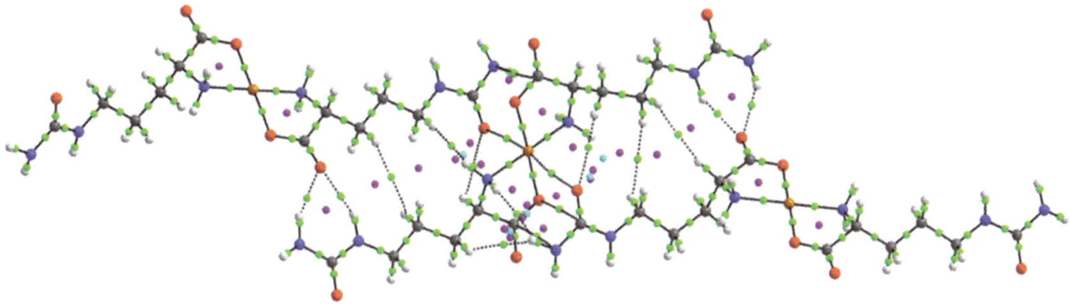

Figure 11. Molecular graphs of compound $[Cu(Citr)_2]_n$ showing the main BCPs of the H-bonds and RCPs between citrulline molecules.

On the other hand, the BCPs between Cu1···N1A and Cu1···N1B of the amine group and Cu1···O1A and Cu1···O1B of the carboxylate group of Citrulline show the highest values of ρ(r) between 0.0749–0.0807 a.u. with $E_{H \cdots Y}$ of 39.28–40.47 kcal mol^{-1} for Cu1···N1A/N1B interaction and 44.43–57.73 kcal mol^{-1} for Cu1···O1A/O1B interaction. It indicates the significant metal-ligand non-covalent interactions in the equatorial positions coordinated of the Cu1 atom, similarly to those found for the interactions Cu···N of metformin and bipyridine groups at the range of 0.0814–0.1012 a.u., Cu···N of imidazol-pyridine and glycine groups at the range of 0.0854–0937 a.u. and Cu···N of phenanthroline and glutamine groups at the range of 0.0774–0.0864 a.u., in the axial positions of the square pyramidal geometry of Cu in V/Cu heterobimetallic compounds [29,73]. Furthermore, RCPs between Cu1··· N1A/N1B and Cu1··· O1A/O1B with ρ(r) of 0.0255–0.0260 a.u. are observed. In addition, several RCPs are observed between the Citrulline chains, indicating the formation of stable rings in the structure. Finally, the BCPs formed between Cu1···O3 of the amide group of Citrulline at the axial positions of the octahedral geometry have values of ρ(r) between 0.0057–0.0082 a.u. with small interaction energy values of 1.26–1.98 kcal mol^{-1}, indicating that the Cu1···O3A and Cu1···O3B of the amide group of Citrulline interactions are due to the distorted octahedral geometry.

The Hirshfeld surfaces (HS) and the 2D-fingerprint plot of compound $[Cu(Citr)_2]_n$ were generated with the function d_{norm}, as shown in Figure 12. Two systems were analyzed: in Figure 12a, the molecules Citr1 and Citr2 are located inside the HS, while Citr3–Citr4 act as ligands outside the HS (see Figure 12a for labeling). In this case, the red spots on the HS are due to the hydrogen bonds between N–H of amide groups inside the HS, acting as a donor, and the oxygen of carboxylate groups of the Citrulline molecules outside the HS, acting as acceptors. In Figure 12b, the molecules Citr3–Citr5 and Citr4–Citr6 are located inside the HS, with Citr1 and Citr2 outside acting as ligands. In this case, the red spots show the same strong hydrogen bond interaction between the oxygen of the carboxylate group inside the HS and the N-H of amide groups outside the HS. In both cases, other red spots are located, indicating the close intermolecular interactions of N–H of amine and carboxylate groups with adjacent Citrulline molecules, and the Cu atom interacting with the oxygen atoms of the amide group coordinated in axial positions of the octahedral geometry. The main hydrogen bonds are indicated with dashed cyan lines in Figure 10. In the 2D-fingerprint plot, the hydrogen bonds N–H··· O connecting the Citrulline molecules inside and outside the HS have contributions of 17.9% and 21.6%, as shown in Figure 10c. The major contribution to the HS with 45.7% is due to the close intermolecular interactions H··· H of H of amine and the $-C_3H_6$ groups with adjacent Citrulline molecules. Other interactions with minor contributions are Cu··· O (1.5%), N··· H (2.4%), and C··· H (2.0%).

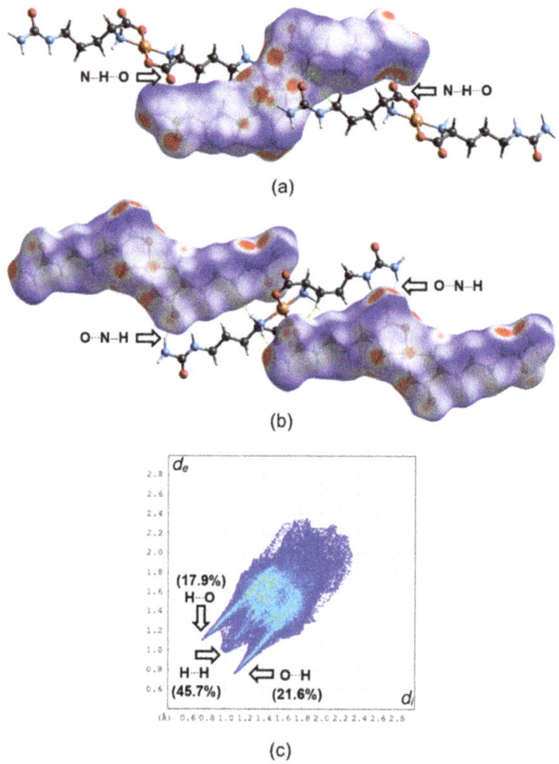

Figure 12. Hirshfeld surfaces mapped with d_{norm} parameter of $[Cu(Citr)_2]_n$ in two equivalent surfaces: (**a**) Citr1 and Citr2 inside the HS with Citr3–Citr4 acting as ligands outside the HS, (**b**) the Citr3–Citr5 and Citr4–Citr6 inside the HS with Citr1 and Citr2 outside the HS, and (**c**) 2D-fingerprint plot of non-covalent interactions.

4. Discussion

Citrulline is a no-coded amino acid that has received significant attention. The number of papers published on it, from its discovery in 1930 till 2011, is now equal to the ones reported in the last decade.

Citrulline is almost fully present in physiological settings as a zwitterion. Both carboxylic and amine groups can act as Lewis donors at alkaline pH, implying they can complex metallic cations. However, the lateral chain is significantly less complexing as the nitrogen atoms' non-binding electrons are conjugated with a double bond. As a result, Citrulline complexes are identical to those generated by other classical amino acids. However, there is only one crystal structure report in the Cambridge Crystallographic Data-Base. The structure of the palladium complex reported by Mascaliovas et al. [25] was obtained by powder X-ray diffraction, and most studies have been carried out in solution. Although Kurtz proposed the structure of the copper compound in 1939, it was considered an intermediate in the synthesis of Citrulline, and due to its insolubility, it was not characterized thoroughly. However, the work of Ganadu et al., in 1991 [24] on the $[Pd(Citr)_2]$ gave some light on the structure using spectroscopic methods. The coordination was typical of an aminoacidato, coordinating through the oxygen of the carboxylate and the amino group. No mention of some role of the ureide moiety was presented. Mascaliovas corroborated this, describing the compound as a bis-citrullinate of palladium. The metal is coordinated by two citrulline molecules acting as bidentate ligands in a square planar arrangement with nitrogens and oxygens in *trans* positions. The compound crystallizes on the P-1 space group, so it has an inversion center that requires the presence of the L and D-Citrulline. Interestingly enough, the authors do not mention this fact.

Our findings indicate that Citrulline is a very adaptable chemical, potentially generating several hydrogen-bonding contacts since seven strong interactions were observed in its crystal structure [74]. This is responsible for the polymeric nature of the compound $[Cu(Citr)_2]_n$, contrary to the case of palladium(II). Palladium prefers square planar geometries; instead, the copper(II) ion could accommodate two monodentate ligands in an axial position in a distorted octahedral environment. Therefore, in this case, the ureide carbonyl group is not-innocent and coordinates the copper(II) ions in the axial positions. The theoretical studies using AIM and Hirshfeld surface analysis also demonstrate that hydrogen bond interactions highly dominate the chemistry of Citrulline. Some complexes based on Cu(II) and amino acids similar to Citrulline have been published. Typically, the synthesis of these compounds uses the L-enantiomer of arginine or ornithine; thus, the complexes obtained contain this enantiomer [75]. In this case, we report a change in the conformation of Citrulline; therefore, due to the coordination of Cu(II), the reactive L-Citrulline is transformed into D-Citrulline. Racemization of amino acids in the presence of metal ions and pyridoxal was discovered by Olivard et al., in 1952 [76]. In this regard, the copper(II) ion was the most effective in the racemization of alanine. Using NMR studies, Abbott and Martell in 1969 showed that aluminum(III) exchange of the α proton of the amino acid in the region pD (pH) 10 [77]. This reaction results in the racemization of the asymmetric center, which is demonstrated by the signals of the racemic complex initially containing one of the amino acid enantiomers.

Recently the term SIST (solubility-induced stereoisomer transformation) was introduced by Fu et al., (2020) [78]. This new concept could be applied to understand $[Cu(Citr)_2]_n$ since, regardless of the source of L-Citrulline, the final product contains the D stereoisomer. In SIST, one stereoisomer that is below saturation is in rapid equilibrium with the other that is above saturation. A continuous transformation of the more soluble stereoisomer into the less soluble stereoisomer is produced. SIST can theoretically provide double the yield of classical resolutions and does not require the development of stereoselective receptors. It has been a challenge to rapidly racemize free unactivated amino acids because of the weak acidity of the α-carbon-hydrogen atom. Two strong hydrogen bonds can be used in concert to deracemize unactivated amino acids by SIST under mild reaction conditions. The same method can convert readily available L-amino acids into D-amino acids. In addition, if crystallization occurs during a process known as reactive crystallization, it augments the equilibrium reaction yields significantly [79]. Thus, crystallization could propel the development of specific crystals of stereoisomers. Crystallization-induced stereoisomer transformation (CIST) describes a stereoisomer produced by crystallization. In the case of $[Cu(Citr)_2]_n$, the rapid formation of polymeric chains induces the crystallization of the compound with the D enantiomer imposed by the symmetry of the crystal. It appears that the conjunction of SIST, CIST, hydrogen bonds, and copper allows the rapid formation of catena-trans-syn-$[Cu(D\text{-}Citr)_2]_n$.

The development of effective methods for manufacturing D-amino acids has attracted considerable attention since they are common pharmaceutical building pieces for drugs, such as saxagliptin [80], tadalafil [81], clopidogrel [82], and cyclo-serine [83], against diabetes, erectile dysfunction, heart disease or bacteria. Moreover, D-amino acids are also important in the field of unnatural-peptide-based-therapies, now in growth, such as telaprevir [84], degarelix [85] or carfilzomib [86], antiviral, prostate anticancer, and against multiple myeloma drugs, respectively. Until now, many interesting chemical [87] and biological [88] methods have been developed for making D-amino acids [89]. Therefore, new and interesting findings with this molecule are expected soon.

5. Conclusions

The synthesis and characterization of the bis-citrullinato copper(II) were achieved. Two methodologies and three copper sources were used. Regardless of the method and sources of copper, the complex $[Cu(Citr)_2]_n$ was obtained as a pale blue microcrystalline solid. The reaction takes 2 h using hydrothermal synthesis; however, the crystals are extremely small. Using electrosynthesis, which takes 6 to 10 h, bigger crystals were obtained and used for X-ray diffraction studies. Surprisingly, the solid has a polymeric nature, which accounts for its high insolubility. As for the aminoacidato complexes, the amino group's nitrogen and the carboxylic group's oxygen are coordinated to the copper atom. In addition, the carbonyl group of the ureide moiety is not-innocent, and it is coordinated in the axial positions of the copper, which due to a Jahn-Teller effect, has long distances. Therefore, the compound has a copper(II) ion in a distorted octahedral geometry. The aminoacidato ligands are disposed of with oxygen and nitrogen in *trans* positions in the equatorial plane. The stereochemistry of the ligands is D, D, which is contrary to the expected L, L, and the positions of the hydrogens of the α-carbons point in the same direction; thus, the structure corresponds to a syn conformer, although both positions are slightly different.

The molecular structure and non-covalent interactions were analyzed thoroughly and allow us to understand the properties of the compound. The interactions Cu1···N1 of the amine group and Cu1···O2 of the carboxylate group of Citrulline showed high interaction energies, indicating the formation of metal-ligand non-covalent interactions in the equatorial positions coordinated by the Cu1 atom. The hydrogen bonds N-H···O connecting the Citrulline molecules are clearly characterized in the Hirshfeld surfaces. So, the theoretical studies using AIM and Hirshfeld surface analysis demonstrated that hydrogen bond interactions highly dominate the chemistry of Citrulline. The calculated spectra also were very useful for the assignments of all bands and peaks of the IR, Raman, and ^{13}C and ^{1}H NMR spectroscopic studies.

As part of our search for Casiopeina® analogs, we have synthesized mixed complexes of diimines with lysine, ornithine, glutamine, and asparagine. Currently, attempts are being carried out for Citrulline-mixed complexes since the bis-D-citrullinato is completely insoluble for such purposes. However, the synthesis provides an easy way to obtain the D-Citrulline stereoisomer, which could serve other applications.

Supplementary Materials: The following supporting information can be downloaded at: https://www.mdpi.com/article/10.3390/cryst12101386/s1, Equations for NMR calculations. Figure S1: Stereoisomers of Citrulline. Figure S2: Structure of the complex and structure of the zwitter ion of L-Citruline from CCDC(FIGOQ). Figure S3: Top and side views of the intermediary of the racemization of Citrulline mediated by copper coordination; Figure S4: ORTEP representation of a fragment of the bis-citrullinato Cu(II) polymer. Table S1: Fractional atomic coordinates and isotropic or equivalent isotropic displacement parameters (Å2) for (Cu_citrulline); Table S2: Atomic displacement parameters (Å2) for (Cu_citrulline); Table S3: Geometric parameters (Å, °) for (Cu_citrulline). Proof of structure deposit and Checkcif.

Author Contributions: D.R.-C. carried out experimental work (synthesis, crystallization, and experimental characterization). FT-Raman measurements were carried out by L.E.S.-d.l.R., F.J.M. and M.E.C. carried out the theoretical characterization. A.G.-G., D.C.-L. and A.R.-D. carried out the X-ray diffraction determination. D.R.-C., E.G.-V., M.E.C., A.G.-G. and B.L.S.-G. wrote and revised the manuscript. D.R.-C. and E.G.-V. conceived and designed this study. All authors contributed extensively to the work presented in this paper. All authors have read and agreed to the published version of the manuscript.

Funding: Projects funded this research: 100108444-VIEP, 100256733-VIEP, and 100233622-VIEP. The PRODEP Academic Group BUAP-CA-263 (SEP, Mexico), and the Ministerio de Universidades and Next Generation for the Margarita Salas contract 401 (Spain).

Data Availability Statement: Data can be obtained directly from authors upon request.

Acknowledgments: Diego Ramírez-Contreras wishes to thank CONACyT (Mexico) M.Sc. fellowship support number 1143792. María Eugenia Castro and Francisco J. Melendez wishes to thank Laboratorio Nacional de Supercómputo del Sureste de México (LNS-BUAP) and the CONACyT network of national laboratories for the computer resources and support provided. Amalia García-García thanks to Ministerio de Universidades and funds Next Generation (Spain).

Conflicts of Interest: The authors declare no conflict of interest.

References

1. Wada, M. On the Occurrence of a New Amino Acid in Watermelon, Citrullus Vulgaris, Schrad. *Bull. Agric. Chem. Soc. Jpn.* **1930**, *6*, 32–34. [CrossRef]
2. Fragkos, K.C.; Forbes, A. Was Citrulline First a Laxative Substance? The Truth about Modern Citrulline and Its Isolation. *Nihon Ishigaku Zasshi* **2011**, *57*, 275–292. [PubMed]
3. Rimando, A.M.; Perkins-Veazie, P.M. Determination of Citrulline in Watermelon Rind. *J. Chromatogr. A* **2005**, *1078*, 196–200. [CrossRef] [PubMed]
4. Kaore, S.N.; Kaore, N.M. Citrulline: Pharmacological Perspectives and Role as a Biomarker in Diseases and Toxicities. In *Biomarkers in Toxicology*; Gupta, R.C., Ed.; Elsevier Inc.: Amsterdam, The Netherlands, 2014; pp. 883–905. ISBN 978-0-12-404630-6.
5. Curis, E.; Crenn, P.; Cynober, L. Citrulline and the Gut. *Curr. Opin. Clin. Nutr. Metab. Care* **2007**, *10*, 620–626. [CrossRef]
6. Aguayo, E.; Martínez-Sánchez, A.; Fernández-Lobato, B.; Alacid, F. L-Citrulline: A Non-Essential Amino Acid with Important Roles in Human Health. *Appl. Sci.* **2021**, *11*, 3293. [CrossRef]
7. Curis, E.; Nicolis, I.; Moinard, C.; Osowska, S.; Zerrouk, N.; Bénazeth, S.; Cynober, L. Almost All about Citrulline in Mammals. *Amino Acids* **2005**, *29*, 177–205. [CrossRef]
8. Papadia, C.; Osowska, S.; Cynober, L.; Forbes, A. Citrulline in Health and Disease. Review on Human Studies. *Clin. Nutr.* **2018**, *37*, 1823–1828. [CrossRef]
9. Endo, F.; Matsuura, T.; Yanagita, K.; Matsuda, I. Clinical Manifestations of Inborn Errors of the Urea Cycle and Related Metabolic Disorders during Childhood. *J. Nutr.* **2004**, *134*, 1605S–1609S. [CrossRef]
10. Khalaf, D.; Krüger, M.; Wehland, M.; Infanger, M.; Grimm, D. The Effects of Oral L-Arginine and L-Citrulline Supplementation on Blood Pressure. *Nutrients* **2019**, *11*, 1679. [CrossRef]
11. Collins, J.K.; Wu, G.; Perkins-Veazie, P.; Spears, K.; Claypool, P.L.; Baker, R.A.; Clevidence, B.A. Watermelon Consumption Increases Plasma Arginine Concentrations in Adults. *Nutrition* **2007**, *23*, 261–266. [CrossRef]
12. Gramaglia, I.; Velez, J.; Chang, Y.-S.; Caparros-Wanderley, W.; Combes, V.; Grau, G.; Stins, M.F.; van der Heyde, H.C. Citrulline Protects Mice from Experimental Cerebral Malaria by Ameliorating Hypoargininemia, Urea Cycle Changes and Vascular Leak. *PLoS ONE* **2019**, *14*, e0213428. [CrossRef]
13. Kaore, S.N.; Amane, H.S.; Kaore, N.M. Citrulline: Pharmacological Perspectives and Its Role as an Emerging Biomarker in Future. *Fundam. Clin. Pharmacol.* **2013**, *27*, 35–50. [CrossRef]
14. Bahri, S.; Zerrouk, N.; Aussel, C.; Moinard, C.; Crenn, P.; Curis, E.; Chaumeil, J.-C.; Cynober, L.; Sfar, S. Citrulline: From Metabolism to Therapeutic Use. *Nutrition* **2013**, *29*, 479–484. [CrossRef]
15. Rashid, J.; Kumar, S.S.; Job, K.M.; Liu, X.; Fike, C.D.; Sherwin, C.M.T. Therapeutic Potential of Citrulline as an Arginine Supplement: A Clinical Pharmacology Review. *Paediatr. Drugs* **2020**, *22*, 279–293. [CrossRef]
16. Alghamdi, M.; Alasmari, D.; Assiri, A.; Mattar, E.; Aljaddawi, A.A.; Alattas, S.G.; Redwan, E.M. An Overview of the Intrinsic Role of Citrullination in Autoimmune Disorders. *J. Immunol. Res.* **2019**, *2019*, 7592851. [CrossRef]
17. Maric, S.; Restin, T.; Muff, J.L.; Camargo, S.M.; Guglielmetti, L.C.; Holland-cunz, S.G.; Crenn, P.; Vuille-dit-bille, R.N. Citrulline, Biomarker of Enterocyte Functional Mass and Dietary Supplement. Metabolism, Transport, and Current Evidence for Clinical Use. *Nutrients* **2021**, *13*, 2794. [CrossRef]
18. Crans, D.C.; Kostenkova, K. Open Questions on the Biological Roles of First-Row Transition Metals. *Commun. Chem.* **2020**, *3*, 104. [CrossRef]
19. Fox, S.W. The Preparation of Citrulline by Hydrolysis of Arginine. *J. Biol. Chem.* **1938**, *123*, 687–690. [CrossRef]
20. Kurtz, A.C. A Simple Synthesis of Dl-Citrulline. *J. Biol. Chem.* **1938**, *122*, 477–484. [CrossRef]
21. Trikha, K.C.; Nair, B.C.; Singh, R.P. Complexation of Bivalent Metal Ions with Amino Acids. I. L-Citrulline Complexes. *Indian J. Chem.* **1968**, *6*, 532.
22. Clarke, E.R.; Martell, A.E. Metal Chelates of Arginine and Related Ligands. *J. Inorg. Nucl. Chem.* **1970**, *32*, 911–926. [CrossRef]
23. Yamauchi, O.; Sakurai, T.; Nakahara, A. Histidine-Containing Ternary Amino Acid-Copper(II) Complexes. Syntheses and Properties. *J. Am. Chem. Soc.* **1979**, *101*, 4164–4172. [CrossRef]
24. Ganadu, M.L.; Leoni, V.; Crisponi, G.; Nurchi, V. An Investigation on the Interaction between Palladium(II) and L-Citrulline by 1H and 13C NMR Spectroscopy and Potentiometry. *Polyhedron* **1991**, *10*, 333–336. [CrossRef]
25. Mascaliovas, B.Z.; Bergamini, F.R.G.; Cuin, A.; Corbi, P.P. Synthesis and Crystal Structure of a Palladium(II) Complex with the Amino Acid L-Citrulline. *Powder Diffr.* **2015**, *30*, 357–361. [CrossRef]
26. Singh, M.; Shankar, V.; Singh, D.; Krishna, V. Chelation and Stabilization Properties of Citrulline and Uracil with Hg(II) as a Heavy Metal Ion in Solution. *Chem. Sci. Trans.* **2017**, *6*, 646–652. [CrossRef]
27. Singh, M.; Sinha, S.; Krishna, V. Computed Distribution of Quaternary Complexes of Cu(II), Zn(II) Co(II) and Ni(II) with Citrulline and Trypthtophan as Primary Ligand and Thymine as Secondary Ligand. *Proc. Natl. Acad. Sci. India Sect. A—Phys. Sci.* **2021**, *91*, 1–7. [CrossRef]
28. Martínez-Valencia, B.; Corona-Motolinia, N.D.; Sánchez-Lara, E.; Noriega, L.; Sánchez-Gaytán, B.L.; Castro, M.E.; Meléndez-Bustamante, F.; González-Vergara, E. Cyclo-Tetravanadate Bridged Copper Complexes as Potential Double Bullet pro-Metallodrugs for Cancer Treatment. *J. Inorg. Biochem.* **2020**, *208*, 111081. [CrossRef]

29. Martínez-Valencia, B.; Corona-Motolinia, N.D.; Sánchez-Lara, E.; Sánchez-Gaytán, B.L.; Cerro-López, M.; Mendoza, A.; Castro, M.E.; Meléndez-Bustamante, F.J.; González-Vergara, E. Synthesis and Experimental-Computational Characterization of a Copper/Vanadium Compound with Potential Anticancer Activity. *Crystals* **2020**, *10*, 492. [CrossRef]
30. Corona-Motolinia, N.D.; Martínez-Valencia, B.; Noriega, L.; Sánchez-Gaytán, B.L.; Mendoza, A.; Meléndez-Bustamante, F.J.; Castro, M.E.; González-Vergara, E. Ternary Copper Complex of L-Glutamine and Phenanthroline as Counterions of Cyclo-Tetravanadate Anion: Experimental-Theoretical Characterization and Potential Antineoplastic Activity. *Metals* **2021**, *11*, 1541. [CrossRef]
31. Su, C.-C.; Tai, T.-Y.; Wu, S.-P.; Wang, S.-L.; Liao, F.-L. Spectroscopic and Electronic Properties of Mixed Ligand Aminoacidatocopper(II) Complexes: Molecular Structure of [Cu(4,7-Dimethyl-1,10-Phenanthroline)(l-Phenylalaninato)](ClO4). *Polyhedron* **1999**, *18*, 2361–2368. [CrossRef]
32. Pérez-Benítez, A.; Méndez-Rojas, M.Á.; Bernès, S.; González-Vergara, E. Hybrid (Electrochemical-Chemical) Single Crystal Synthesis of Copper Aspirinate Starting from an Aspirin Tablet: An Undergraduate Bioinorganic Experiment. *Chem. Educ. J.* **2008**, *11*, 11–18.
33. Schotten, C.; Nicholls, T.P.; Bourne, R.A.; Kapur, N.; Nguyen, B.N.; Willans, C.E. Making Electrochemistry Easily Accessible to the Synthetic Chemist. *Green Chem.* **2020**, *22*, 3358–3375. [CrossRef]
34. Bruker. *APEX3, SAINT, and SADABS*; Bruker AXS Inc.: Madison, WI, USA, 2016.
35. Sheldrick, G.M. SHELXT—Integrated Space-Group and Crystal-Structure Determination. *Acta Crystallogr. Sect. A Found. Adv.* **2015**, *71*, 3–8. [CrossRef] [PubMed]
36. Dolomanov, O.V.; Bourhis, L.J.; Gildea, R.J.; Howard, J.A.K.; Puschmann, H. OLEX2: A Complete Structure Solution, Refinement and Analysis Program. *J. Appl. Crystallogr.* **2009**, *42*, 339–341. [CrossRef]
37. Hohenberg, P.; Kohn, W. Inhomogeneous Electron Gas. *Phys. Rev.* **1964**, *136*, B864–B871. [CrossRef]
38. Adamo, C.; Jacquemin, D. The Calculations of Excited-State Properties with Time-Dependent Density Functional Theory. *Chem. Soc. Rev.* **2013**, *42*, 845–856. [CrossRef]
39. Adamo, C.; Barone, V. Exchange Functionals with Improved Long-Range Behavior and Adiabatic Connection Methods without Adjustable Parameters: The MPW and MPW1PW Models. *J. Chem. Phys.* **1998**, *108*, 664–675. [CrossRef]
40. Rassolov, V.A.; Pople, J.A.; Ratner, M.A.; Windus, T.L. 6-31G * Basis Set for Atoms K through Zn. *J. Chem. Phys.* **1998**, *109*, 1223–1229. [CrossRef]
41. Hay, P.J.; Wadt, W.R. Ab Initio Effective Core Potentials for Molecular Calculations. Potentials for the Transition Metal Atoms Sc to Hg. *J. Chem. Phys.* **1985**, *82*, 270–283. [CrossRef]
42. Marenich, A.V.; Cramer, C.J.; Truhlar, D.G. Universal Solvation Model Based on Solute Electron Density and on a Continuum Model of the Solvent Defined by the Bulk Dielectric Constant and Atomic Surface Tensions. *J. Phys. Chem. B* **2009**, *113*, 6378–6396. [CrossRef]
43. Wolinski, K.; Hinton, J.F.; Pulay, P. Efficient Implementation of the Gauge-Independent Atomic Orbital Method for NMR Chemical Shift Calculations. *J. Am. Chem. Soc.* **1990**, *112*, 8251–8260. [CrossRef]
44. Frisch, M.J.; Trucks, G.W.; Schlegel, H.B.; Scuseria, G.E.; Robb, M.A.; Cheeseman, J.R. *Gaussian 16, Revision, B.01*; Gaussian Inc.: Pittsburgh, PA, USA, 2016.
45. Dennington, R.D.; Keith, T.A.; Millam, J.M. *Gauss View*; Version 6.0.16; Semichem, Inc.: Shawnee, KS, USA, 2016.
46. Keith, T.A. *TK Gristmill Software*; Version 19.02.13; AIMAll; TK Gristmill Software: Overland Park, KS, USA, 2019.
47. Spackman, P.R.; Turner, M.J.; McKinnon, J.J.; Wolff, S.K.; Grimwood, D.J.; Jayatilaka, D.; Spackman, M.A. CrystalExplorer: A Program for Hirshfeld Surface Analysis, Visualization and Quantitative Analysis of Molecular Crystals. *J. Appl. Crystallogr.* **2021**, *54*, 1006–1011. [CrossRef]
48. Halcrow, M.A. Interpreting and Controlling the Structures of Six-Coordinate Copper(II) Centres—When Is a Compression Really a Compression? *Dalt. Trans.* **2003**, *23*, 4375–4384. [CrossRef]
49. Hemissi, H.; Nasri, M.; Abid, S.; Al-Deyab, S.S.; Dhahri, E.; Hlil, E.K.; Rzaigui, M. Crystal Structure, Spectroscopic, Magnetic and Electronic Structure Studies of a Novel Cu(II) Amino Acid Complex [Cu(l-Arg)2(H2O)]2(P4O12)·8H2O. *J. Solid State Chem.* **2012**, *196*, 489–497. [CrossRef]
50. Stevanović, N.; Zlatar, M.; Novaković, I.; Pevec, A.; Radanović, D.; Matić, I.Z.; Đorđić Crnogorac, M.; Stanojković, T.; Vujčić, M.; Gruden, M.; et al. Cu(II), Mn(II) and Zn(II) Complexes of Hydrazones with a Quaternary Ammonium Moiety: Synthesis, Experimental and Theoretical Characterization and Cytotoxic Activity. *Dalt. Trans.* **2022**, *51*, 185–196. [CrossRef]
51. Suktanarak, P.; Watchasit, S.; Chitchak, K.; Plainpan, N.; Chainok, K.; Vanalabhpatana, P.; Pienpinijtham, P.; Suksai, C.; Tuntulani, T.; Ruangpornvisuti, V.; et al. Tuning the Reactivity of Copper Complexes Supported by Tridentate Ligands Leading to Two-Electron Reduction of Dioxygen. *Dalt. Trans.* **2018**, *47*, 16337–16349. [CrossRef]
52. Halcrow, M.A. Jahn-Teller Distortions in Transition Metal Compounds, and Their Importance in Functional Molecular and Inorganic Materials. *Chem. Soc. Rev.* **2013**, *42*, 1784–1795. [CrossRef]
53. Jezowska-Trzebiatowska, B.; Gerega, K.; Vogt, A. Cobalt(II) Complexes Reversibly Uptaking Nitric Oxide in Aqueous Solution. *Inorg. Chim. Acta* **1978**, *31*, 183–185. [CrossRef]
54. Mofidabadi, A.H.J.; Dehghani, A.; Ramezanzadeh, B. Investigating the Effectiveness of Watermelon Extract-Zinc Ions for Steel Alloy Corrosion Mitigation in Sodium Chloride Solution. *J. Mol. Liq.* **2022**, *346*, 117086. [CrossRef]

55. Etter, M.C. Encoding and Decoding Hydrogen-Bond Patterns of Organic Compounds. *Acc. Chem. Res.* **1990**, *23*, 120–126. [CrossRef]
56. Humphrey, W.; Dalke, A.; Schulten, K. VMD: Visual Molecular Dynamics. *J. Mol. Graph.* **1996**, *14*, 33–38. [CrossRef]
57. Smith, D.W. Ligand Field Splittings in Copper (II) Compounds. In *Structure and Bonding*; Springer: Berlin/Heidelberg, Germany, 1972; Volume 12, pp. 49–112.
58. Hathaway, B.J. A New Look at the Stereochemistry and Electronic Properties of Complexes of the Copper(II) Ion. In *Complex Chemistry. Structure and Bonding*; Springer: Berlin/Heidelberg, Germany, 1984; pp. 55–118. ISBN 978-3-540-13411-4.
59. Stanila, A.; Marcu, A.; Rusu, D.; Rusu, M.; David, L. Spectroscopic Studies of Some Copper(II) Complexes with Amino Acids. *J. Mol. Struct.* **2007**, *834*, 364–368. [CrossRef]
60. Reddy, S.L.; Endo, T.; Reddy, G.S. Electronic (Absorption) Spectra of 3d Transition Metal Complexes. In *Advanced Aspects of Spectroscopy*; Farrukh, M.A., Ed.; IntechOpen: London, UK, 2012; pp. 3–48.
61. Wojciechowska, A.; Szuster-Ciesielska, A.; Sztandera, M.; Bregier-Jarzębowska, R.; Jarząb, A.; Rojek, T.; Komarnicka, U.K.; Bojarska-Junak, A.; Jezierska, J. L-Argininato Copper(II) Complexes in Solution Exert Significant Selective Anticancer and Antimicrobial Activities. *Appl. Organomet. Chem.* **2020**, *34*, e5698. [CrossRef]
62. Sciortino, G.; Maréchal, J.-D.; Fábián, I.; Lihi, N.; Garribba, E. Quantitative Prediction of Electronic Absorption Spectra of Copper(II)-Bioligand Systems: Validation and Applications. *J. Inorg. Biochem.* **2020**, *204*, 110953. [CrossRef] [PubMed]
63. Cuevas, A.; Viera, I.; Torre, M.H.; Kremer, E.; Etcheverry, S.B.; Baran, E.J. Infrared Spectra of the Copper (II) Complexes of Amino Acids with Hydrophobic Residues. *Acta Farm. Bonaer.* **1998**, *17*, 213–218.
64. Baran, E.J.; Wagner, C.C.; Torre, M.H.; Kremer, E.; Kögerler, P. Vibrational Spectra of the Cu(II) Complexes of Aspartic and Glutamic Acids. *Acta Farm. Bonaer.* **2000**, *19*, 231–234.
65. Jenkins, A.L.; Larsen, R.A.; Williams, T.B. Characterization of Amino Acids Using Raman Spectroscopy. *Spectrochim. Acta A. Mol. Biomol. Spectrosc.* **2005**, *61*, 1585–1594. [CrossRef]
66. Nakamoto, K. *Infrared and Raman Spectra of Inorganic and Coordination Compounds: Part B: Applications in Coordination, Organometallic, and Bioinorganic Chemistry*, 6th ed.; John Wiley and Sons: Hoboken, NJ, USA, 2009; ISBN 9780471743392.
67. Sreevalsa, V.G.; Jayalekshmi, S. Investigations on the Growth and Characterization of L-Citrulline Oxalate Monohydrate Single Crystal. *J. Cryst. Growth* **2011**, *324*, 172–176. [CrossRef]
68. Freire, P.T.C.; Barboza, F.M.; Lima, J.A.; Melo, F.E.A.; Filho, J.M. Raman Spectroscopy of Amino Acid Crystals. In *Raman Spectroscopy and Applications*; Khan, M., Ed.; IntechOpen: London, UK, 2017; pp. 201–223.
69. Fernández, L.E.; Delgado, G.E.; Maturano, L.V.; Tótaro, R.M.; Varetti, E.L. Experimental and Theoretical Vibrational Study of N-Carbamoyl-L-Proline. *J. Mol. Struct.* **2018**, *1168*, 84–91. [CrossRef]
70. Vusak, D.; Pejic, J.; Jurkovic, M.; Szalontai, G.; Sabolovic, J. Coordination Polymers of Paramagnetic Bis(Leucinato)Copper(Ii) Diastereomers: Experimental and Computational Study of the Stereoisomerism and Conformations. *CrystEngComm* **2020**, *22*, 5587–5600. [CrossRef]
71. Mao, J.; Zhang, Y.; Oldfield, E. Nuclear Magnetic Resonance Shifts in Paramagnetic Metalloporphyrins and Metalloproteins. *J. Am. Chem. Soc.* **2002**, *124*, 13911–13920. [CrossRef]
72. Patil, B.H.; Peraje, P.; Naik, D.; Rajaramakrishna, R.; Dittmer, J.; Kumar, S.; Swamy, K. Experimental 1H and 13C Solid-State NMR Signal Assignment of Paramagnetic Copper (II) 2-Pyrazine-Carboxylate Complex Using Density Functional Theory Calculations. *J. Phys. Conf. Ser.* **2021**, *1819*, 012032. [CrossRef]
73. Corona-Motolinia, N.D.; Martínez-Valencia, B.; Noriega, L.; Sánchez-Gaytán, B.L.; Méndez-Rojas, M.Á.; Melendez, F.J.; Castro, M.E.; González-Vergara, E. Synthesis, Crystal Structure, and Computational Methods of Vanadium and Copper Compounds as Potential Drugs for Cancer Treatment. *Molecules* **2020**, *25*, 4679. [CrossRef]
74. Caruso, A.; Rossi, M.; Gahn, C.; Caruso, F. A Structural and Computational Study of Citrulline in Biochemical Reactions. *Struct. Chem.* **2017**, *28*, 1581–1589. [CrossRef]
75. Jia, G.-G.; Liu, Y.; Zheng, Y.-G.; Zhang, L.; Liu, L.; Ling, X.; Gou, L.-S.; Yin, C.; Zhuang, X.-M. A Novel Preparation of L-Citrulline and L-Homocitrulline. *J. Chem. Soc. Pak.* **2012**, *34*, 451–454.
76. Olivard, J.; Metzler, D.E.; Snell, E.E. Catalytic Racemization of Amino Acids by Pyridoxal and Metal Salts. *J. Biol. Chem.* **1952**, *199*, 669–674. [CrossRef]
77. Abbott, E.H.; Martell, A.E. Mechanism of Formation, Structure, Stereochemistry, and Racemization of Bis[Pyridoxylidene (Amino Acidato)]Aluminum(III) Complexes. *J. Am. Chem. Soc.* **1970**, *92*, 5845–5851. [CrossRef]
78. Fu, R.; So, S.M.; Lough, A.J.; Chin, J. Hydrogen Bond Assisted L to D Conversion of α-Amino Acids. *Angew. Chemie Int. Ed.* **2020**, *59*, 4335–4339. [CrossRef]
79. Kolarovič, A.; Jakubec, P. State of the Art in Crystallization-Induced Diastereomer Transformations. *Adv. Synth. Catal.* **2021**, *363*, 4110–4158. [CrossRef]
80. Augeri, D.J.; Robl, J.A.; Betebenner, D.A.; Magnin, D.R.; Khanna, A.; Robertson, J.G.; Wang, A.; Simpkins, L.M.; Taunk, P.; Huang, Q.; et al. Discovery and Preclinical Profile of Saxagliptin (BMS-477118): A Highly Potent, Long-Acting, Orally Active Dipeptidyl Peptidase IV Inhibitor for the Treatment of Type 2 Diabetes. *J. Med. Chem.* **2005**, *48*, 5025–5037. [CrossRef]
81. Daugan, A.; Grondin, P.; Ruault, C.; Le Monnier De Gouville, A.C.; Coste, H.; Kirilovsky, J.; Hyafil, F.; Labaudinière, R. The Discovery of Tadalafil: A Novel and Highly Selective PDE5 Inhibitor. 1: 5,6,11,11a-Tetrahydro-1H-Imidazo[1′,5′:1,6]Pyrido[3,4-b]Indole-1,3(2H)-Dione Analogues. *J. Med. Chem.* **2003**, *46*, 4525–4532. [CrossRef]

82. Herbert, J.M.; Frehel, D.; Vallee, E.; Kieffer, G.; Gouy, D.; Berger, Y.; Necciari, J.; Defreyn, G.; Maffrand, J.P. Clopidogrel, A Novel Antiplatelet and Antithrombotic Agent. *Cardiovasc. Drug Rev.* **1993**, *11*, 180–198. [CrossRef]
83. Kuehl, F.A.; Wolf, F.J.; Trenner, N.R.; Peck, R.L.; Howe, E.; Hunnewell, B.D.; Downing, G.; Newstead, E.; Folkers, K.; Buhs, R.P.; et al. D-4-Amino-3-Isoxazolidone, A New Antibiotic. *J. Am. Chem. Soc.* **1955**, *77*, 2344–2345. [CrossRef]
84. Kwong, A.D.; Kauffman, R.S.; Hurter, P.; Mueller, P. Discovery and Development of Telaprevir: An NS3-4A Protease Inhibitor for Treating Genotype 1 Chronic Hepatitis C Virus. *Nat. Biotechnol.* **2011**, *29*, 993–1003. [CrossRef]
85. Anderson, J. Degarelix: A Novel Gonadotropin-Releasing Hormone Blocker for the Treatment of Prostate Cancer. *Futur. Oncol.* **2009**, *5*, 433–443. [CrossRef]
86. Khan, M.L.; Stewart, A.K. Carfilzomib: A Novel Second-Generation Proteasome Inhibitor. *Futur. Oncol.* **2011**, *7*, 607–612. [CrossRef]
87. So, S.M.; Kim, H.; Mui, L.; Chin, J. Mimicking Nature to Make Unnatural Amino Acids and Chiral Diamines. *Eur. J. Org. Chem.* **2012**, *2012*, 229–241. [CrossRef]
88. Breuer, M.; Ditrich, K.; Habicher, T.; Hauer, B.; Keßeler, M.; Stürmer, R.; Zelinski, T. Industrial Methods for the Production of Optically Active Intermediates. *Angew. Chem. Int. Ed. Engl.* **2004**, *43*, 788–824. [CrossRef]
89. Blaskovich, M.A.T. Unusual Amino Acids in Medicinal Chemistry. *J. Med. Chem.* **2016**, *59*, 10807–10836. [CrossRef]

crystals

MDPI

Article

Magnesium Coordination Chemistry: A Case Study of Magnesium Carboxylate Complexes with Hexamethylenetetramine

Tomasz Sierański *, Agata Trzęsowska-Kruszyńska *, Marcin Świątkowski, Marta Bogdan and Paulina Sobczak

Institute of General and Ecological Chemistry, Lodz University of Technology, Zeromskiego 116, PL-90924 Lodz, Poland

* Correspondence: tomasz.sieranski@p.lodz.pl (T.S.); agata.trzesowska-kruszynska@p.lodz.pl (A.T.-K.)

Abstract: Three magnesium coordination compounds were obtained to explore the influence of carboxylate anions on the pattern of the formed hydrogen bonds. For their synthesis, various salts of magnesium carboxylic acid (formate, acetate, and propionate) were utilized. As an *N*-donor ligand, hexamethylenetetramine was employed. The supramolecular structures of the obtained compounds were determined and evaluated in Hirshfeld analysis. The length of the carbon chain of the used carboxylate anions has been proven to have a considerable impact on the self-organization of the supramolecular system by altering the three-dimensional net of the created hydrogen bonds. IR spectroscopy was used to characterize the obtained compounds, revealing significant differences between distinct systems. The thermal analysis of the investigated compounds also shows noticeable differences, demonstrating better stability of the systems containing formate anions.

Keywords: magnesium; formate; acetate; propionate; hexamethylenetetramine; crystal structure; thermal analysis; infrared spectroscopy; hydrogen bonds

Citation: Sierański, T.; Trzęsowska-Kruszyńska, A.; Świątkowski, M.; Bogdan, M.; Sobczak, P. Magnesium Coordination Chemistry: A Case Study of Magnesium Carboxylate Complexes with Hexamethylenetetramine. *Crystals* **2022**, *12*, 1434. https://doi.org/10.3390/cryst12101434

Academic Editor: Maria Gazda

Received: 16 September 2022
Accepted: 8 October 2022
Published: 11 October 2022

1. Introduction

The design of the distinctive metal–organic materials attracts growing interest [1–3]. However, synthesizing a solid-state structure with desired properties is a challenging task that demands understanding the influence of many factors, such as characteristics of the used cation, ligands, and solvent(s); and the formation of intermolecular interactions in such a selected environment [4–6]. Currently, the most prevalent strategies that are used in crystal engineering engage hydrogen/halogen-bonding and coordination bonding [7–10]. Achievements in this field contribute to developing many new materials and compounds that might be used, among others, in electronics and the medical and pharmaceutical industries [11–13]. Yet, developing and constructing coordination systems with the requisite topologies and features is still difficult. This requires deep knowledge of the impact of the factors mentioned above on the hierarchy of interactions occurring in the emerging coordination system. One of the cations for which coordination compounds are of interest and for which there is a need for their synthesis is the magnesium cation. The coordination chemistry of this cation is of particular importance due to the unique role of Mg^{2+} in many biological processes [14–17]. At the molecular level, magnesium cations participate in homeostasis, are essential in electrolyte pathophysiology, and activate many enzyme systems such as alkaline phosphatase, peptidases, and enzymes transferring phosphate [18,19]. In addition, enzymes that contribute to vitamin D metabolism are magnesium-dependent [20–22]. Since divalent magnesium is classified as a hard acid, it tends to bind with ligands that contain a hard oxygen atom, such as water molecules or carboxylate, phosphate, and hydroxylate ions. It may be coordinated directly to a macromolecular binding site or indirectly through water molecules. The first type of binding concerns proteins, while binding through coordinating water molecules dominates the interactions with nucleic acids. Even though the coordination chemistry of magnesium is widely explored, the factors that make such systems (including "outer-sphere" coordination mode or "inner-sphere" mode) the most energetically beneficial are still not fully understood [23–25]. Hence, the investigation of magnesium coordination chemistry, especially related to *N*-donor ligand systems, may help to learn the binding mechanism of this metal to many bioactive molecules (including proteins and nucleic acids) and have vital importance due to the need for biologically active or biologically important magnesium compounds for clinical applications.

In order to broaden the knowledge on crystal engineering of coordination compounds containing an important ion, which is a magnesium cation, a series of its coordination compounds, of general formula $[Mg(H_2O)_6]A_2{\cdot}2$ hmta$\cdot 4H_2O$ (A–carboxylate anion), was obtained. Hexamethylenetetramine (hmta) was selected as an *N*-donor ligand. Carboxylic anions, differing in chain length in each compound, were selected as counterions. Hmta serves as a model ligand for many bioactive molecules (including amino acids, proteins, and nucleotides) due to possessing unshielded nitrogen atoms [26–28]. This ligand also forms many coordination links. It can be present in both the outer and/or inner coordination sphere, where it acts as a bridging and/or terminal ligand [28,29]. Hmta was also a potent agent used for treating urinary tract infections [30]. For all the above reasons, some coordination aspects of the presented magnesium complex compounds can be extended to other bioactive molecules. Additionally, hmta is a vital agent used in the production of powdery or in synthesizing phenolic resins and their molding compounds [31,32]. Usage of modified, hmta-containing compounds as hardening components in these processes can increase the beneficial properties of the mentioned resins. In general, the salts of carboxylic acids are well soluble in water and the carboxylic anions with short carbon chains form associations with water molecules through hydrogen bonds. Using different anions of carboxylic acids (formates, acetates, and propionates) will allow concluding the influence of the length of the carbon chains occurring in these anions on the type of created intermolecular interactions and the geometry of the formed coordination entity. The presented study may be essential not only in obtaining the new materials for potential medical uses, but may also provide further conceptual benefits, and help to understand the relationship between the structure and compound properties (for instance, thermolysis).

2. Materials and Methods

2.1. Materials and Synthesis

All the reagents (hexamethylenetetramine, magnesium carbonate, formic, acetic, and propionic acids) were analytically pure and obtained from POCh S.A (Gliwice, Poland). The samples of magnesium carbonate (0.002 mmol) were suspended in a possibly small amount of water (about 5 cm^3) and then mixed with the formic, acetic, and propionic acid (0.002 mmol), respectively. The mixtures were stirred on a magnetic stirrer for about 15 min, and next, they were filtrated to remove unreacted excess magnesium carbonate. The residues on the filters were washed three times with 5 cm^3 of cold water (the magnesium ions were not detected in the last portion of filtrates). To each solution of combined filtrates (only filtrates relating to the particular magnesium carbonate sample were combined), the 3 cm^3 of a water solution containing 0.1402 g (0.001 mmol) of hmta was added. The obtained solutions were stirred on the magnetic stirrer for 15 min and were placed in a refrigerator at 5 °C. All syntheses were repeated with 1:1 and 1:2 Mg:hmta ratios (0.002 mmol and 0.004 mmol of hmta, respectively). After 3 weeks, the colorless crystals started to grow. The crystals were isolated from the solutions shortly after they were grown. Therefore, a few fractions of crystals were collected from each synthesis solution. The IR spectrum was conducted for each fraction, and the fractions were combined if their spectra were the same. For syntheses performed with 2:1 and 1:1 ratios, the last few fractions were pure magnesium carboxylate salts. Independently of the substrate ratio, the same product was formed for each used magnesium carboxylate. The reaction yields (in relation to hmta), respectively for the syntheses performed with 2:1, 1:1, and 1:2 ratios were: compound 1 (magnesium formate and hmta) 98%, 95%, and 94%; compound 2 (magnesium acetate and hmta) 98%, 94%, and 90%; compound 3 (magnesium propionate and hmta) 97%, 92%, and 89%. Elemental analysis for the obtained compounds 1–3 (calculated/found) [%] (1): C:29.27%/29.21; H:8.01%/7.99%; Mg:4.23%/4.31%; N:19.50%/19.53%; O:38.99%/39.04; (2): C:31.85%/32,05%; H:8.38%/8.50%; Mg:4.03%/3.89%; N:18.59%/18.71%; O:37.15%/37.01; (3): C:34.28%/34.35; H:8.56%/8.60%; Mg:3.86%/3.77%; N:17.77%/17.89%; O:35.52%/35.60%.

2.2. Crystal Structure Determination

X-ray diffraction data of 1–3 were collected at temperature 291.0(3) K, on a KM-4-CCD automatic diffractometer equipped with CCD detector and fine-focus sealed X-ray tubes generated monochromatic MoK_α and CuK_α radiations. Lorentz, polarization, and numerical absorption [33] corrections were applied. The structures were solved with the SHELXT [34] using intrinsic phasing and refined with the SHELXL [35] using least squares minimization. All the non-hydrogen atoms were refined anisotropically. All the hydrogen atoms were found from the Fourier difference map and refined using the "riding" model. The structure of **2** could not be fully refined due to the disorder of the outer coordination sphere species, which were impossible to localize on the Fourier difference map, despite splitting molecules into parts. The application of the solvent mask function only worsened the refinement. Details concerning crystal data and refinement of **1** and **3** are given in Table 1.

Table 1. Crystal data and structure refinement details for **1** and **3**.

Compound	1	3
Empirical formula	$C_{14}H_{46}MgN_8O_{14}$	$C_{18}H_{54}MgN_8O_{14}$
Formula weight	574.90	631.00
Crystal system, space group	Triclinic, *P*-1	Triclinic, *P*-1
Temperature (K)	291.0 (3)	291.0 (3)
Radiation	MoKα (λ = 0.71073 Å)	CuKα (λ = 1.54178)
a (Å)	9.2503 (3)	8.3713 (4)
b (Å)	9.2802 (4)	9.1600 (7)
c (Å)	9.3256 (4)	12.0634 (6)
α (°)	76.621 (4)	94.210 (6)
β (°)	60.933 (4)	100.128 (3)
γ (°)	79.673 (3)	114.963 (2)
Volume (Å^3)	678.63 (5)	814.18 (8)
Z, Calculated density (g/cm^3)	1, 1.407	1, 1.287
Absorption coefficient (mm^{-1})	0.142	1.094
Min. and max. transmission	0.946 and 0.968	0.811 and 0.819
F(000)	310	342
Crystal size (mm)	0.381 × 0.299 × 0.267	0.187 × 0.183 × 0.180
2θ range for data collection (°)	4.526 to 50.04	3.77 to 67.64
Index ranges	$-10 \leq h \leq 10, -11 \leq k \leq 10, -11 \leq l \leq 10$	$-10 \leq h \leq 9, -10 \leq k \leq 10, -14 \leq l \leq 14$
Reflections collected	6794	9286
Independent reflections	2386 [R_{int} = 0.0202, R_{sigma} = 0.0190]	2800 [R_{int} = 0.0287, R_{sigma} = 0.0275]
Data/restraints/parameters	2386/0/169	2800/0/189
Goodness-of-fit on F^2	1.084	1.059
Final R indexes [$I > 2\sigma (I)$]	$R_1 = 0.0329$, $wR_2 = 0.1012$	$R_1 = 0.0372$, $wR_2 = 0.0979$
Final R indexes [all data]	$R_1 = 0.0379$, $wR_2 = 0.1033$	$R_1 = 0.0377$, $wR_2 = 0.0984$
Largest diff. peak and hole (e•Å^{-3})	0.47 and −0.19	0.32 and −0.26

2.3. Other Measurements

The thermal analyses were carried out in a TG/DTA-SETSYS-16/18 thermoanalyzer. The samples were heated in corundum crucibles up to 1000 °C at a heating rate of 10 °C min^{-1} in the air atmosphere. Elemental analyses were carried out using a Vario EL III CHNOS elemental analyzer (C, H, N, O). Magnesium content was determined by complexometric titration with the 0.01 mol/dm^3 water EDTA solution as a complexing agent [36]. IR spectra of the investigated compound were recorded as KBr disc on a Nicolet Magna 560 spectrophotometer over the range 4000−400 cm^{-1}.

3. Results and Discussion

3.1. Structural Analysis

The syntheses of magnesium carboxylates (formate, acetate, and propionate) with hmta led to obtaining three coordination compounds of the same formula $[Mg(H_2O)_6]A_2 \cdot 2$ hmta$\cdot 4H_2O$ (A–carboxylate anion). They were formed regardless of the substrate stoichiometry applied in the synthesis reactions. These are discrete compounds with the hexaaquamagnesium coordination entities (Figure 1). In all cases, the central atoms are located on the inversion centers (special positions *h*, *c*, and *e* of the $P2_1$ space group, respectively for **1**–**3**). The coordination polyhedra are almost perfect octahedrons (Table 2). The carboxylate anions are located in the outer coordination sphere together with hmta and water molecules. The Mg:hmta ratio is 1:2. In the case of **2**, the outer coordination sphere species, except hmta, could not be refined due to unsolvable disorder (Figure S1) (see supplementary materials). In consequence, compound 2 was not included in the following structural analysis concerning intermolecular interactions. The composition of **2** was established based on elemental and thermal analyses.

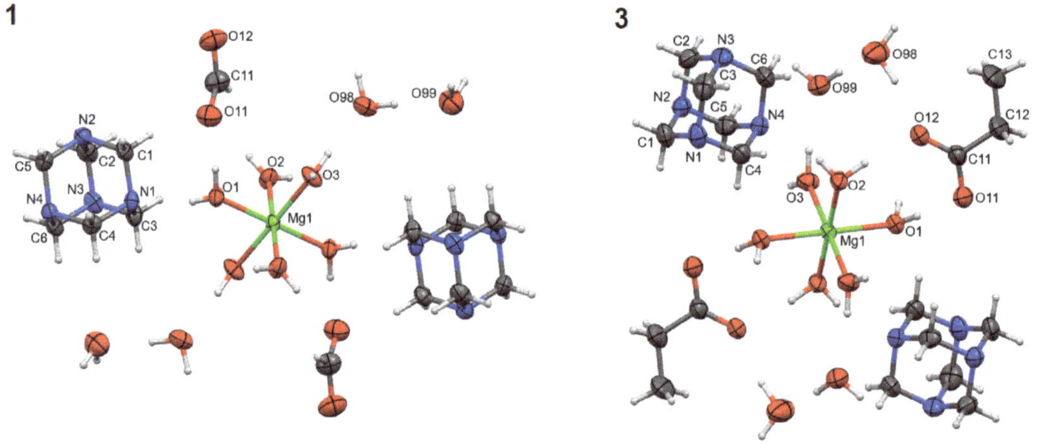

Figure 1. Molecular structures of **1** and **3**, plotted with a 50% probability of displacement ellipsoids of nonhydrogen atoms and as spheres of arbitrary radii for hydrogen atoms. The equivalent atoms (without labels) were generated according to transformation: (**1**) −x + 1, −y + 1, −z + 1; (**3**) −x + 1, −y + 1, −z + 2.

Table 2. Selected structural data for **1** and **3** (Å, °).

Compound 1		Compound 3	
Mg1—O1	2.0350 (9)	Mg1—O1	2.0732 (9)
Mg1—O2	2.0695 (10)	Mg1—O2	2.0587 (9)
Mg1—O3	2.0349 (9)	Mg1—O3	2.0716 (9)
O1—Mg1—O2	87.01 (4)	O1—Mg1—O2	90.59 (4)
O1—Mg1—O2 ^{i}	92.99 (4)	O1—Mg1—O2 ^{ii}	89.41 (4)
O1—Mg1—O3	91.06 (4)	O1—Mg1—O3	90.02 (4)
O1—Mg1—O3^{i}	88.94 (4)	O1—Mg1—O3 ^{ii}	89.98 (4)
O2—Mg1—O3	87.93 (4)	O2—Mg1—O3	90.42 (4)
O2—Mg1—O3 ^{i}	92.07 (4)	O2—Mg1—O3 ^{ii}	89.58 (4)

Symmetry transformations used to generate equivalent atoms: (i) −x + 1, −y + 1, −z + 1; (ii) −x + 1, −y + 1, −z + 2.

Despite the same general formula, the supramolecular structures of the studied systems show significant differences. The following hydrogen bonds stabilize their three-dimensional crystal nets: O—H•••O, O—H•••N, and C—H•••O (Table 3). They comprise carboxylate anions, water, and hmta molecules (Figure 2), creating several different graph set patterns [37,38]. Considering the unitary graph set level, only D motifs are observed. Chain and ring patterns appear in the binary graph set. In the case of **1**, $N_2C_2^2(8)$, $N_2C_4^4(16)$, $N_2R_4^4(12)$, and $N_2R_4^4(16)$ patterns are found. The $C_2^2(8)$ pattern is composed of O1—H1O•••O11 and O2—H2O•••O12 hydrogen bonds, and in this case, the water molecules (acting as hydrogen bond donors) coordinating magnesium ions (each molecule to a different ion) are linked to oxygen atoms of the formate ion (hydrogen bond acceptors). This $C_2^2(8)$ pattern shares the graph paths with the $N_2R_4^4(16)$ ring pattern. One of the $N_2R_4^4(16)$ ring patterns contains coordinated water and hmta molecules linked by O1—H1P•••N1 and O2—H2P•••N4 intermolecular hydrogen bonds. The next two (of a total four) different $N_2R_4^4(16)$ ring patterns also contain hmta molecules. The four existing $N_2C_4^4(16)$ patterns can be expressed as *abba*, *cddc*, *ceec*, and *deed* (where a, b, c, d, and e denote the O1—H1O•••O11, O2—H2O•••O12, O1—H1P•••N1, O2—H2P•••N4, and O3—H3P•••N3 hydrogen bonds, respectively). The solely $N_2R_4^4(12)$ ring is created by O99—H99O•••O11 and O99—H99P•••O12 hydrogen bonds, and it exists between the formate ions and water molecules located in the outer coordination sphere. Unitary graph set level of **3** is created by $C_2^2(8)C_2^2(8)C_2^2(8)R_2^2(8)C_4^4(16)C_4^4(16)C_4^4(16)R_4^4(16)R_4^4(16)R_4^4(16)$ motifs. $C_2^2(8)$, $C_4^4(16)$, and $R_4^4(16)$ patterns are formed by O—H•••N hydrogen bonds between magnesium coordinated water molecules and hmta molecules. $C_2^2(8)$ patterns can be expressed as *ab*, *ac*, and *bc*, whilst $C_4^4(16)$ as *abba*, *acca*, and *bccb* (where a, b, and c refer to O1—H10•••N1, O2—H20•••N4, and O3—H30•••N2 hydrogen bonds, respectively). The same hydrogen bonds create $R_4^4(16)$ patterns and the particular rings can be written as *abab*, *acac*, and *bcbc*. The propionic ions, unlike in **1**, are engaged only in the solely $R_2^2(8)$ pattern created by hydrogen bonds appearing between these ions and magnesium water coordinated molecules (O1—H1P•••O11 and O2—H2P•••O12).

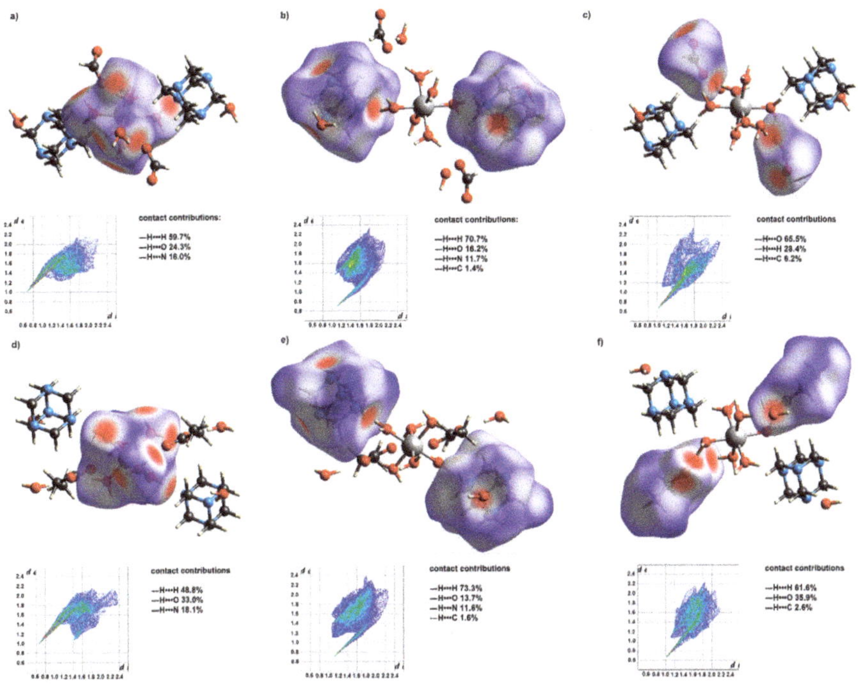

Figure 2. Hirshfeld surfaces of 1 and 3 for their coordination entities (**a**,**d**), hmta ligands (**b**,**e**), and carboxylate anions (**c**,**f**). Two-dimensional fingerprint maps of the mentioned species, together with the contact contributions, are located below the respective surfaces. The Hirshfeld surfaces were plotted over d_{norm}.

Table 3. Hydrogen bonds in **1** and **3** (Å, °).

D-H•••A	d(D-H)	d(H•••A)	d(D•••A)	<(DHA)
		Compound 1		
O1—H1O•••O11	0.89	1.80	2.6885(15)	173.5
O1—H1P•••N1 i	0.76	2.07	2.8226(15)	166.7
O2—H2O•••O12 ii	0.85	1.93	2.7683(16)	169.9
O2—H2P•••N4 iii	0.80	2.07	2.8467(15)	164.1
O3—H3O•••O98 iv	0.86	1.80	2.6634(14)	173.0
O3—H3P•••N3 v	0.88	1.94	2.8182(16)	170.8
O98—H98O•••N2	0.91	1.90	2.8010(16)	174.8
O98—H98P•••O99 vi	0.92	1.80	2.7251(18)	177.8
O99—H99O•••O11 vii	0.85	1.97	2.7560(18)	154.3
O99—H99P•••O12 viii	0.90	1.86	2.729(2)	161.4
		Compound 3		
O1—H1O•••N1	0.86	2.00	2.8386(17)	163.6
O1—H1P•••O11 ix	0.93	1.75	2.6716(15)	171.6
O2—H2O•••N4 x	0.81	2.06	2.8496(18)	163.2
O2—H2P•••O12	0.90	1.81	2.7053(16)	173.0
O3—H3O•••N2 xi	0.85	1.98	2.8271(17)	177.4
O3—H3P•••O99 xii	0.88	1.89	2.7618(16)	170.0
O98—H98O•••N3 xiii	0.91	2.09	2.9609(19)	160.0
O98—H98P•••O12	0.98	1.78	2.751(2)	170.1
O99—H99O•••O11 xiv	0.85	1.91	2.756(2)	173.8
O99—H99P•••O98 xii	0.93	1.88	2.800(2)	170.0

Symmetry transformations used to generate equivalent atoms: (i) −x + 1, −y + 1, −z + 1; (ii) +x, +y − 1, +z; (iii) +x + 1, +y, +z; (iv) +x, +y + 1, +z; (v) −x + 1, −y + 1, −z; (vi) +x, +y − 1, +z; (vii) +x − 1, +y, +z; (viii) −x + 1, −y + 2, −z; (ix) −x + 1, −y + 1, −z + 2; (x) −x + 1, −y + 2, −z + 2; (xi) x−1, y−1, z; (xii) −x, −y + 1, −z + 1; (xiii) x − 1, y − 1, z − 1; (xiv) −x + 1, −y + 1, −z + 1.

The resulting systems' Hirshfeld surfaces and 2D fingerprints are comparable, yet changes in supramolecular structure may be detected. Such differences demonstrate the influence of the carbon chain length of the present carboxylic acid anions on their participation in the formation of intermolecular interactions. The Hirshfeld surface of the formate anions contains a much greater proportion of the red area (Figure 2), which corresponds to the strong —H•••O interactions. In the case of propionate anions, due to their longer carbon chain, —H•••H contacts account for a more significant contribution to intermolecular interactions. Thus, water molecules of the outer coordination sphere are more actively engaged in creating hydrogen bonds with formate anions. It is reflected in hydrogen bond patterns engaging uncoordinated water molecules and carboxylate anions. In these patterns, the ratio of formate anions to water molecules is 1 to 1, while for propionate anions, it is 1 to 2.

3.2. FT-IR Analysis

The IR spectra of the studied compounds show many similarities (Figures S2–S4). The most noticeable is the strongest band corresponding to the stretching vibration of CN bonds (υ CN) of hmta molecules (Table 4). In the spectra of the coordination compounds, it is slightly shifted towards higher frequencies. Another strong band, corresponding to the NCN bending vibration of hmta (δ NCN), is also shifted toward higher frequencies; however, this shift is stronger (by about 15 cm^{-1}). These shifts are due to the involvement of the nitrogen atoms of the hmta ligand in forming O-H•••N hydrogen bonds that engage water molecules directly coordinated to the magnesium cation. The mentioned band of υ CN and the band assigned to rocking vibrations of CH_2 (ρ CH_2), which serve together to recognize a binding mode of hmta, are not split; thus, confirming uncoordinated hmta [39]. The spectra of all the compounds also contain strong bands corresponding to the asymmetric and symmetric stretching vibrations of the COO groups of carboxylate anions. The difference between their wavenumbers, known as a separation parameter Δυ, is used to establish the COO coordination behavior [40,41]. For the studied compounds, $\Delta\upsilon$ is 246, 160, and 141 cm^{-1}. These values are in agreement with those for known compounds containing uncoordinated formate, acetate, and propionate anions [41–43]. The significantly greater value of $\Delta\upsilon$ for formate, regardless of the same binding mode, results from its different structure. The OCO angle is larger for formate than for acetate and propionate due to a substituent bonded to the COO group (hydrogen versus aliphatic group). It was proven that the larger the OCO angle, the greater $\Delta\upsilon$ is [41]; thus, $\Delta\upsilon$ is generally greater for formate than for acetate and propionate. The most significant differences between the spectra of the studied compounds are noted for **1**. Its spectrum contains several weak bands occurring at the wavenumbers in the range of 2700–2800 cm^{-1}. These bands correspond to the stretching vibrations of the CH_2 groups of hmta molecules, and, in the given range, they are not present in the spectra of **2** and **3** (Table 4). Moreover, in the spectrum of **1**, one cannot see a separate band corresponding to the bending vibrations of water molecules. The respective band is seen in the spectra of the other compounds (**2** and **3**) at around 1680 cm^{-1} (Table 4). In the case of all the spectra, the typical broad band associated with the stretching vibrations of the O-H group of water molecules is centered at around 3420–3450 cm^{-1}.

Table 4. Vibrational frequencies (cm^{-1}) and their assignments for the studied compounds.

1	2	3	hmta [44]	$Mg(HCOO)_2$ [45]	$Mg(CH_3COO)_2$ [46]	$NaCH_3CH_2COO$ [43]	Assignment
3427 br	3444 br	3422 br					υ OH (H_2O)
2975 w	2974 w	2977 w	2966			2973	υ_{as} CH_2, υ_{as} CH_3
2937 w	2938 w	2941 w	2955	2907	2930	2937	υ_s CH_2, υ CH, υ_s CH_3
	2891 w	2888 w	2874				υ_s CH_2
2796 w							υ_s $CH_2(NCH_2N)$
2742 w							υ_s $CH_2(NCH_2N)$
2719 w							υ_s $CH_2(NCH_2N)$
2477 w	2477 w	2477 w					υ_s $CH_2(NCH_2N)$
	1683 m	1679 m					δ OH (H_2O)
1596 s	1568 s	1558 s		1615	1554	1563	υ_{as} COO
1466 m	1464 s	1465 s	1458		1450	1461	σ CH_2, δ_{as} CH_3
	1408 s	1417 s			1430	1428	υ_s COO
1383 s	1380 m	1379 m	1370	1392		1376	ω CH_2, δ-α CH, δ_s CH_3
1350 s				1365			υ_s COO
	1343 w				1351		δ_s CH_3
		1299 s				1301	ω CH_2
1241 s	1240 s	1240 s	1234				ρ CH_2
		1078 w				1077	ρ-α CH_3
1009 s	1010 s	1010 s	1007				υ CN
	924 m	877 w			949	881	υ CC
814 m	814 w	817 w	825				υ CN
765 m	752 br	758 br		761	671	646	σ COO
688 s	687 s	690 s	673				δ NCN
508 m	507 m	509 m	512				ω NCN

Vibrations symbols: w—weak, m—medium, s—strong, br—broadened, υ—stretching, δ—bending, τ—twisting, ω—wagging, σ—scissoring, ρ—rocking, α—in-plane, γ—out-of-plane, s—symmetric, as—asymmetric.

3.3. Thermal Analysis

The thermal decomposition of the studied compounds is a gradual process (Figures S5–S7). In the case of all the compounds, the first state is associated with the endothermic removal of water molecules. In **1** and **3**, this is a one-step process; in 2 at the beginning, only water molecules of the outer coordination sphere are lost. The mass loss indicates four water molecules, which next to elemental analysis, is an additional confirmation of the water content in **2** (Table 5). The six remaining water molecules (belonging to the inner coordination sphere) are removed during the second stage. At this point, the mass spectra show ion current signals m/z = 17 and m/z = 18, which correspond to the OH^+ and H_2O^+ species. The second stage of thermal decomposition is the removal of hmta molecules. At the beginning, the hmta is lost during its sublimation, as indicated by an endothermic process registered on DTA curves (Figures S5–S7), as well as the most characteristic fragmentation ions of hmta; i.e., $C_2H_4N^+$ (m/z = 42) is detected in the mass spectra. As the decomposition continues, the non-sublimated hmta molecules are combusted, and the mass spectrum shows ion current signals corresponding to C^+, N^+, OH^+, H_2O^+, CO^+, NO^+, and CO_2^+ species. For **2** and **3**, the carboxylate anions decompose before the hmta degradation is completed (third stage). The mass spectra show the same ion current signal as stated above (excluding $m/z = 42$). In all cases, the process finishes with the formation of magnesium oxide as the final product.

Table 5. Thermal analysis data: temperature ranges, DTA extrema (exo—exothermic, endo—endothermic), mass losses (experimental/calculated), and mz/signals; for the studied compounds.

	1	2	3	m/z
I stage	120–202 °C, 135 °C endo 31.6%/31.3% $-10\ H_2O$	43–132 °C, 90 °C endo 11.8%/11.9% $-4\ H_2O$	45–133 °C, 90 °C endo 26.8%/28.5% $-10\ H_2O$	17, 18
II stage	202–313 °C, 250 °C endo 48.4%/48.7% −2 hmta	132–278 °C, 215 °C endo 59.1%/−	133–260 °C, 225 °C endo 38.6%/−	12, 14, 17, 18, 30, 42, 44
III stage	313–423 °C, 380 °C exo 12.9%/12.9% $-2\ HCOO^-$, +0.5 O_2	278–450 °C, 315 °C exo, 355 °C exo 22.2%/−	260–455 °C, 305 °C endo, 365 °C exo 28.1%/−	12, 14 *, 17, 18, 30 *, 44
		II and III stages totally: 81.3%/81.4% $-6\ H_2O$, −2 hmta, $-2\ CH_3COO^-$, +0.5 O_2	II and III stages totally: 66.7%/65.0% −2 hmta, $-2\ C_2H_5COO^-$, +0.5O_2	
Final product	7.1 %/7.1% MgO	6.9%/6.7% MgO	6.5%/6.5% MgO	-

* m/z signals present only in the thermal decomposition of **2** and **3**.

4. Conclusions

The studied compounds were formed regardless of the applied substrate stoichiometry during the synthesis. The location of the hmta molecules in the outer coordination sphere and the presence of $Mg(H_2O)_6^{2+}$ ions are beneficial for the optimal pharmaceutical activity of both species. Due to the uncoordinated nature of the carboxylate ions, which contrasts with the majority of magnesium carboxylate complexes, the studied compounds are structurally uncommon [47]. The carbon chain length of the carboxylate anions is a key factor in supramolecular assembling. In the studied systems, the propioniate anions engage twice as many water molecules of the outer coordination sphere to form hydrogen bonds as do formate anions. Thus, a slight change in the carbon chain length significantly affects the type of created supramolecular interactions. This, in turn, considerably affects the thermal stability of the investigated compounds. The compound containing formate anions occurred to be notably more thermally stable.

Supplementary Materials: The following supporting information can be downloaded at: https://www.mdpi.com/article/10.3390/cryst12101434/s1, Figure S1: Unit cell of 2 showing the unsolvable disorder in the outer coordination sphere title; Figures S2–S4: IR spectra of 1–3; Figures S5–S7: TG and DTA curves of 1–3.

Author Contributions: Conceptualization, T.S. and A.T.-K.; investigation, visualization, T.S., M.Ś., M.B. and P.S.; writing—original draft preparation, T.S., A.T.-K. and M.Ś.; writing—review and editing, M.Ś., M.B. and P.S.; supervision, T.S. and A.T.-K. All authors have read and agreed to the published version of the manuscript.

Funding: This research received no external funding.

Institutional Review Board Statement: Not applicable.

Informed Consent Statement: Not applicable.

Data Availability Statement: CCDC 2207826 and 2207827 contain the supplementary crystallographic data for 1 and 3. These data can be obtained free of charge via http://www.ccdc.cam.ac.uk/conts/retrieving.html (accessed on 16 September 2022) or from the CCDC, 12Union Road, Cambridge CB2 1EZ, UK (fax: +44−1223−336033; e-mail: deposit@ccdc.cam.ac.uk).

Acknowledgments: This article has been completed while the fifth author (Paulina Sobczak) was the Doctoral Candidate in the Interdisciplinary Doctoral School at the Lodz University of Technology, Poland.

Conflicts of Interest: The authors declare no conflict of interest.

References

1. Koroteev, P.S.; Ilyukhin, A.B.; Gavrikov, A.V.; Babeshkin, K.A.; Efimov, N.N. Mononuclear Transition Metal Cymantrenecarboxylates as Precursors for Spinel-Type Manganites. *Molecules* **2022**, *27*, 1082. [CrossRef] [PubMed]
2. Krejner, E.; Sierański, T.; Świątkowski, M.; Bogdan, M.; Kruszyński, R. Physicochemical Insight into Coordination Systems Obtained from Copper(II) Bromoacetate and 1,10-Phenanthroline. *Molecules* **2020**, *25*, 5324. [CrossRef] [PubMed]
3. Liu, J.; Xie, D.; Xu, X.; Jiang, L.; Si, R.; Shi, W.; Cheng, P. Reversible Formation of Coordination Bonds in Sn-Based Metal-Organic Frameworks for High-Performance Lithium Storage. *Nat. Commun.* **2021**, *12*, 3131. [CrossRef] [PubMed]
4. Malinowski, J.; Zych, D.; Jacewicz, D.; Gawdzik, B.; Drzeżdżon, J. Application of Coordination Compounds with Transition Metal Ions in the Chemical Industry—A Review. *Int. J. Mol. Sci.* **2020**, *21*, 5443. [CrossRef] [PubMed]
5. Seetharaj, R.; Vandana, P.V.; Arya, P.; Mathew, S. Dependence of Solvents, PH, Molar Ratio and Temperature in Tuning Metal Organic Framework Architecture. *Arab. J. Chem.* **2019**, *12*, 295–315. [CrossRef]
6. Chen, X.-M. Chapter 10-Assembly Chemistry of Coordination Polymers. In *Modern Inorganic Synthetic Chemistry*; Xu, R., Pang, W., Huo, Q., Eds.; Elsevier: Amsterdam, The Netherlands, 2011; pp. 207–225, ISBN 978-0-444-53599-3.
7. Mukherjee, A.; Tothadi, S.; Desiraju, G.R. Halogen Bonds in Crystal Engineering: Like Hydrogen Bonds yet Different. *Acc. Chem. Res.* **2014**, *47*, 2514–2524. [CrossRef]
8. Biradha, K. Crystal Engineering: From Weak Hydrogen Bonds to Co-Ordination Bonds. *CrystEngComm* **2003**, *5*, 374–384. [CrossRef]
9. Desiraju, G.R. Crystal Engineering: A Holistic View. *Angew. Chem. Int. Ed. Engl.* **2007**, *46*, 8342–8356. [CrossRef]
10. Metrangolo, P.; Resnati, G.; Pilati, T.; Liantonio, R.; Meyer, F. Engineering Functional Materials by Halogen Bonding. *J. Polym. Sci. Part A Polym. Chem.* **2007**, *45*, 1–15. [CrossRef]
11. Wang, H.-N.; Meng, X.; Dong, L.-Z.; Chen, Y.; Li, S.-L.; Lan, Y.-Q. Coordination Polymer-Based Conductive Materials: Ionic Conductivity vs. Electronic Conductivity. *J. Mater. Chem. A* **2019**, *7*, 24059–24091. [CrossRef]
12. Gul, Z.; Khan, S.; Ullah, S.; Ullah, H.; Khan, M.U.; Ullah, M.; Altaf, A.A. Recent Development in Coordination Compounds as a Sensor for Cyanide Ions in Biological and Environmental Segments. *Crit. Rev. Anal. Chem.* 2022, 1–21, *online ahead of print*. [CrossRef]
13. Medici, S.; Peana, M.; Crisponi, G.; Nurchi, V.M.; Lachowicz, J.I.; Remelli, M.; Zoroddu, M.A. Silver Coordination Compounds: A New Horizon in Medicine. *Coord. Chem. Rev.* **2016**, *327–328*, 349–359. [CrossRef]
14. Yamagami, R.; Sieg, J.P.; Bevilacqua, P.C. Functional Roles of Chelated Magnesium Ions in RNA Folding and Function. *Biochemistry* **2021**, *60*, 2374–2386. [CrossRef] [PubMed]
15. Rezaei Behbehani, G.; Saboury, A.A. A Thermodynamic Study on the Binding of Magnesium with Human Growth Hormone. *J. Anal. Calorim.* **2007**, *89*, 857–861. [CrossRef]
16. Grubbs, R.D.; Maguire, M.E. Magnesium as a Regulatory Cation: Criteria and Evaluation. *Magnesium* **1987**, *6*, 113–127.
17. Ayuk, J.; Gittoes, N.J. Contemporary View of the Clinical Relevance of Magnesium Homeostasis. *Ann. Clin. Biochem* **2014**, *51*, 179–188. [CrossRef]
18. Mildvan, A.S. Role of Magnesium and Other Divalent Cations in ATP-Utilizing Enzymes. *Magnesium* **1987**, *6*, 28–33.
19. Jahnen-Dechent, W.; Ketteler, M. Magnesium Basics. *Clin. Kidney J.* **2012**, *5*, i3–i14. [CrossRef]
20. Uwitonze, A.M.; Razzaque, M.S. Role of Magnesium in Vitamin D Activation and Function. *J. Am. Osteopath Assoc.* **2018**, *118*, 181–189. [CrossRef]
21. Gong, R.; Liu, Y.; Luo, G.; Yang, L. Dietary Magnesium Intake Affects the Vitamin D Effects on HOMA-β and Risk of Pancreatic β-Cell Dysfunction: A Cross-Sectional Study. *Front. Nutr.* **2022**, *9*, 849747. [CrossRef]
22. Dai, Q.; Zhu, X.; Manson, J.E.; Song, Y.; Li, X.; Franke, A.A.; Costello, R.B.; Rosanoff, A.; Nian, H.; Fan, L.; et al. Magnesium Status and Supplementation Influence Vitamin D Status and Metabolism: Results from a Randomized Trial. *Am. J. Clin. Nutr* **2018**, *108*, 1249–1258. [CrossRef]
23. Petrov, A.S.; Pack, G.R.; Lamm, G. Calculations of Magnesium−Nucleic Acid Site Binding in Solution. *J. Phys. Chem. B* **2004**, *108*, 6072–6081. [CrossRef]
24. Dudev, T.; Cowan, J.A.; Lim, C. Competitive Binding in Magnesium Coordination Chemistry: Water versus Ligands of Biological Interest. *J. Am. Chem. Soc.* **1999**, *121*, 7665–7673. [CrossRef]
25. Piovesan, D.; Profiti, G.; Martelli, P.L.; Casadio, R. The Human "Magnesome": Detecting Magnesium Binding Sites on Human Proteins. *BMC Bioinform.* **2012**, *13*, S10. [CrossRef] [PubMed]

26. Sieranski, T.; Kruszynski, R. Magnesium Sulphate Complexes with Hexamethylenetetramine and 1,10-Phenanthroline. *J. Anal. Calorim* **2012**, *109*, 141–152. [CrossRef]
27. Yufanyi, D.M.; Ondoh, A.M.; Foba-Tendo, J.; Mbadcam, K.J. Effect of Decomposition Temperature on the Crystallinity of α-Fe_2O_3 (Hematite) Obtained from an Iron(III)-Hexamethylenetetramine Precursor. *Am. J. Chem.* **2015**, *5*, 1–9.
28. Kirillov, A.M. Hexamethylenetetramine: An Old New Building Block for Design of Coordination Polymers. *Coord. Chem. Rev.* **2011**, *255*, 1603–1622. [CrossRef]
29. Czubacka, E.; Kruszynski, R.; Sieranski, T. The Structure and Thermal Behaviour of Sodium and Potassium Multinuclear Compounds with Hexamethylenetetramine. *Struct. Chem.* **2012**, *23*, 451–459. [CrossRef]
30. Chwa, A.; Kavanagh, K.; Linnebur, S.A.; Fixen, D.R. Evaluation of Methenamine for Urinary Tract Infection Prevention in Older Adults: A Review of the Evidence. *Adv. Drug Saf.* **2019**, *10*, 2042098619876749. [CrossRef]
31. Hirano, K.; Asami, M. Phenolic Resins—100 years of Progress and Their Future. *React. Funct. Polym.* **2013**, *73*, 256–269. [CrossRef]
32. Choi, M.H.; Chung, I.J.; Lee, J.D. Morphology and Curing Behaviors of Phenolic Resin-Layered Silicate Nanocomposites Prepared by Melt Intercalation. *Chem. Mater.* **2000**, *12*, 2977–2983. [CrossRef]
33. STOE & Cie GmbH. *X-RED Version 1.18*; STOE & Cie GmbH: Darmstadt, Germany, 1999.
34. Sheldrick, G.M. SHELXT–Integrated Space-Group and Crystal-Structure Determination. *Acta Cryst. A* **2015**, *71*, 3–8. [CrossRef] [PubMed]
35. Sheldrick, G.M. Crystal Structure Refinement with SHELXL. *Acta Cryst. C* **2015**, *71*, 3–8. [CrossRef] [PubMed]
36. Welcher, F.J. *Analityczne Zastosowanie Kwasu Wersenowego (Eng. The Analytical Uses of Etylenediamineteraacetic Acid)*; WNT: Warsaw, Poland, 1963.
37. Bernstein, J.; Davis, R.E.; Shimoni, L.; Chang, N.-L. Patterns in Hydrogen Bonding: Functionality and Graph Set Analysis in Crystals. *Angew. Chem. Int. Ed. Engl.* **1995**, *34*, 1555–1573. [CrossRef]
38. Shimoni, L.; Glusker, J.P.; Bock, C.W. Energies and Geometries of Isographic Hydrogen-Bonded Networks. 1. The (8) Graph Set. *J. Phys. Chem.* **1996**, *100*, 2957–2967. [CrossRef]
39. Ahuja, I.S.; Singh, R.; Yadava, C.L. Infrared Spectral Evidence for Mono-, Bi- and Tetra-Dentate Behaviour of Hexamethylenetetramine. *Proc. Indian Acad. Sci. (Chem. Sci.)* **1983**, *92*, 59–63. [CrossRef]
40. Deacon, G.B.; Phillips, R.J. Relationships between the Carbon-Oxygen Stretching Frequencies of Carboxylato Complexes and the Type of Carboxylate Coordination. *Coord. Chem. Rev.* **1980**, *33*, 227–250. [CrossRef]
41. Sutton, C.C.R.; da Silva, G.; Franks, G.V. Modeling the IR Spectra of Aqueous Metal Carboxylate Complexes: Correlation between Bonding Geometry and Stretching Mode Wavenumber Shifts. *Chem.–A Eur. J.* **2015**, *21*, 6801–6805. [CrossRef]
42. Swiatkowski, M.; Kruszynski, R. Structurally Diverse Coordination Compounds of Zinc as Effective Precursors of Zinc Oxide Nanoparticles with Various Morphologies. *Appl. Organomet. Chem.* **2019**, *33*, e4812. [CrossRef]
43. Kakihana, M.; Akiyama, M. Vibrational Analysis of the Propionate Ion and Its Carbon-13 Derivatives: Infrared Low-Temperature Spectra, Normal-Coordinate Analysis, and Local-Symmetry Valence Force Field. *J. Phys. Chem.* **1987**, *91*, 4701–4709. [CrossRef]
44. Jensen, J.O. Vibrational Frequencies and Structural Determinations of Hexamethylenetetraamine. *Spectrochim. Acta A Mol. Biomol. Spectrosc.* **2002**, *58*, 1347–1364. [CrossRef]
45. Stoilova, D.; Koleva, V. IR Study of Solid Phases Formed in the Mg(HCOO)2–Cu(HCOO)2–H2O System. *J. Mol. Struct.* **2000**, *553*, 131–139. [CrossRef]
46. Koleva, V.; Stoilova, D. Infrared and Raman Studies of the Solids in the Mg(CH3COO)2–Zn(CH3COO)2–H2O System. *J. Mol. Struct.* **2002**, *611*, 1–8. [CrossRef]
47. Groom, C.R.; Bruno, I.J.; Lightfoot, M.P.; Ward, S.C. The Cambridge Structural Database. *Acta Cryst. B* **2016**, *72*, 171–179. [CrossRef] [PubMed]

crystals

MDPI

Article

Novel Organotin(IV) Complexes of 2-[4-Hydroxy-3-((2-hydroxyethylimino)methyl)phenylazo]benzoic Acid: Synthesis, Structure, Noncovalent Interactions and In Vitro Antibacterial Activity

Pratima Debnath [1], Paresh Debnath [1], Manojit Roy [1,*], Lesław Sieroń [2], Waldemar Maniukiewicz [2,*], Tamanna Aktar [3], Debasish Maiti [3], Alexander S. Novikov [4,5,*] and Tarun Kumar Misra [1,*]

1 Department of Chemistry, National Institute of Technology Agartala, Jirania 799046, Tripura, India
2 Institute of General and Ecological Chemistry, Lodz University of Technology, Zeromskiego 116, 90-924 Lodz, Poland
3 Department of Human Physiology, Tripura University, Agartala 799022, Tripura, India
4 Institute of Chemistry, Saint Petersburg State University, Universitetskaya Nab., 7/9, 199034 Saint Petersburg, Russia
5 Research Institute of Chemistry, Peoples' Friendship University of Russia (RUDN University), Miklukho-Maklaya Street, 6, 117198 Moscow, Russia
* Correspondence: mroychem@gmail.com (M.R.); waldemar.maniukiewicz@p.lodz.pl (W.M.); a.s.novikov@spbu.ru (A.S.N.); tkmisra70@yahoo.com (T.K.M.)

Abstract: Three new organotin(IV) complexes, $[Me_3Sn(H_2L)]_2$ (**1**), $Bu_3Sn(H_2L)$ (**2**), and $[(Bu_2Sn(H_2L))_2O]_2$ (**3**) were synthesized by the reaction of 2-[4-hydroxy-3-((2-hydroxyethylimino)methyl)phenylazo]benzoic acid (**H_3L**) with appropriate alkyltin(IV) precursors. The complexes were characterized by elemental analysis, IR, and multinuclear (^{1}H, ^{13}C and ^{119}Sn) NMR spectroscopy. Further, the complex **1** was analyzed by single-crystal X-ray analysis. It displays a 24-membered cyclic dimeric $Me_3Sn^{IV}(H_2L)$ unit where the ligand act as a bridging framework using its carboxylate-O and phenoxy-O atoms. The Sn(IV) adopts distorted trigonal-bipyramidal geometry. In the solution state, the structures were determined by ^{119}Sn-NMR spectroscopy, and the complexes **1** and **2** have distorted tetrahedral geometry, whereas complex **3** shows distorted trigonal-bipyramidal geometry around the tin centres. The Hirshfeld surface analysis and DFT calculations, together with a topological analysis of the electron density distribution in the crystal structure of complex **1**, indicate that its molecular packing determined by various noncovalent interactions, including stacking and hydrogen bonding. The antibacterial studies of the ligand and the complexes (**1–3**) against gram-negative bacteria *viz. Klebsiella pneumoniae (A), Vibrio cholerae (M)* and *Shigella boydii (Q)* and gram-positive bacteria *viz. Staphylococcus aureus* (J), *Streptococcus pneumonia* (K) are promising and the compounds can be treated as potential common antibacterial materials.

Keywords: organotin(IV) compounds; NMR spectroscopy; crystal structure; cyclic dimeric; noncovalent interactions; antibacterial activity

Citation: Debnath, P.; Debnath, P.; Roy, M.; Sieroń, L.; Maniukiewicz, W.; Aktar, T.; Maiti, D.; Novikov, A.S.; Misra, T.K. Novel Organotin(IV) Complexes of 2-[4-Hydroxy-3-((2-hydroxyethylimino)methyl)phenylazo]benzoic Acid: Synthesis, Structure, Noncovalent Interactions and In Vitro Antibacterial Activity. *Crystals* **2022**, *12*, 1582. https://doi.org/10.3390/cryst12111582

Academic Editor: Kil Sik Min

Received: 24 October 2022
Accepted: 4 November 2022
Published: 7 November 2022

1. Introduction

The development of organotin(IV) compounds for use in biomedical, industrial, and agricultural applications is attracting significant interest from researchers [1]. They have been commercially used as a homogeneous catalyst for polymeric materials, such as PVC, polyurethanes, silicone polymers, etc. [2]. Because of their specific interactions with the phosphate group of DNA, they have been considered advanced anti-cancer agents [3–7] and the agents for other diseases of proliferation [8–10]. They are distinctive from a stereochemical perspective, displaying monomeric, polymeric or cyclic motifs [11,12]. Nevertheless, coordinated ligands play a vital role in the structural variety of organotin(IV) compounds. Ligands bearing carboxylate/hydroxo groups with Schiff base imine, azo, or azo-imine functionalities have been extensively investigated [13–17] for their stereochemical motifs and biological applications [18–25].

It has been observed that the cyclic compounds formed by organotin(IV) with ligands containing hydroxy-azo/imino-carboxylate groups at the specific positions are either tetrameric or dimeric forms of Sn(IV) [25–31]. Tetrameric Sn(IV) complexes are found to be 36- to 48-members. They adopt distorted trigonal-bipyramidal geometry regardless of the organo-groups linked to Sn(IV). The ring size and multimeric character of Sn(IV) units are thus controlled by the location of the coordinating sites that are present in a ligand frame. Roy M. et al. [25] reported the 48-membered macrocyclic tetrameric trimethyltin(IV) compound with 4-(2,4-dihydroxy-phenylazo)-benzoic acid, where in respect of the azo group the carboxylate is at 4- and the hydroxyl is at 2′-positions and the ligand is in zwitterionic form. The complex showed very good antidiabetic effectiveness, which overcomes the effectiveness of standard compound acarbose. The ligand, 3-(salicylideneamino)benzoate where the carboxylate is at 3- and the hydroxyl is at 2′-positions in respect of the imino function yielded the 44-membered macrocyclic tetrameric triphenyltin(IV) compound [26]. The compound has shown effective anticancer activity as well. Basu Baul T.S. et al. [27] designed the ligand, 5-[(*E*)-2-(3-pyridyl)-1-diazenyl]-2-hydroxybenzoic acid in which the carboxylate and the hydroxyl groups at the 3- and 4-positions of a phenyl ring are linked with 3′-pyridine through the azo-group. With the triphenyltin(IV) fragment, the ligand produced a 44-membered macrocyclic tetrameric structure employing carboxylate-O and pyridine-N. The research group framed a different ligand in which phenylazo group (having no coordination site) attached at 4-position of the hydroxyl group and 3-position of the (methylidene)aminoacetate group. The ligand in its zwitterionic form produced a 36-membered macrocyclic tetrameric structure with the triphenyltin(IV) fragment [28]. Thus, the ligands with coordination sites (carboxylate, hydroxy) at (4,2′), (3,3′), (3,2′) positions yield tetrameric organotin(IV) complexes. On the other hand, the ligands of type (2,4′), i.e., the carboxylate group at 2-position and the hydroxyl group at 4′-position, with $SnPh_3$ and $SnMe_3$ fragments yielded centrosymmetric dimeric 24-membered macrocyclic organotin(IV) compounds [29,30]. In both cases, there are (methylidene)amino fragments at the ortho-position of the hydroxy-group and unlike the (methylidene)aminoacetate group [28], these were found to be uncoordinated. All of the compounds are bioactive and it could be stated that the polynuclear organotin(IV) compounds are supposed to have higher antitumor activity than that of monomeric ones [31].

In this present contribution, we report synthesis of a new ligand of type (2,4′) with a pendent arm of (methylidene)aminoethyl alcohol at the *ortho*-position and its Sn(IV) complexes with trimethyl-, tributyl- and dibutyl-organotin(IV) precursors. The report includes the crystal structure of the complex of $SnMe_3$ fragment which is a 24-membered macrocyclic centrosymmetric dimer of organotin(IV), establishing our observations of dimeric form with the (2,4′) systems from the literature [29,30]. The Hirshfeld surface analysis and DFT calculations, together with a topological analysis of the electron density distribution, were used for theoretical study noncovalent interactions in the obtained crystal structure. Antibacterial properties of all the compounds against both gram-positive and gram-negative bacteria are also highlighted.

2. Materials and Methods

2.1. Materials

Ethanolamine, salicylaldehyde, trimethyltin(IV) chloride, bis-tri-n-butyltin(IV) oxide, *o*-amino benzoic acid, and dibutyltin(IV) oxide were taken from Merck by purchase process and then used without further purification. Nutrient Agar (Hi-media) was used as culture media for the growth of bacterial species. The analysis for the detection of elemental Carbon, hydrogen and nitrogen analyses was performed on a PerkinElmer 2400 series II instrument. The IR spectra were obtained from Perkin Elmer FTIR spectrophotometer in the range of 4000–400 cm^{-1} using KBr discs. ^{1}H, ^{13}C and ^{119}Sn NMR spectra of ligand and the complexes were recorded on a Bruker AMX 400 spectrometer measured at 400, 100 and 149 MHz, respectively. Me_4Si was employed as a reference compound set at 0.00 ppm for ^1H- and ^{13}C-chemical shifts while Me_4Sn was used as a reference for ^{119}Sn-chemical shifts set at 0.00 ppm.

2.2. Methods

2.2.1. Synthesis of 2-{4-Hydroxy-3-[(2-hydroxyethylimino)methyl]phenylazo}benzoic Acid (H_3L)

The diazotization of 2-aminobenzoic acid, followed by coupling with salicylaldehyde led 2-(3-formyl-4-hydroxyphenylazo)benzoic acid [30]. It (1.00 g, 3.70 mmol) was then dissolved in toluene (50 mL) and heated for 30 min. To this hot toluene solution, ethanolic solution of ethanolamine (0.226 g, 3.70 mmol) was added dropwise in stirring condition, and the mixture was made to reflux for 3 hrs. After the solid product fully formed, it was filtered out, vacuum-dried, and then again recrystallized from anhydrous methanol to produce the light yellow H_3L product.

Yield: 0.967 g, 83.3%; m.p.: 210–214 °C. Anal.Calcd. for $C_{16}H_{15}N_3O_4$: C, 61.34; H, 4.83; N, 13.41%. Found: C, 61.75; H, 5.01; N, 13.23%. UV-visible (DMF) λmax (nm): 267, 370. IR (KBr, cm^{-1}): 3450 ν(OH), 2929 ν(C-H str. of aliphatic -CH_2), 1698 ν(COO)asy, 1387 $\nu(COO)_{sym}$, 1613 ν(C=N), 1494 ν(N=N), 1273 ν(aromatic C-O), 1167 ν(aliphatic C-O). ^1H NMR (DMSO-*d6*, 400 MHz) δ_H: 10.33[s, 1H, Ar-OH], 8.66 [s, 1H, H-7], 8.02 [d, 1H, H-6, J = 2.8 Hz], 7.84 [dd, 1H, H-4, J = 9.2 and 2.8 Hz], 7.75 [d, 1H, H-3′], 7.60 [m, 1H, H-5′], 7.50 [m, 1H, H-4′ and H-6′], 6.78 [d, 1H, H-3, J = 9.2 Hz], 3.69 [m, 4H, H-8 & H-9] ppm. ^{13}C NMR (DMSO-*d6*, 100 MHz) δ_C: 174.24 [COO], 168.45 [C-7], 166.78 [C-2], 150.45 [C-1′], 141.37 [C-5], 134.29 [C-5′], 131.34 [C-2′], 129.94 [C-4′], 129.00 [C-4], 125.76 [C-6], 122.25 [C-6′], 117.58 [C-1], 115.22 [C-3], 59.65 & 55.78 [C-8 & C-9] ppm.

2.2.2. Preparation of Cyclic Dimeric Trimethyltin(IV) Complex of H_3L, $[Me_3Sn(H_2L)]_2$ (**1**)

The cyclic dimeric trimethyltin(IV) complex, $[Me_3Sn(H_2L)]_2$ (**1**) was synthesised by reacting 2-{4-hydroxy-3-[(2-hydroxyethylimino)methyl]phenylazo}benzoic acid (H_3L) with trimethyltin(IV) chloride in the stoichiometric molar ratio 1:1 (L:M) with triethylamine as a base in a refluxing condition. The ligand H_3L (0.4716 g, 1.505 mmol) in 30 mL of distilled toluene in a round bottom flask was first treated with a triethylamine base (0.1522 g, 1.505 mmol) and the mixture was then refluxed for 30 min. After that, trimethyltin(IV) chloride (0.3 g, 1.505 mmol) was added to the mixture and continually refluxed for a further 6–7 h. The reaction mixture was eventually filtered off and a yellow colored product was extracted from the solution. The product was washed thoroughly with hexane, dried and recrystallized from anhydrous toluene. The pure yellow crystals of compound **1** were finally obtained.

Yield: 0.62 g, 87.32%; m.p.: 102–103 °C. Anal.Calcd. for $C_{38}H_{46}N_6O_8Sn_2$: C, 47.93; H, 4.87; N, 8.83%. Found: C, 47.87; H, 4.50; N, 8.90%. UV-visible (DMF) λmax (nm): 268, 372. IR (KBr, cm^{-1}): 3389 ν(O-H), 2921 ν(C-H str. of Me), 1654 $\nu(COO)_{asy}$, 1493 ν(N=N), 1372 $\nu(COO)_{sym}$, 1281 ν(Aromatic C-O),1183 ν(Aliphatic C-O), 772 ν(C-H oop bending), 579 & 538 (Sn-C), 492 ν(Sn-O).^1H NMR ($CDCl_3$, 400 MHz) δ_H, Ligand skeleton: 8.48 [s, 1H, H-7], 7.91 [m, 1H, H-6], 7.70 [m, 1H, H-4], 7.63 [m, 1H, H-3′], 7.58 [m, 3H, H-4′,5′and 6′], 7.05 [d, 1H, H-3, J = 8.8 Hz], 4.07 [m, 1H, Aliphatic-OH], 3.98 [m, 2H, H-9, J = 4.4 Hz], 3.89 [m, 2H, H-8, J = 4.4 Hz]; Sn-CH_3 Skeleton: 0.61 [s, 9H, (Sn-CH_3)] ppm.^{13}C NMR ($CDCl_3$, 100 MHz) δ_C: Ligand skeleton: ^{13}C NMR ($CDCl_3$, 100 MHz) δ_C: 173.97 [COO], 169.32 [C-7], 151.76 [C-2], 145.70 [C-1′], 133.89 [C-5], 132.48 [C-5′], 131.14 [C-2′], 130.36 [C-4′], 128.74 [C-4], 127.75 [C-6], 119.65[C-6′], 118.13 [C-1], 117.50 [C-3]; Sn-CH_3 skeleton: −1.73 [Sn-CH_3] ppm. ^{119}Sn NMR ($CDCl_3$, 149 MHz): +139.8 ppm.

2.2.3. Preparation of Tributyltin(IV) Complex of H_3L, $Bu_3Sn(H_2L)$ (**2**)

The tri-butyltin(IV) compound (**2**) was synthesized by the reaction of the ligand, 2-{4-hydroxy-3-[(2-hydroxyethylimino)methyl]phenylazo}benzoic acid with *bis*-tributyltin

(IV) oxide by using the Dean–Stark apparatus under refluxing condition in (L:M) 1:1 molar ratio. The whole reaction was carried out in a toluene solvent as described for compound **1**. Yield: 0.26 g, 85.76%; m.p.: 86–89 °C. Anal.Calcd. for $C_{28}H_{41}N_3O_4Sn$: C, 55.83; H, 6.86; N, 6.98%. Found: C, 55.67; H, 6.85; N, 7.01%. UV-visible (DMF) λ_{max} (nm): 365, 372, 495. IR (KBr, cm^{-1}): 3450 ν(O-H), 2957 ν(C-H str. of ^{n}Bu), 1640 $\nu(COO)_{asy}$, 1494 ν(N=N), 1391 $\nu(COO)_{sym}$, 1281 ν(Aromatic C-O),1183 ν(Aliphatic C-O), 598 & 585 ν(Sn-C), 507 ν(Sn-O).^{1}H NMR ($CDCl_3$, 400 MHz) δ_H, Ligand skeleton: 8.41 [s, 1H, H-7], 7.93 [dd, 1H, H-4, *J* = 9.2 Hz & 2 Hz], 7.87 [s, 1H, H-6], 7.80 [d, 1H, H-3′, *J* = 7.2 Hz], 7.48 [s, 1H, H-5′], 7.40 [m, 1H, H-4′ and H-6′], 6.99[d, 1H, H-3, *J* = 9.2 Hz], 3.92 [m, 2H, H-9], 3.77 [m, 2H, H-8] ppm; Sn-^{n}Bu Skeleton: 1.65[m,6H, H-α], 1.25 [m, 12H, H-β, H-γ], 0.89 [t, 9H, H-δ] ppm. ^{13}C NMR ($CDCl_3$, 100 MHz) δ_C: 172.74 [COO-], 166.46 [C-7],151.78 [C-2], 145.04 [C-1′], 131.37 [C-5], 130.89 [C-5′], 129.77 [C-2′], 129.50 [C-4′], 129.11 [C-4], 126.50 [C-6], 118.70 [C-6′], 117.65 [C-1], 117.58 [C-3], 61.77 & 60.47 [C-8 & C-9] ppm; Sn-^{n}BuSkeleton: 27.83 [C-β] ^2J [^{119}Sn-^{13}C(19.9 Hz)], 27.07 [C-γ] ^3J [^{119}Sn-^{13}C(64.4 Hz)], 16.67 [C-α] ^1J [^{119}Sn-^{13}C(352.6 Hz)], 13.65 [C-δ] ppm. ^{119}Sn NMR ($CDCl_3$, 149 MHz): +115.97 ppm. The numbering scheme of Sn-Bu skeletal in the triorganotin complex is shown below:

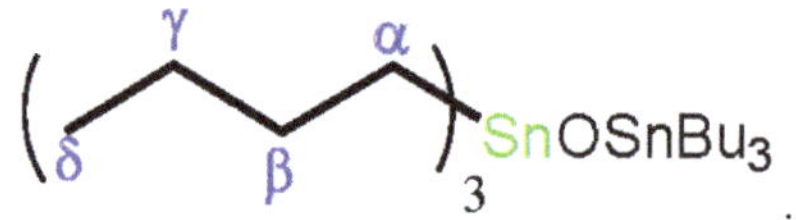

2.2.4. Preparation of Dibutyltin (IV) Complex of H_3L, $[(Bu_2Sn(H_2L))_2O]_2$ (**3**)

The dibutyltin(IV) complex $[(Bu_2Sn(H_2L))_2O]_2$ (**3**) was synthesized in a toluene solvent by the reaction of the ligand 2-{4-hydroxy-3-[(2-hydroxyethylimino)methyl]phenylazo}benzoic acid with dibutlytin(IV) oxide with the use of Dean-Stark apparatus under refluxing condition at 1:1 (L:M) stoichiometric ratio. Yield: 0.50 g, 74.86%; m.p.: 201–203 °C. Anal.Calcd. for $C_{96}H_{132}N_{12}O_{18}Sn_4$: C, 52.01; H, 6.00; N, 7.58%. Found: C, 51.65; H, 5.61; N, 7.13%. UV-visible (DMF) λ_{max} (nm): 265, 273, 483. IR (KBr, cm^{-1}): 3442 ν(O-H), 2957 ν(C-H str. of ^{n}Bu), 1630 $\nu(COO)_{asy}$, 1470 ν(N=N), 1394 $\nu(COO)_{sym}$, 1311 ν(Aromatic C-O),1193 ν(Aliphatic C-O), 595 & 581 ν(Sn-C), 538 ν(Sn-O). ^{1}H NMR ($CDCl_3$, 400 MHz) δ_H, Ligand skeleton: 9.91 [s, 1H, H-7], 8.31 [dd, 1H, H-6], 8.19 [m, 1H, H-4], 7.85 [m, 1H, H-3′], 7.54 [m, 1H, H-5′], 7.06 [m, H-4′ and H-6′], 6.95 [m, 1H, H-3], 4.07 [m, 1H, aliphatic-OH], 3.83 [m, 4H, H-8 & H-9] ppm; Sn-^{n}Bu Skeleton: 1.46 [m,12H, H-α, H-β, H-γ], 0.91 [m, 6H, H-δ] ppm. ^{13}C NMR ($CDCl_3$, 100 MHz) δ_C: 171.76 [COO-], 167.23 [C-7],150.51 [C-2], 147.23 [C-1′ and C-9], 133.79 [C-5 and C-8], 131.40 [C-5′], 129.74 [C-2′], 128.89 [C-4′], 128.07 [C-4], 125.75 [C-6], 119.63 [C-6′], 117.45 [C-1], 116.06 [C-3]; Sn-^{n}Bu Skeleton: 29.70[C-β], 27.29 [C-γ], 26.47 [C-α], 13.58 [C-δ] ppm. ^{119}Sn NMR ($CDCl_3$, 149 MHz): −93.1 and −140.4 ppm. The numbering scheme of Sn-Bu skeletal in the diorganotin complex is shown below:

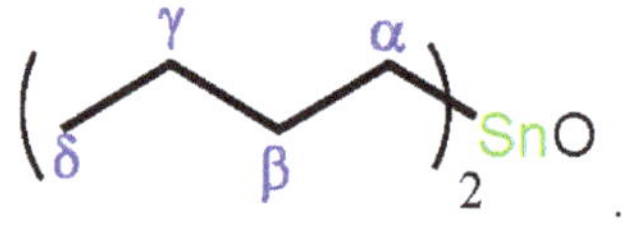

2.3. *Crystallographic Data Collection and Structure Refinement*

Single crystal X-ray diffraction data were collected using the ω-scan technique using MoKα (λ = 0.71073Å) radiation. The title compound was studied at 100 K using a RIGAKU XtaLAB Synergy, Dualflex, Pilatus 300 K diffractometer [32] with Photon Jet micro-focus X-ray Source. Data collection, cell refinement, data reduction and absorption correction were carried out using CrysAlis PRO software [32]. The crystal structures were solved by using direct methods with the SHELXT 2018/3 program [33]. Atomic scattering factors were taken from the International Tables for X-ray Crystallography. Positional parameters of non-H-atoms were refined by a full-matrix least-squares method on F2 with anisotropic thermal parameters by using the SHELXL 2018/3 program [34]. Hydrogen atoms participating in hydrogen bonding were found on the Fourier map and freely refined while the others were placed in calculated positions (C–H = 0.93–0.98 Å) and included as riding contributions with isotropic displacement parameters set to 1.2-times the Ueq of the parent atom. Crystal data and structure refinement parameters are shown in Table 1.

Table 1. Crystal data and structure refinement parameters for complex **1**.

Parameters	1
Empirical formula	$C_{38}H_{46}N_6O_8Sn_2$
Formula weight	952.23
Temperature (K)	100 (3)
Wavelength (Å)	0.71073
Crystal system	triclinic
Space group	P-1
a (Å)	7.1138 (1)
b (Å)	10.4253 (2)
c (Å)	14.0461 (2)
α (°)	94.753 (1)
β (°)	93.448 (1)
γ (°)	108.603 (1)
Volume (Å^3)	979.74 (3)
Z	1
Density (Mg/m^3)	1.614
Absorp. coeff. (mm^{-1})	1.333
F (000)	480
Crystal size (mm^3)	0.04 × 0.05 × 0.33
Theta range for data collection	2.9, 25.0
Index ranges	−8<= h <=8; −12<= k <=12; −16<= l <=16
Reflection collected	26,309
Data completeness	99.9%
Independent reflections	3559 [R(int) = 0.022]
Goodness of fit on F^2	1.11
Final R indices [I > 2sigma(I)]	0.0152
R indices (all data)	0.0159
Largest diff. peak and hole (e Å^{-3})	0.58 and −0.22

2.4. Computational Details

The Hirshfeld molecular surface was generated by the CrystalExplorer program (version 17.5) [35,36]. The normalized contact distances, d_{norm} [37], based on Bondi's van der Waals radii [38], were mapped into the Hirshfeld surface. The DFT calculations based on the experimental X-ray geometry of **1** were carried out using the dispersion-corrected hybrid functional ωB97XD [39] with the help of Gaussian-09 [40] program package. The Douglas–Kroll–Hess 2nd order scalar relativistic calculations requesting relativistic core Hamiltonian were carried out using the DZP-DKH basis sets [41–44] for all atoms. The topological analysis of the electron density distribution with the help of the quantum theory of atoms-in-molecules (QTAIM) method, electron localization function (ELF), and reduced density gradient (RDG) analyses were performed using the Multiwfn program (version 3.7) [45].

2.5. Antibacterial Studies

The antibiotic activities of the ligand (**H_3L**) and the compounds (**1–3**) were studied against the three gram-negative bacterial strains viz. *Klebsiella pneumoniae* (A), *Vibrio cholera* (M), *Shigella boydii* (Q) andtwo gram-positive bacterial strains *viz. Staphylococcus aureus* (J) and *Streptococcus pneumonia* (K) according to the Kirby–Bauer Disk Diffusion Susceptibility Test Protocol [46]. The standard antibiotics such as Polymyxin B and Gentamycin were used as positive control against gram-negative bacteria and Vancomycin for gram-positive bacteria. They were treated at various concentrations. In order to determine the zone of inhibition, discs of 5 mm in diameter were used on pre-spread bacterial strains in a nutrient agar plate. Strains were treated with the compounds at different concentrations (1 μg/disc, 100 μg/disc and 500 μg/disc). After 18–24 h of incubation at 37 °C, the diameters (mm) of clear zone of inhibitions were measured using a scale.

3. Results and Discussion

3.1. Synthesis

The ligand, an azo-imine carboxylate Schiff base, was synthesized by condensation of 2-(3-formyl-4-hydroxyphenylazo) benzoic [30] with ethanolamine in toluene at warm condition. The newly obtained ligand, 2-{4-hydroxy-3-[(2-hydroxyethylimino)methyl]phenylazo}benzoic acid (**H_3L**) was subjected to develop Sn(IV) compounds. Thus, the three new organotin(IV) complexes (**1–3**) were synthesized by the reaction of **H_3L** with appropriate alkyl tin(IV) precursors such as: trimethyltin(IV) chloride, *bis*-tributyltin(IV) oxide and dibutyltin(IV) oxide in refluxing condition by the use of distilled toluene. The single crystals of the complex **1** were obtained from the reaction mixture, whereas the complexes **2** and **3** were recrystallized from toluene. The three complexes were obtained in good yield and all the three complexes **1–3** were found to be soluble in all common organic solvents. The reaction scheme for the synthesis of the complexes **1–3** is shown in Scheme 1.

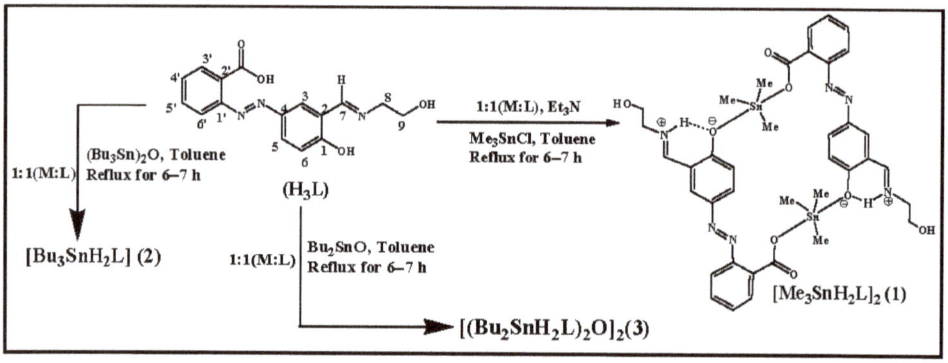

Scheme 1. The reaction scheme for the synthesis of complexes **1**, **2** and **3**.

3.2. IR-Spectroscopy

The IR spectroscopic data for the ligand (**H_3L**) and the complexes (**1–3**) are given in the Experimental section, and their spectra are shown in Figures S1–S4 (see Supplementary Materials). The ligand has various functional groups such as carboxylate, aliphatic and phenolic-OH, carbonyl (-C=O), azo (-N=N-), imine (-C=N-), aromatic -C=C-, aliphatic and aromatic-C-O. These appeared in the spectra in the usual positions for the ligand and its Sn(IV) complexes. The overlapping stretching frequencies of all the –OH groups appear as strong and broad band at 3450 cm^{-1} in H_3L. The band is quite sharp and shifts to lower frequency at 3389 cm^{-1}, indicating involvement of carboxylate-O and phenolic-O coordination to Sn(IV) in the complex **1**. However, there is not much change observed in complexes **2** and **3**. The weak band for aliphatic –CH_2- groups can be seen at 2929 cm^{-1} for H_3L and at 2921 cm^{-1} for the complex **1**. However, in complexes **2** and **3**, it is very prominent and multiplet in form, indicating coordination of the ligand to Sn(IV) that has a butyl substituent. The asymmetric ($\nu_{(COO)asy}$) and symmetric ($\nu_{(COO)sy}$) stretching frequencies of the carboxylate group in the ligand appear at 1698 cm^{-1} and 1385 cm^{-1}, respectively. As a result of COO-coordination to Sn(IV), these bands shift to the frequency range at 1630–1654 cm^{-1} for asymmetric -COO and 1372–1394 cm^{-1} for symmetric-COO [10,15–17,21,30]. The difference between the asymmetric and symmetric frequencies, i.e., $\Delta\nu(_{\nu(COO)asy} - \nu_{(COO)sym})$ is 282, 248 and 236 cm^{-1} respectively for the complexes **1**, **2** and **3**, which are greater than 200 cm^{-1}, indicating that the carboxylate group coordinates to Sn(IV) in monodentate mode in each complex [15,16]. The -C=N-, -C=C-, and -N=N- bands appear in the usual positions in the spectra of the ligand and the complexes, respectively ranging from 1645–1494 cm^{-1} (**H_3L**), 1615–1493 cm^{-1} (**1**), 1620–1472 cm^{-1} (**2**), and 1610–1470 cm^{-1} (**3**). The bands for aromatic and aliphatic-C-O groups can be seen at 1273 and 1167 cm^{-1} for **H_3L**, 1281 and 1183 cm^{-1} for **1**, **2**, and 1311 and 1193 for **3** cm^{-1}, respectively. All the compounds exhibit a strong band for out-of-plane bending of ring C-H bonds in the range of 750 (complex **3**) to 772 (complex **1**). The Sn-C band of the complex **1** is doublet in nature (579 and 538 cm^{-1}), which means that the Sn-C bonds are non-linear and the Sn-O bond is sharp at 492 cm^{-1}. For complexes **2** and **3**, these are assigned in a similar way (for **2**: 598, 585 cm^{-1} and 507 cm^{-1}; for **3**: 595, 581 cm^{-1} and 538 cm^{-1}) to Sn-C and Sn-O bonds, respectively [10,21,30].

3.3. Multinuclear (^{1}H, ^{13}C and ^{119}Sn) NMR-Spectroscopy

The NMR spectra of the ligand (**H_3L**) were recorded in DMSO-d_6 solvent, whilst these were recorded in the $CDCl_3$ solvent for the complexes **1–3**; their spectral data are provided in the experimental section. The ^{1}H NMR spectra of the ligand and the complex **1** are given in Figures 1 and 2; all other spectra including ^{1}H NMR and ^{13}C NMR are placed in Figures S5–S10 (Supplementary Materials). While the aromatic protons of the azo-imino ligand **H_3L** are seen in the range of 6.78–8.02 ppm (Figure 1), the aromatic protons in all the complexes **1–3** are found to be in the range of 6.94–8.43 ppm. The imine protons of all the compounds are found at 8.67 ppm for **H_3L** and at 8.48, 8.41 and 9.91 ppm, respectively, for **1–3** compounds. The phenolic-OH at 10.33 ppm for the ligand is virtually absent in the complexes. The carboxylate-OH, which is absent in the ligand spectra due to the DMSO solvent, is absent in complexes **1** and **3**, but present in complex **2** spectra. For the ligand **H_3L**, the four methylene protons attached with imino/OH functions appear as a triplet at 3.69 ppm, while the same protons appear in the range of 4.11–3.76 ppm for the three complexes **1–3**. Based on the Sn-alkyl protons signals, the complexes could be distinguished from the uncoordinated ligand, as well as from one another. The complex **1** has Sn-methyl groups, which is detected as a singlet peak at 0.61 ppm [21]; the complex **2** has Sn-Bu_3, which results in two distinct peaks as a triplet peak at 0.89 ppm and two multiplets in the range of 1.25–1.66 ppm for the methyl and the three methylene protons, respectively; similarly, the complex **3** with Sn-Bu_2 shows two multiplet peaks in the range of 0.91–1.46 ppm for tin-butyl protons [10,21].

The ^{13}C-NMR spectrum of the azo-imino carboxylate ligand **H_3L** is shown as a δ(COO) signal at 174.25 ppm, whereas the signal in the spectra of complexes is shifted downfield and appears at 173.97, 172.74, and 171.76 ppm for **1–3**, respectively. The imine carbons can be found at 168.46 (**H_3L**), 169.32 (**1**), 166.46 (**2**), and 167.24 (**3**) ppm. The aromatic ring carbons of the ligand **H_3L** show signals in the range of 166.79–115.22 ppm (Figure S7), but the same signals are found in the range of 116.07–151.78 ppm in the spectra of the complexes **1–3** (Figures S8–S10). In the ligand spectrum, the ethylene carbons are found at 59.66 and 55.79 ppm, and in the complex **2** it is at 61.77 and 60.47 ppm. In complex **1**, the tin-methyl-carbon shows a highly shielded signal at −1.73 ppm. Moreover, the complexes **2** and **3** have tin-butyl carbons (c-α, c-β, c-γ, c-δ), which appear in the range of 27.93–13.65 ppm for **2** and of 31.93–13.58 ppm for **3**.

Moreover, the geometry and coordination number of the complex **2** can be predicted from its carbon satellites (Figure S9) by determining ^{n}J (^{119}Sn-^{13}C) coupling constant values [15,16,21]. The complex **2** has 352.6 Hz, 19.9 Hz and 64.4 Hz coupling constant for ^{1}J (^{119}Sn-^{13}C), ^{2}J (^{119}Sn-^{13}C) and ^{3}J (^{119}Sn-^{13}C) coupling satellites, respectively, indicating that the complex has four coordinate quasi tetrahedral geometry around tin centres [15,16,21] in the solution state. It was further confirmed by its ^{119}Sn NMR spectral signal [15], which is found at +115.98 ppm (Figure 3). The ^{119}Sn NMR spectral signal at +139.86 ppm (Figure 3) for the complex **1** also reveals that it has a similar structure in the solution to complex **2**, whereas in the case of complex **3**, two sharp signals of ^{119}Sn NMR spectrum are found at −93.1 and −140.4 ppm, which indicate that it must have two types of tin centres with a distorted trigonal bipyramidal geometry [10].

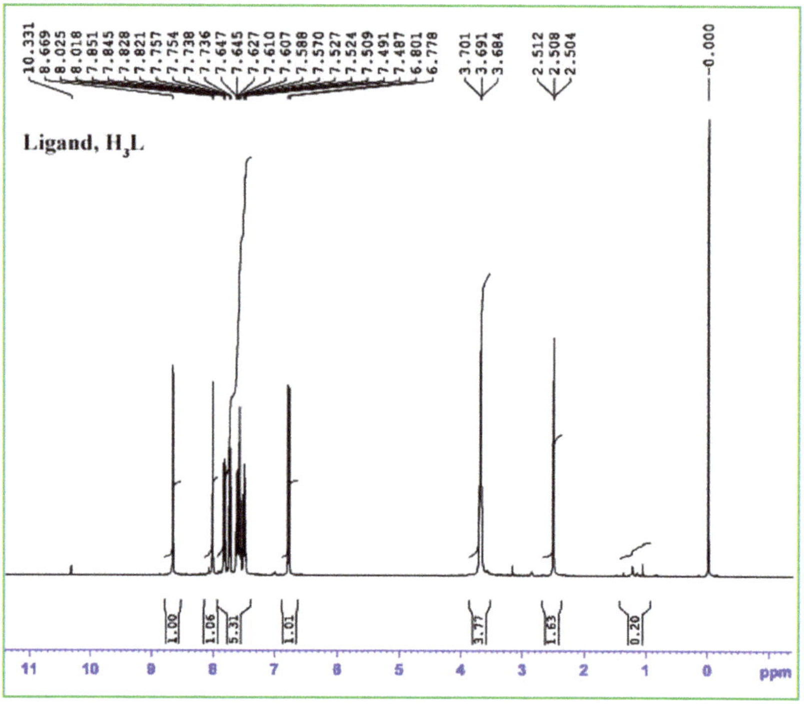

Figure 1. ^{1}H NMR spectrum of the Ligand, H_3L.

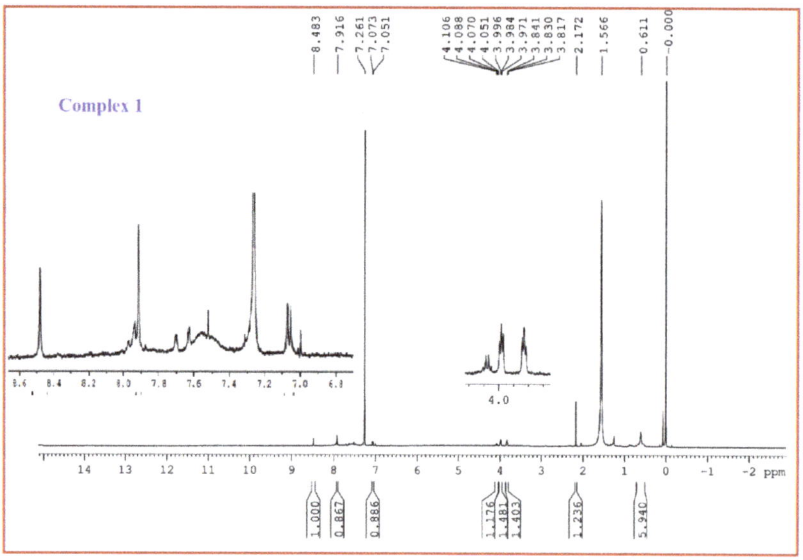

Figure 2. ^{1}H NMR spectrum of the complex **1**.

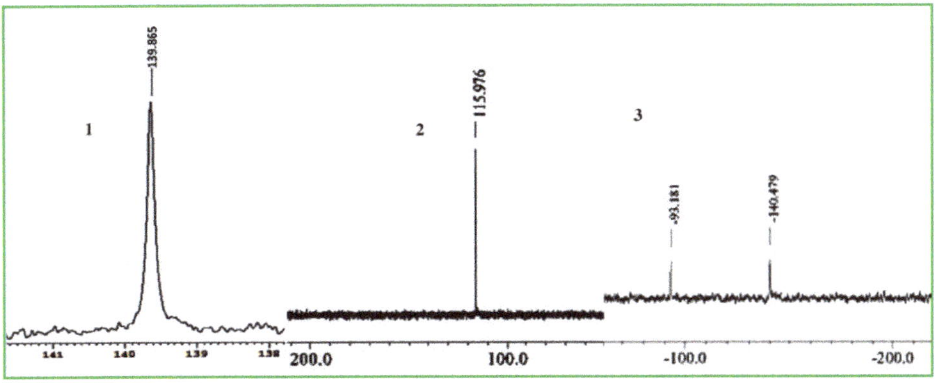

Figure 3. ^{119}Sn NMR spectra of the complexes **1**, **2** and **3**.

3.4. Crystal Structure of 1

The ORTEP view with atom numbering of the complex **1** is shown in Figure 4. It is a cyclic dimer of triorganotin(IV) ($C_{38}H_{46}N_6O_8Sn_2$) with the asymmetric unit Me3Sn^{IV}(H_2L). It is a triclinic crystal system with the space group P-1. The coordination environment about Sn(IV) is trigonal-bipyramidal, constituted by the three methyl group at the equatorial positions and the two axial positions occupied by the one carboxylate group via unidentate fashion and one phenoxo group from the ligand moiety. The coordinate ligand is in the zwitterionic state, where phenolic H-atom shifts to the imino-N atom, forming a 6-membered ring through H-bonding. The distortion of coordination sphere of Sn(IV) can be characterized quantitatively by parameter τ, defined by Addison et al. [47] (the τ values for the idealized geometries are $\tau = 0$, square planar, $\tau = 1$, trigonal bipyramidal). The value of $\tau = (b - a)/60$ where b is the largest and a is the second largest basal angle around the tin atom] for complex **1** is 0.94. The selected bond lengths, bond angles and hydrogen bonds geometry are presented in Tables 2 and 3, respectively. The coordinated ligand is in azo-form with N1-N2 bond length of 1.261(2) Å, which is comparable with literature [21,25,30].

Table 2. The selected bond lengths (Å) and bond angles (°) for **1**.

Atoms	Bond Length (Å)
Sn1-O1	2.2126 (11)
Sn1-C1	2.1270 (18)
Sn1-C2	2.1306 (15)
Sn1-C3	2.1303 (19)
Sn1-O3 ^{i}	2.3091 (11)
N1-N2	1.261 (2)
Atoms	**Bond Angle (°)**
O1-Sn1-C1	95.52 (6)
O1-Sn1-C2	93.02 (6)
O1-Sn1-C3	86.09 (6)
O1-Sn1-O3 ^{i}	177.68 (5)
C1-Sn1-C2	117.20 (7)
C1-Sn1-C3	121.47 (8)
O3 ^{i}-Sn1-C1	84.61 (5)
C2-Sn1-C3	121.13 (7)
O3 ^{i}-Sn1-C2	88.97 (6)
O3 ^{i}-Sn1-C3	91.86 (6)

Symmetrycode: ^{i} 1 − x, 2 − y, 1 − z.

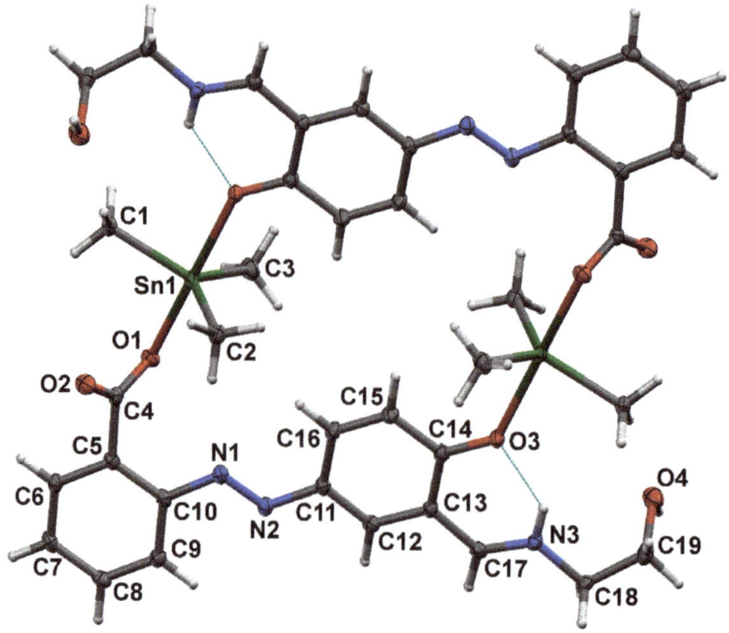

Figure 4. The crystal structure of the complex **1**. Displacement ellipsoids are drawn at 50% probability level.

Table 3. The hydrogen bonding parameters for compound **1**.

Compound	D–H . . . A	d (D–H)	d (H . . . A)	d (D . . . A)	<(DHA)
	N3-H3 . . . O3	0.83 (2)	1.97 (2)	2.6316 (18)	136 (2)
1	O4-H4 . . . O2 ^{ii}	0.85 (3)	1.96 (3)	2.7829 (18)	159 (3)
	C18-H18B . . . O2 ^{iii}	0.99	2.56	3.529(2)	167

Symmetrycode: ^{ii} x, y, 1 + z ^{iii} −x, 1 − y, 1 − z.

The imine bond length (N3-C17 of 1.293(2) Å) is also comparable with the literature value [21,22,30]. The complex is viewed as end-to-end bridged by the ligand, OOC-H_2L-O to Me_3Sn^{IV} fragment. The C-Sn-C angles are of 117.20(7), 121.47(8), and 121.13(7)° and the O-Sn-O angle is 177.68(5)°. The Sn-C bonds are almost equivalent (Sn1-C1, 2.1270(18) Å; Sn1-C2, 2.1306(15) Å Sn1-C3, 2.1303(19) Å). In the case of Sn-O bonds, the Sn1-O1(carboxylate-O) (2.2126(11) Å) is shorter than that of Sn1-O3(phenolate-O) (2.3091(11) Å). Thus, the complex possesses somewhat distorted trigonal bipyramidal geometry about Sn(IV), where axial bonds are longer than the equatorial bonds. The distance between Sn1 and other carboxylate oxygen is 3.303 Å, which is in line with the sum of their Van der Waals radii. Thus, it can be treated as a strong secondary interaction. The complex has a centre of symmetry with plane-centroid-Sn distance, 4.00 Å. The complex holds an intramolecular and an intermolecular H-bonding between N3-H3 . . . O3 and O4-H4 . . . .O2 ^{i} (^{i} +x, 1 + y, 1 + z), respectively (Table 3, Figure S11). Another similar type, but with longer H-bonding, occurs between C18-H18B . . . .O2 ^{ii} (^{ii} −x, 1 − y, 1 − z).

3.5. Theoretical Study of Intermolecular Interactions

The molecular Hirshfeld surface (visualization of short interatomic contacts using sums of appropriate vdW radii) represents an area where molecules come into contact, and its analysis gives the possibility of an additional insight into the nature of intermolecular interactions in the crystal state. We carried out the Hirshfeld surface analysis for the X-ray structure of **1** to understand what kind of intermolecular contacts gives the largest contributions in crystal packing (Figure 5). For the visualization, we have used a mapping of the normalized contact distance (d_{norm}); its negative value enables identification of molecular regions of substantial importance for the detection of short contacts. In the Hirshfeld surface, the regions of shortest intermolecular contacts are visualized by red circle areas (corresponding to hydrogen bonds O–H···O and stacking interactions). The fingerprint plots from the Hirshfeld surface analysis for the X-ray structure of **1** are given in Figure S12. Results of the Hirshfeld surface analysis for the X-ray structure of **1** reveal that intermolecular contacts involving hydrogen atoms (viz. H···H, O···H, and C···H) give the largest contributions in the crystal packing.

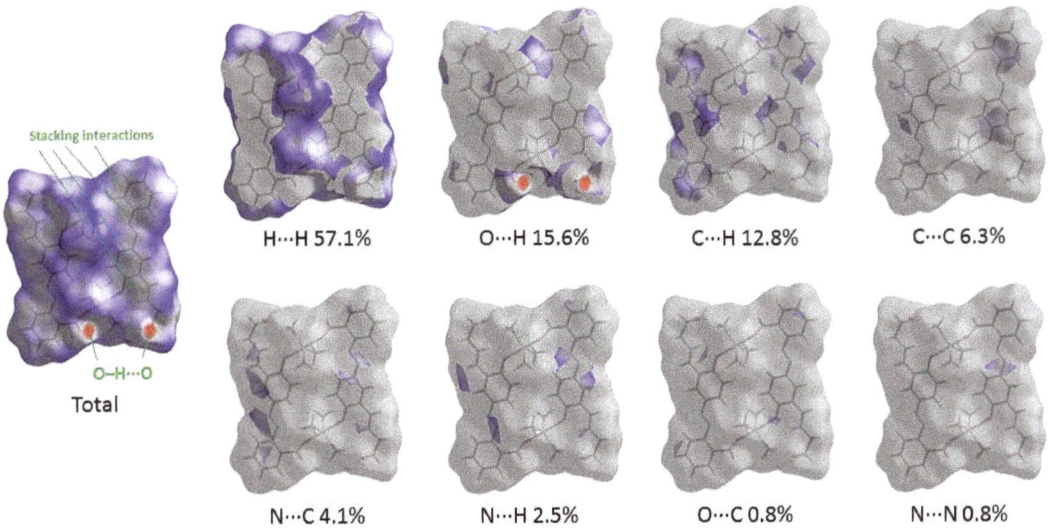

Figure 5. Results of a Hirshfeld surface analysis for the X-ray structure **1**.

We carried out DFT calculations, followed by the topological analysis of the electron density distribution, at the ωB97XD/DZP-DKH level of theory for model supramolecular associates to more deeply study the phenomenon of hydrogen bonds O–H···O and stacking interactions in the X-ray structure of **1**. Results of topological analysis of the electron density distribution are summarized in Table 4; the contour line diagrams of the Laplacian of electron density distribution $\nabla^2\rho(\mathbf{r})$, visualization of electron localization function (ELF) and reduced density gradient (RDG) analyses for hydrogen bonds O–H···O and stacking interactions in the X-ray structure of **1** are shown in Figure 6.

Table 4. Values of the density of all electrons—$\rho(\mathbf{r})$, Laplacian of electron density—$\nabla^2\rho(\mathbf{r})$ and appropriate λ_2 eigenvalues, energy density—H_b, potential energy density—$V(\mathbf{r})$, Lagrangiankinetic energy—$G(\mathbf{r})$, and electron localization function—ELF (a.u.) at the bond critical points (3, −1), corresponding to hydrogen bonds O–H···O and stacking interactions in the X-ray structure of **1**, and estimated strength for these contacts E_{int} (kcal/mol).

Contact	$\rho(\mathbf{r})$	$\nabla^2\rho(\mathbf{r})$	λ_2	H_b	$V(\mathbf{r})$	$G(\mathbf{r})$	ELF	E_{int} *
O–H···O 1.958 Å	0.024	0.094	−0.024	0.003	−0.018	0.021	0.067	5.6
C···C 3.310 Å	0.006	0.019	−0.006	0.001	−0.003	0.004	0.022	0.9

* $E_{int} \approx -V(\mathbf{r})/2$.

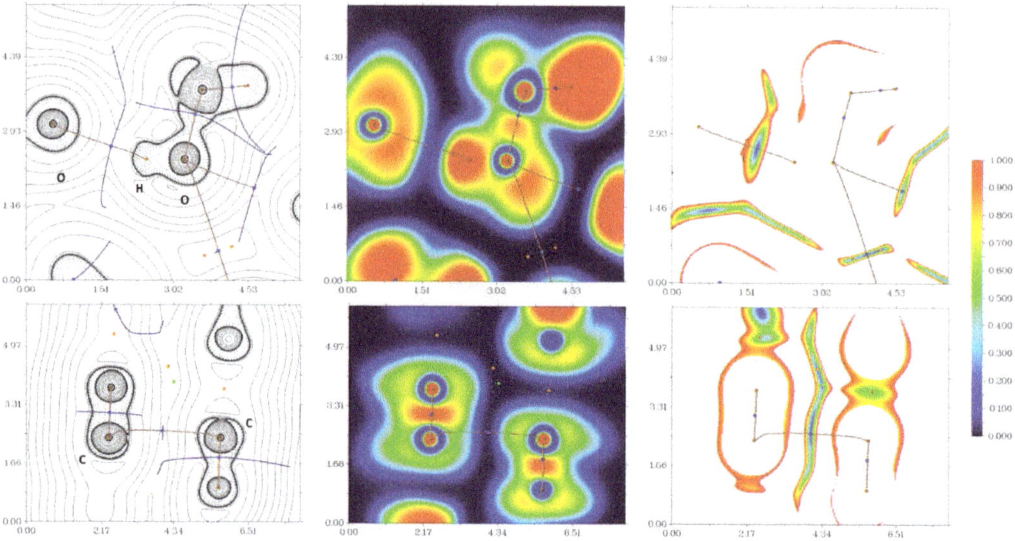

Figure 6. Contour line diagrams of the Laplacian of electron density distribution $\nabla^2\rho(\mathbf{r})$ and selected zero-flux surfaces (left panels), visualization of electron localization function (ELF, center panels) and reduced density gradient (RDG, right panels) analyses for hydrogen bond O–H···O (top) and stacking interactions (bottom) in the X-ray structure of **1**. Bond critical points (3, −1) are shown in blue, nuclear critical points (3, −3)—in pale brown, ring critical points (3, +1)—in orange, cage critical points (3, +3)—in light green, length units—Å, and the color scale for the ELF and RDG maps is presented in a.u.

The topological analysis of the electron density distribution in model supramolecular associates demonstrates the presence of bond critical points (3, −1) for hydrogen bonds O–H···O and stacking interactions in the X-ray structure of **1** (Table 4 and Figure 6). The low magnitude of the electron density, positive values of the Laplacian of electron density, and very close to zero positive energy density in these bond critical points (3, −1) and estimated strength for appropriate short contacts are typical for weak noncovalent interactions. The balance between the Lagrangian kinetic energy G(**r**) and potential energy density V(**r**) at the bond critical points (3, −1) for hydrogen bonds O–H···O and stacking interactions in the X-ray structure of **1** (viz. $-G(\mathbf{r})/V(\mathbf{r}) > 1$) reveals that covalent contribution in these short contacts is negligible [48]. The Laplacian of electron density is typically decomposed into the sum of contributions along the three principal axes of maximal variation, giving the three eigenvalues of the Hessian matrix (λ_1, λ_2 and λ_3), and the sign of λ_2 can be utilized to distinguish bonding (attractive, $\lambda_2 < 0$) weak interactions from nonbonding ones (repulsive, $\lambda_2 > 0$) [49,50]. Thus, hydrogen bonds O–H···O and stacking interactions in the X-ray structure of **1** are attractive. Note that estimated cumulative strengths of hydrogen bonds O–H···O and stacking interactions responsible for the formation of formal dimers in the X-ray structure of **1** by conventional supermolecule method (as the difference between the total electronic energies of appropriate dimer and two monomers) are 44.8 and 46.3 kcal/mol, respectively.

3.6. Antibacterial Activities

Antibacterial activities of the ligand (**H_3L**) and the compounds (**1–3**) were evaluated at different concentrations (1 μg/disc, 100 μg/disc, 500 μg/disc) and compared with the standard antibiotics. The *Klebsiella pneumoniae* (A), *Vibrio cholerae* (M) and *Shigella boydii* (Q) (gram-negative) and *Staphylococcus aureus* (J), *Streptococcus pneumonia* (K) (gram-positive) bacteria were included as bacterial strains. The *Polymyxin B* and Gentamicin as standard antibiotics against gram-negative bacteria and *Vancomycin* against gram-positive bacterial strains were used as positive control at four different concentrations (10 µg/disc, 30 µg/disc, 100 µg/disc, 500 µg/disc) in order to compare with the activity of the tested compounds. The calculated zones of inhibitions were measured and the data are given in Table 5. The data are represented in bar-diagram in Figure 7 for the ligand and Figure 8 for the complex **1**. The zone diameter is the measure of activity; the zone diameter activity is higher and stronger. The *Vancomycin,* an antibiotic drug for gram (+) bacteria only, shows lethal activity at 30 µg/disc concentration against gram-positive bacterial strains *viz. Staphylococcus aureus* (J) (Zone diameter, 14 ± 0 mm), *Streptococcus pneumonia* (K) (Zone diameter, 12 ± 0 mm) and remains inactive at higher concentrations (100 µg/disc and 500 µg/disc). The activities of the compounds are less or comparable (see Table 5) with the standard drug. Most importantly, they show activity even at a lower concentration—1 µg/disc (e.g., 6.3 ± 0.5 and 10.7 ± 0.5 for **3**)—and increases with an increase in concentrations. Their activities are more or less the same. On the other hand, the standard gram (−) ve antibiotic drugs, *Polymyxin B* for *Klebsiella pneumoniae* (A) and Gentamicin for *Vibrio cholerae* (M) and *Shigella boydii* (Q) bacteria show lethal activity at concentrations 10 µg/disc, 100 µg/disc and 500 µg/disc but not at 30 µg/disc (see Table 5). However, the compounds (**H_3L** and complexes, **1–3**) under investigation show lethal activity against all kinds of bacterial strains at all concentrations. The *Polymyxin B* show zone of inhibition ranges from 5.0 ± 0.0 to 12.0 ± 0.6 mm against *Klebsiella pneumoniae* (A), which is exclusively comparable with the compounds (**H_3L**: 6.0 ± 1.0–14.0 ± 1.0; **1**: 5.0 ± 0.0–13.7 ± 0.6 mm). The activities of Gentamicin range from 15 ± 0.0–25 ± 0.0 against *Vibrio cholerae* (M) and 17.0 ± 0.6–36.7 ± 0.6 against *Shigella boydii* (Q) bacteria. The activities of the compounds against these bacteria are less or comparable (e.g., **H_3L**: 5.0 ± 0.0–13.7 ± 0.6 against *Vibrio cholerae* (M) and 5.7 ± 1.0–15.3 ± 1.0 against *Shigella boydii* (Q); **2**: 8.3 ± 1.0–18.7 ± 1.0 against *Vibrio cholerae* (M) and 6.3 ± 1.0–18.0 ± 0.0 against *Shigella boydii* (Q)). The study highlights two important characteristic points regarding their potential antibacterial properties; first, they are active evenly against both types of bacteria; second, their activities are comparable against the corresponding standard drugs. Thus, the compounds could be treated as common potential antibacterial materials.

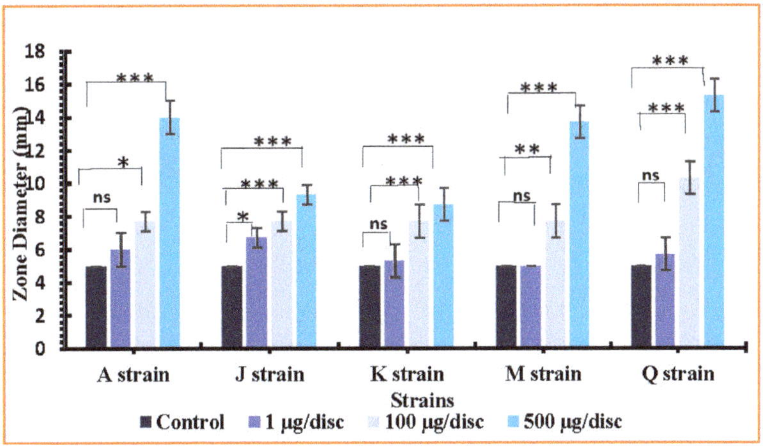

Figure 7. The zone diameter at different concentrations of **H_3L** on Strains, *Klebsiellapneumoniae* (A), *Staphylococcus aureus* (J), Streptococcus *pneumonia* (K), *Vibrio cholerae* (M) & *Shigella boydii* (Q). Data are represented as mean ± SD of three concentration of doses (* $p < 0.05$, ** $p < 0.01$, *** $p < 0.001$, ns—no significance $p > 0.05$).

Table 5. Antibacterial activities of the compounds and standard drugs against three (−) ve and two (+) ve bacteria.

Compounds	Conc.	Zone of Inhibition at Different Concentration (mm)				
		Gram (−) ve Bacteria			Gram (+) ve Bacteria	
		K. pneumoniae (A)	*V. cholerae* (M)	*Shigella boydii* (Q)	*S. Aureus* (J)	*S. Pneumoniae* (K)
Control		5.0 ± 0.00	5.0 ± 0.00	5.0 ± 0.00	5.0 ± 0.00	5.0 ± 0.00
Polymyxin B	10 µg	5.0 ± 0.0	-	-	-	-
	100 µg	7.0 ± 0.0	-	-	-	-
	500 µg	12.0 ± 0.6	-	-	-	-
Gentamicin	10 µg	-	15 ± 0.0	17 ± 0	-	-
	100 µg	-	22 ± 0.0	23.3 ± 0.6	-	-
	500 µg	-	25 ± 0.0	36.7 ± 0.6	-	-
Vancomycin	30 µg	-	-	-	14 ± 0	12 ± 0
H_3L	1 µg	6.0 ± 1.0	5.0 ± 0.0	5.7 ± 1.0	6.7 ± 0.6	5.3 ± 1.0
	100 µg	7.7 ± 0.6	7.7 ± 1.0	10.3 ± 1.0	7.7 ± 0.6	7.7 ± 1.0
	500 µg	14.0 ± 1.0	13.7 ± 1.0	15.3 ± 1.0	9.3 ± 0.6	8.7 ± 1.0
1	1 µg	5.0 ± 0.0	5.0 ± 0.0	5.0 ± 0.0	6.0 ± 0.0	6.0 ± 0.0
	100 µg	6.7 ± 0.6	9.0 ± 0.0	11.7 ± 1.0	7.3 ± 0.6	8.7 ± 0.6
	500 µg	13.7 ± 0.6	13.7 ± 1.0	5.0 ± 0.0	10.3 ± 0.6	14.7 ± 0.6
2	1 µg	5.0 ± 0.0	8.3 ± 1.0	6.3 ± 1.0	6.3 ± 0.6	6.0 ± 0.0
	100 µg	5.0 ± 0.0	17.7 ± 1.0	14.7 ± 1.0	7.0 ± 0.0	7.3 ± 0.6
	500 µg	6.3 ± 0.6	18.7 ± 1.0	18.0 ± 0.0	8.7 ± 0.6	8.0 ± 1.0
3	1 µg	5.0 ± 0.0	5.0 ± 0.0	6.0 ± 0.0	6.3 ± 0.5	10.7 ± 0.5
	100 µg	5.0 ± 0.0	11.7 ± 0.5	14.7 ± 0.5	7.0 ± 0.0	6.0 ± 0.0
	500 µg	5.0 ± 0.0	8.3 ± 0.5	19.7 ± 0.5	8.7 ± 0.5	9.3 ± 0.5

'-' means no activity; Polymyxin B and Gentamicin shows no activity at conc. 30 µg but Vancomycin only at conc. 30 µg.

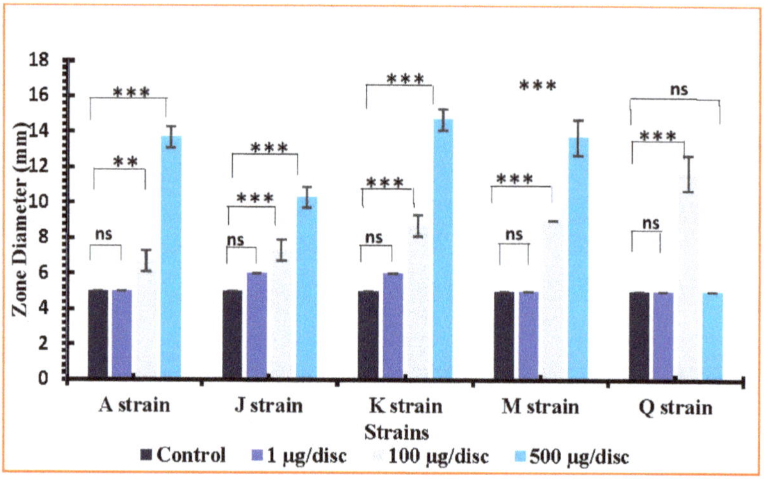

Figure 8. The zone diameter at different concentrations of complex (**1**) on Strains *Klebsiella pneumoniae* (**A**), *Staphylococcus aureus* (**J**), *Streptococcus pneumoniae* (**K**), *Vibrio cholerae* (**M**) & *Shigella boydii* (**Q**). Data are represented as mean ± SD of three concentration of doses (** $p < 0.01$, *** $p < 0.001$, ns—no significance $p > 0.05$).

4. Conclusions

A new ligand containing versatile functional groups, including azo-, imine-, phenolic-OH, alcoholic-OH and carboxylate functions, was synthesized. The ligand generates versatile organotin(IV) complexes. With trimethyltin(IV), the ligand forms a 24-membered cyclic dimeric organotin(IV) complex. In the solid state, various noncovalent interactions, including stacking and hydrogen bonding, determine molecular packing. The solid state TBP geometry about Sn(IV) changes into a distorted tetrahedral geometry in the solution. Again, with tributyltin(IV), the ligand forms distorted tetrahedral geometry in the solution, but with dibutyltin(IV), the ligand shows distorted trigonal bipyramidal geometry with a two tin centre. The compounds possess considerable antibacterial activity against both gram-negative and gram-positive bacteria. Thus, the compounds can be an effective potential common antibacterial material.

Supplementary Materials: The following supporting information can be downloaded at: https://www.mdpi.com/article/10.3390/cryst12111582/s1, Supplementary material 1: The CCDC number 2203973 for the compound **1** has crystallographic supplementary data and can be obtained without paying any charge from http://www.ccdc.cam.ac.uk/conts/retrieving.html (accessed on 1 October 2022), or the Cambridge Crystallographic Data Center, 12 Union Road, Cambridge CB2 1EZ, UK; fax: (+44) 1223-336-033; or e-mail: deposit@ccdc.cam.ac.uk. The supplementary data include IR spectra in Figures S1–S4 respectively for the ligand and the complexes **1–3**; ^{1}H NMR and ^{13}C NMR spectra of compounds in Figures S5–S10; molecular packing and fingerprint plots from Hirshfeld surface analysis for the X-ray structure **1** in Figure S11 and S12, respectively.

Author Contributions: Conceptualization and visualization, P.D. (Pratima Debnath) and M.R.; investigation, P.D. (Pratima Debnath), M.R. and T.K.M.; data curation, P.D. (Pratima Debnath), P.D. (Paresh Debnath), M.R., T.K.M., L.S., W.M., T.A., D.M. and A.S.N.; writing-original draft preparation, P.D. (Pratima Debnath), P.D. (Paresh Debnath), M.R., T.K.M., D.M., W.M., and A.S.N.; writing-review and editing, P.D., M.R., T.K.M., W.M., and A.S.N.; supervision, M.R. and T.K.M. All authors have read and agreed to the published version of the manuscript.

Funding: This research received no external funding.

Data Availability Statement: Not applicable.

Acknowledgments: We thank Department of Chemistry, NIT Agartala for providing research facilities. PB is grateful to NIT Agartala for receiving institutional fellowship from MHRD, Govt. of India. The DFT calculations, topological analysis of the electron density distribution, and Hirshfeld surface analysis were supported by the RUDN University Strategic Academic Leadership Program. We would like to thank SAIF, IISC Bangalore for NMR studies. Moreover, all authors in this section have consented to the acknowledgements.

Conflicts of Interest: All authors declare that there are no conflict of interests on this research work.

References

1. Saxena, A.K. Organotin compounds: Toxicology and biomedicinal applications. *Appl. Organomet. Chem.* **1987**, *1*, 39–56. [CrossRef]
2. Ghazi, D.; Rasheed, Z.; Yousif, E. Review of organotin compounds: Chemistry and applications. *Int. J. Res. Eng. Innov.* **2018**, *2*, 340–348.
3. Tabassum, S.; Khan, R.A.; Arjmand, F.; Sen, S.; Kayal, J.; Juvekar, A.S.; Zingde, S.M. Synthesis and characterization of glycoconjugate tin (IV) complexes: In vitro DNA binding studies, cytotoxicity, and cell death. *J. Organomet. Chem.* **2011**, *696*, 1600–1608. [CrossRef]
4. Arjmand, F.; Jamsheera, A. Synthesis, characterization and in vitro DNA binding studies of tin (IV) complexes of tert-butyl 1-(2-hydroxy-1-phenylethylamino)-3-methyl-1-oxobutan-2-yl carbamate. *J. Organomet. Chem.* **2011**, *696*, 3572–3579. [CrossRef]
5. Paul, A.; Hazra, S.; Guedes da Silva, M.F.C.; Pombeiro, A.J. Biological Evaluation of Azo-and Imino-Based Carboxylate Triphenyltin (IV) Compounds. *Eur. J. Inorg. Chem.* **2020**, *11–12*, 930–941. [CrossRef]
6. Amir, M.K.; Khan, S.; Shah, A.; Butler, I.S. Anticancer activity of organotin (IV) carboxylates. *Inorg. Chim. Acta* **2014**, *423*, 14–25. [CrossRef]
7. Devi, J. Pachwania, S. Recent advancements in DNA interaction studies of organotin (IV) complexes. *Inorg. Chem. Commun.* **2018**, *91*, 44–62. [CrossRef]

8. Dahmani, M.; Harit, T.; Et-Touhami, A.; Yahyi, A.; Eddike, D.; Tillard, M.; Benabbes, R. Two novel macrocyclic organotin (IV) carboxylates based on bipyrazoledicarboxylic acid derivatives: Syntheses, crystal structures and antifungal activities. *J. Organomet. Chem.* **2021**, *948*, 121913. [CrossRef]
9. Shoaib Ahmad Shah, S.; Ashfaq, M.; Waseem, A.; Mehboob Ahmed, M.; Najam, T.; Shaheen, S.; Rivera, G. Synthesis and biological activities of organotin (IV) complexes as antitumoral and antimicrobial agents. *A Rev. Mini Rev. Med. Chem.* **2015**, *15*, 406–426. [CrossRef]
10. Debnath, P.; Singh, K.S.; Singh, K.K.; Singh, S.S.; Sieroń, L.; Maniukiewicz, W. Di-butyltin (IV) complexes with azo-carboxylates: Synthesis, characterization, crystal structures and their anti-diabetic assay. *New J. Chem.* **2020**, *44*, 5862–5872. [CrossRef]
11. Tiekink, E.R. Structural chemistry of organotin carboxylates: A review of the crystallographic literature. *Appl. Organomet. Chem.* **1991**, *5*, 1–23. [CrossRef]
12. Abbas, S.M.; Ali, S.; Hussain, S.T.; Shahzadi, S. structural diversity in organotin(IV) dithiocarboxylates and carboxylates. *J. Coord. Chem.* **2013**, *66*, 2217–2234. [CrossRef]
13. Nath, M.; Saini, P.K. Chemistry and applications of organotin (IV) complexes of Schiff bases. *Dalton Trans.* **2011**, *40*, 7077–7121. [CrossRef] [PubMed]
14. Basu, S.; Masharing, C.; Das, B. Diorganotin (IV) complexes of polyaromatic azo-azomethine ligand derived from salicylaldehyde and ortho-aminophenol: Synthesis, characterization, and molecular structures. *Heteroat. Chem.* **2012**, *23*, 457–465. [CrossRef]
15. Debnath, P.; Das, A.; Singh, K.S.; Yama, T.; Singh, S.S.; Butcher, R.J.; Maniukiewicz, W. Synthesis, structural characterization and antimicrobial activities of triorganotin (IV) azo-carboxylates derived from ortho/para-amino benzoic acids and β-naphthol. *Inorg. Chim. Acta.* **2019**, *498*, 119172. [CrossRef]
16. Debnath, P.; Singh, K.S.; Devi, T.S.; Singh, S.S.; Butcher, R.J.; Sieroń, L.; Maniukiewicz, W. Synthesis, characterization, crystal structures and anti-diabetic activity of organotin (IV) complexes with 2-(4-hydroxynaphthylazo)-benzoic acid. *Inorg. Chim. Acta.* **2020**, *510*, 119736. [CrossRef]
17. Debnath, P.; Singh, K.S.; Sharma, S.; Debnath, P.; Singh, S.S.; Sieroń, L.; Maniukiewicz, W. Synthesis, structural characterization, Hirshfeld surface analysis and in vitro-antimicrobial activities of triphenyltin (IV) compounds of azo-carboxylates derived from 2-or 4-amino benzoic acids and naphthalen-1 or 2-ol. *J. Mol. Struct.* **2021**, *1223*, 128971. [CrossRef]
18. Yin, H.D.; Chen, S.W. Synthesis and characterization of di-and tri-organotin (IV) complexes with Schiff base ligand pyruvic acid 3-hydroxy-2-naphthoyl hydrazone. *Inorg. Chim. Acta.* **2006**, *359*, 3330–3338. [CrossRef]
19. BasuBaul, T.S.; Basu, S.; de Vos, D.; Linden, A. Amino acetate functionalized Schiff base organotin (IV) complexes as anticancer drugs: Synthesis, structural characterization, and in vitro cytotoxicity studies. *Investig. New Drugs.* **2009**, *27*, 419–431. [CrossRef]
20. Yin, H.; Liu, H.; Hong, M. Synthesis, structural characterization and DNA-binding properties of organotin (IV) complexes based on Schiff base ligands derived from 2-hydroxy-1-naphthaldy and 3-or 4-aminobenzoic acid. *J. Organomet. Chem.* **2012**, *713*, 11–19. [CrossRef]
21. Roy, M.; Roy, S.; Devi, N.M.; Singh, C.B.; Singh, K.S. Synthesis, structural characterization and antimicrobial activities of diorganotin (IV) complexes with azo-imino carboxylic acid ligand: Crystal structure and topological study of a doubly phenoxide-bridged dimeric dimethyltin (IV) complex appended with free carboxylic acid groups. *J. Mol. Struct.* **2016**, *1119*, 64–70.
22. Baul, T.S.B.; Addepalli, M.R.; Lyčka, A.; van Terwingen, S.; da Silva, M.F.C.G. Synthesis and structural characterization of diorganotin (IV) complexes with heteroditopic pyridyl-ONO′-ligands. *Inorg. Chim. Acta.* **2020**, *512*, 119892. [CrossRef]
23. BasuBaul, T.S.; Chaurasiya, A.; Rabha, M.; Khatua, S.; Lyčka, A.; Schollmeyer, D.; Jurkschat, K. Diorganotin Compounds Containing α-Aminoacidato Schiff Base Ligands Derived from Functionalized 2-Hydroxy-5-(aryldiazenyl) benzaldehyde. Syntheses, Structures and Sensing of Hydrogen Sulfide. *Eur. J. Inorg. Chem.* **2020**, *18*, 1803–1813. [CrossRef]
24. BasuBaul, T.S.; Addepalli, M.R.; Duthie, A.; Singh, P.; Koch, B.; Gildenast, H.; Englert, U.; Rojas-León, I.; Höpfl, H. Triorganotin (IV) derivatives with semirigid heteroditopic hydroxo-carboxylato ligands: Synthesis, characterization, and cytotoxic properties. *Appl. Organomet. Chem.* **2021**, *35*, E608.
25. Roy, M.; Roy, S.; Singh, K.S.; Kalita, J.; Singh, S.S. Synthesis, characterisation and anti-diabetic activities of triorganotin (IV) azo-carboxylates derived from amino benzoic acids and resorcinol: Crystal structure and topological study of a 48 membered macrocyclic-tetrameric trimethyltin (IV) complex. *Inorg. Chim. Acta.* **2016**, *439*, 164–172. [CrossRef]
26. Chen, L.; Wang, Z.; Qiu, T.; Sun, R.; Zhao, Z.; Tian, L.; Liu, X. Synthesis, structural characterization, and properties of triorganotin complexes of Schiff base derived from 3-aminobenzoic acid and salicylaldehyde or 2, 4-pentanedione. *Appl. Organomet. Chem.* **2020**, *34*, E5790. [CrossRef]
27. BasuBaul, T.S.; Chaurasiya, A.; Duthie, A.; Montes-Tolentino, P.; Höpfl, H. Coordination-driven self-assembly of macrocycles and 1D or 2D coordination polymers using heteroditopic pyridyl-carboxylate ligands: The case study of 5-[(E)-2-(3-pyridyl)-1-diazenyl]-2-hydroxybenzoate in combination with {R_nSn}(n= 2 and 3). *Cryst. Growth Des.* **2019**, *19*, 6656–6671. [CrossRef]
28. BasuBaul, T.S.; Das, P.; Eng, G.; Linden, A. Synthesis and Characterization of Some Triphenyltin (IV) Complexes from Sterically Crowded [((E)-1-{2-Hydroxy-5-[(E)-2-(aryl)-1-diazenyl] phenyl} methylidene) amino] acetate Ligands and Crystal Structure Analysis of a Tetrameric Triphenyltin (IV) Compound. *J. Inorg. Organomet. Polym. Mater.* **2010**, *20*, 134–141. [CrossRef]
29. Baul, T.S.B.; Singh, K.S.; Holčapek, M.; Jirásko, R.; Rivarola, E.; Linden, A. Synthesis, characterization and crystal structures of polymeric and dimeric triphenyltin (IV) complexes of 4-[((E)-1-{2-hydroxy-5-[(E)-2-(2-carboxyphenyl)-1-diazenyl] phenyl} methylidene) amino] aryls. *J. Organomet. Chem.* **2005**, *690*, 4232–4242. [CrossRef]

30. Singh, K.S.; Roy, M.; Roy, S.; Ghosh, B.; Devi, N.M.; Singh, C.B.; Mun, L.K. Synthesis, characterization and antimicrobial activities of triorganotin (IV) complexes with azo-azomethine carboxylate ligands: Crystal structure of a tributyltin (IV) and a trimethyltin (IV) complex. *J. Coord. Chem.* **2017**, *70*, 361–380. [CrossRef]
31. Shang, X.; Cui, J.; Wu, J.; Pombeiro, A.J.L.; Li, Q. Polynuclear diorganotin (IV) complexes with arylhydroxamates: Syntheses, structures and in vitro cytotoxic activities. *J. Inorg. Biochem.* **2008**, *102*, 901–909. [CrossRef] [PubMed]
32. Rigaku, O.D. *CrysAlis PRO*; Rigaku Oxford Diffraction Ltd.: Yarnton, UK, 2019.
33. Sheldrick, G.M. SHELXT—Integrated space-group and crystal-structure determination. *Acta Cryst.* **2015**, *A 71*, 3–8. [CrossRef]
34. Sheldrick, G.M. Crystal structure refinement with SHELXL. *Acta Cryst.* **2015**, *C 71*, 3–8.
35. Wolff, S.K.; Grimwood, D.J.; McKinnon, J.J.; Turner, M.J.; Jayatilaka, D.; Spackman, M.A. *Crystal Explorer (Version 3.1)*; University of Western Australia: Nedland, Australia, 2012.
36. Spackman, M.A.; Jayatilaka, D. Hirshfeld surface analysis. *Cryst. Eng. Comm.* **2009**, *11*, 19–32. [CrossRef]
37. McKinnon, J.J.; Jayatilaka, D.; Spackman, M.A. Towards quantitative analysis of intermolecular interactions with Hirshfeld surfaces. *Chem. Commun.* **2007**, 3814–3816. [CrossRef] [PubMed]
38. Bondi, A. Van der Waals volumes and radii of metals in covalent compounds. *J. Phys. Chem.* **1966**, *70*, 3006–3007. [CrossRef]
39. Chai, J.D.; Head-Gordon, M. Long-range corrected hybrid density functionals with damped atom–atom dispersion corrections. *Phys. Chem. Chem. Phys.* **2008**, *10*, 6615–6620. [CrossRef]
40. Frisch, M.J.; Trucks, G.W.; Schlegel, H.B.; Scuseria, G.E.; Robb, M.A.; Cheeseman, J.R.; Scalmani, G.; Barone, V.; Mennucci, B.; Petersson, G.A.; et al. *Fox, in Gaussian 09, Revision C.01*; Gaussian, Inc.: Wallingford, CT, USA, 2010.
41. Barros, C.L.; De Oliveira, P.J.P.; Jorge, F.E.; Canal Neto, A.; Campos, M. Gaussian basis set of double zeta quality for atoms Rb through Xe: Application in non-relativistic and relativistic calculations of atomic and molecular properties. *Mol. Phys.* **2010**, *108*, 1965–1972. [CrossRef]
42. Jorge, F.E.; Canal Neto, A.; Camiletti, G.G.; Machado, S.F. Contracted Gaussian basis sets for Douglas–Kroll–Hess calculations: Estimating scalar relativistic effects of some atomic and molecular properties. *J. Chem. Phys.* **2009**, *130*, 064108. [CrossRef]
43. Neto, A.C.; Jorge, F.E. All-electron double zeta basis sets for the most fifth-row atoms: Application in DFT spectroscopic constant calculations. *Chem. Phys. Lett.* **2013**, *582*, 158–162. [CrossRef]
44. De Berrêdo, R.C.; Jorge, F.E. All-electron double zeta basis sets for platinum: Estimating scalar relativistic effects on platinum (II) anticancer drugs. *J. Mol. Struc-THEOCHEM.* **2010**, *961*, 107–112. [CrossRef]
45. Lu, T.; Chen, F. Multiwfn: A multifunctional wavefunction analyzer. *J. Comput. Chem.* **2012**, *33*, 580–592. [CrossRef] [PubMed]
46. Bauer, A.W.; Kirby, W.M.M.; Sherris, J.C.; Turck, M. Antibiotic susceptibility testing by a standardized single disk method. *Am. J. Clin. Pathol.* **1966**, *45*, 493–496. [CrossRef] [PubMed]
47. Addison, A.W.; Rao, T.N.; Reedijk, J.; van Rijn, J.; Verschoor, G.C. Synthesis, structure, and spectroscopic properties of copper(II) compounds containing nitrogen–sulphur donor ligands; the crystal and molecular structure of aqua [1,7-bis(N-methylbenzimidazol-2′-yl)-2,6-dithiaheptane]copper(II) perchlorate. *J. Chem. Soc. Dalt. Trans.* **1984**, 1349–1356. [CrossRef]
48. Espinosa, E.; Alkorta, I.; Elguero, J.; Molins, E. From weak to strong interactions: A comprehensive analysis of the topological and energetic properties of the electron density distribution involving X–H· · · F–Y systems. *J. Chem. Phys.* **2002**, *117*, 5529–5542. [CrossRef]
49. Johnson, E.R.; Keinan, S.; Mori-Sánchez, P.; Contreras-García, J.; Cohen, A.J.; Yang, W. Revealing noncovalent interactions. *J. Am. Chem. Soc.* **2010**, *132*, 6498–6506. [CrossRef]
50. Contreras-García, J.; Johnson, E.R.; Keinan, S.; Chaudret, R.; Piquemal, J.P.; Beratan, D.N.; Yang, W. NCIPLOT: A program for plotting noncovalent interaction regions. *J. Chem. Theory Comput.* **2011**, *7*, 625–632. [CrossRef]

crystals

MDPI

Article

Crystal Structure, Hirshfeld Analysis, and DFT Calculations of Three Trinuclear Cu(II) Polymorphs

Kelly L. Rue [1], Logesh Mathivathanan [1], Gellert Mezei [2], Alexander M. Mebel [1] and Raphael G. Raptis [1,*]

[1] Department of Chemistry and Biochemistry, Florida International University, Miami, FL 33199, USA
[2] Department of Chemistry, Western Michigan University, Kalamazoo, MI 49008, USA
* Correspondence: rraptis@fiu.edu

Abstract: The crystal structure of the ferromagnetically-coupled Cu^{II}_3−pyrazolato complex, $(Bu_4N)_2[Cu_3(\mu_3\text{-}Cl)_2(\mu\text{-}4\text{-}NO_2\text{-}pz)_3Cl_3]$ (**1a**, pz = pyrazolato anion), was originally determined in the triclinic *P*-1 space group. By varying the recrystallization solvent and temperature, two additional true polymorphs were crystallized in the monoclinic $P2_1/n$ (**1b**) and orthorhombic *Pbca* (**1c**) space groups. Comparison of the metric parameters of the three polymorphs revealed only minor variations in their bond lengths and angles but clearly distinguishable packing patterns. The DFT calculations showed that, in vacuum, **1a** had the lowest energetic minimum (also the densest of three polymorphs), whereas **1b** and **1c** lay at 6.9 kcal/mol and 7.8 kcal/mol higher energies. The existence of isolable **1b** and **1c** is, therefore, attributed to the intermolecular interactions analyzed by the Hirshfeld methods.

Keywords: polymorphism; Hirshfeld analysis; DFT calculations; trinuclear copper; pyrazolato ligands

Citation: Rue, K.L.; Mathivathanan, L.; Mezei, G.; Mebel, A.M.; Raptis, R.G. Crystal Structure, Hirshfeld Analysis, and DFT Calculations of Three Trinuclear Cu(II) Polymorphs. *Crystals* **2022**, *12*, 1611. https://doi.org/10.3390/cryst12111611

Academic Editors: Valerian Dragutan, Ileana Dragutan, Fu Ding and Ya-Guang Sun

Received: 28 October 2022
Accepted: 8 November 2022
Published: 11 November 2022

1. Introduction

Polymorphism [1] is a well-established phenomenon in organic compounds, resulting in the manifestation of critical differences in the pharmaceutical properties of compounds crystallized as different polymorphs [2,3]. Polymorphism is also common among minerals—calcium carbonate and the 14 known polymorphs of silica are good examples—and binary or ternary solid-state materials with often strikingly different properties manifested by their various forms [4,5]. Metal organic frameworks (MOFs) are also commonly encountered [6], as are materials undergoing a phase transition to a new polymorph under pressure. In contrast, transition metal complexes are rarely encountered as true polymorphs at ambient conditions [7–11]; a few more examples were included in a review article [12].

Trinuclear copper pyrazolato complexes of the general formula, $[Cu_3(\mu_3\text{-}E)(\mu\text{-}4\text{-}R\text{-}pz)_3X_3]^z$, where E = OH, O, $(Cl)_2$, $(Br)_2$, and OMe; R = H, Cl, Br, I, Me, NO_2, -CHO, and -COOEt; X = Cl, Br, PhCOO-, py, SCN-, and MeCOO-, and z = 2−, 1−, and 2+, have been studied in our laboratory for over two decades for their interesting magnetic, structural, and electrochemical properties [13–21]. The capping ligand E forces the Cu_3E motif into one of three geometries: planar (Cu_3O), trigonal pyramidal ($Cu_3(OH)$), or trigonal bipyramidal (Cu_3X_2).

We have previously published a ferromagnetically coupled Cu^{II}_3-pyrazolato complex, $(Bu_4N)_2[Cu_3(\mu_3\text{-}Cl)_2(\mu\text{-}4\text{-}NO_2\text{-}pz)_3Cl_3]$ (**1a**), which crystallized in the triclinic *P*-1 space group [22]. Herein, we report its two true polymorphs in the $P2_1/n$ (**1b**) and *Pbca* (**1c**) space groups and discuss the analyses of their Hirshfeld surfaces and energies calculated by DFT methods.

2. Materials and Methods

2.1. Materials

All reagents, except 4-nitro-pyrazole were purchased from commercial sources and used as received. 4-Nitro-pyrazole was synthesized following procedures in the literature [23]. The solvents were purified and dried using standard techniques [24].

2.2. Synthesis

The synthetic method for $(Bu_4N)_2[Cu_3(\mu_3\text{-}Cl)_2(\mu\text{-}4\text{-}NO_2\text{-}pz)_3Cl_3]$ has been previously published along with the crystal structure of **1a** [22]. Whereas the trinuclear complex was synthesized in the manner reported, the crystallization of **1a–c** differed as follows: saturated solutions of the complex in boiling n-propyl alcohol (**1a**), n-butyl alcohol (**1b**), and methanol (**1c**) were filtered while hot and allowed to cool down under ambient conditions, resulting in X-ray quality crystals. Except for the methanol, a small amount of acetonitrile was added to the alcoholic solution during crystallization to prevent the formation of oily products on cooling.

2.3. X-ray Crystallography and Data Collection

X-ray diffraction data were collected with a Bruker AXS SMART 1K CCD diffractometer [25] with graphite-monochromated Mo Kα radiation (λ = 0.71073 Å) at ambient temperature. The data were corrected for Lorentz and polarization effects [26]. The structures were solved using the SHELXTL-direct methods program, then refined by full-matrix least squares methods on F^2 [27]. The crystal data and structure refinement parameters are listed in Table 1. CCDC 2215557 (**1b**) and 2215558 (**1c**) contain the supplementary crystallographic data for this paper and can be obtained free of charge from the Cambridge Crystallographic Data Center via www.ccdc.cam.ac.uk/data_request/cif.

Table 1. Crystal data and structure refinement parameters for **1a–c**.

	1a [22]	**1b**	**1c**
Formula	$C_{41}H_{78}Cl_5Cu_3N_{11}O_6$	$C_{41}H_{78}Cl_5Cu_3N_{11}O_6$	$C_{41}H_{78}Cl_5Cu_3N_{11}O_6$
Formula Weight	1189.01	1189.01	1189.01
Temperature (K)	299(2)	298(2)	298(2)
Crystal System	Triclinic	Monoclinic	Orthorhombic
Space Group	*P*-1 (No. 2)	$P2_1/n$ (No. 14)	*Pbca* (No. 61)
a/Å	13.121(1)	15.526(2)	21.944(2)
b/Å	15.183(1)	20.938(3)	15.783(2)
c/Å	15.625(1)	18.904(3)	33.886(4)
α/°	108.778(2)	90	90
β/°	102.082(2)	108.746(2)	90
γ/°	95.916(2)	90	90
V/Å^3	2832.5(5)	5819(1)	11,736(2)
Z	2	4	8
D_{calc}/g cm^{-3}	1.394	1.357	1.346
μ/mm^{-1}	1.402	1.365	1.354
Refl. Collected	12,635	30,317	62,961
Unique Refl.	8156	10,274	10,358
Obs. Refl. (I > 2σ(I))	4654	6065	4571
θ range/°	1.42–23.30	1.50–25.03	1.20–25.04
Data	8156	10,274	10,358
Restraints/param.	0/603	0/603	6/603
R(F); Rw(F) (I > 2σ(I))	0.0401; 0.0914	0.0476; 0.1294	0.0795; 0.2176
GooF	0.914	1.000	1.003

2.4. DFT Calculations

DFT calculations were carried out, with geometry optimization, without symmetry restrictions using the B3LYP [28,29] hybrid density functional with the 6-31G* basis set for all atoms [30] using Gaussian-09 [31] software. Correctness of the calculated electronic states was ensured by checking the stability of the SCF solutions; the calculations confirmed that all the considered structures had the quartet ground electronic state. Geometry optimizations were carried out in the gas phase.

2.5. Software

Geometric calculations and visual representations of the molecules were obtained from both *Olex2* [32] and Mercury *2020.3.0* [33]. Hirshfeld surfaces, fingerprint plots, and the associated images were calculated and obtained using *CrystalExplorer17* [34]. Hirshfeld surface analysis and the capabilities of *CrystalExplorer* are well discussed in the literature [35–38].

3. Results and Discussion

3.1. Crystal Structure Descriptions

3.1.1. General Structure Description

A representative structure of the $[Cu_3(\mu_3\text{-}Cl)_2(\mu\text{-}4\text{-}NO_2\text{-}pz)_3Cl_3]^{2-}$ unit of **1a**, **1b**, and **1c** is shown in Figure 1, and selected bond lengths of each polymorph are listed in Table 2. The charge in each dianionic complex is balanced by two crystallographically independent tetrabutylammonium cations. **1a** was previously crystallized from boiling n-propyl alcohol and reported in the *P*-1 space group, while **1b** was crystallized from boiling n-butyl alcohol in the $P2_1/n$ space group, and **1c** was crystallized from boiling methanol in the *Pbca* space group. As the symmetry of each polymorph increased from the triclinic to monoclinic to orthorhombic crystal systems, the crystal density (as reported in Table 1) decreased from 1.394 g cm^{-3} to 1.357 g cm^{-3} to 1.346 g cm^{-3}.

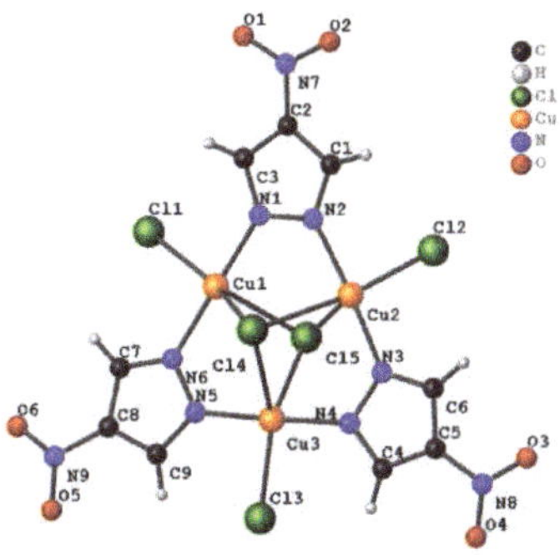

Figure 1. Representative structure and atom numbering scheme of the $[Cu_3(\mu_3\text{-}Cl)_2(\mu\text{-}4\text{-}NO_2\text{-}pz)_3Cl_3]^{2-}$ unit for **1(a–c)**.

Table 2. Selected bond lengths (Å) for **1a–1c**.

	1a	1b	1c
Cu1–Cl1	2.234(2)	2.275(1)	2.256(2)
Cu1–Cl4	2.501(2)	2.659(1)	2.680(3)
Cu1–Cl5	2.628(1)	2.510(1)	2.480(2)
Cu1–N1	1.939(4)	1.948(4)	1.950(6)
Cu1–N6	1.950(4)	1.944(3)	1.942(6)
Cu2–Cl2	2.255(2)	2.232(1)	2.267(2)
Cu2–Cl4	2.558(1)	2.594(1)	2.679(2)
Cu2–Cl5	2.614(2)	2.560(1)	2.490(2)
Cu2–N2	1.947(4)	1.950(4)	1.937(6)
Cu2–N3	1.946(4)	1.951(3)	1.926(7)
Cu3–Cl3	2.276(2)	2.271(1)	2.255(3)
Cu3–Cl4	2.584(2)	2.513(1)	2.465(2)
Cu3–Cl5	2.501(1)	2.615(1)	2.642(2)
Cu3–N4	1.960(4)	1.945(3)	1.964(7)
Cu3–N5	1.966(4)	1.953(3)	1.950(7)
Cu1 … Cu2	3.381(1)	3.4194(9)	3.420(2)
Cu1 … Cu3	3.389(1)	3.4042(8)	3.386(2)
Cu2 … Cu3	3.433(1)	3.397(1)	3.423(2)

3.1.2. Tau Parameter Determination

In all cases, the Cu centers are in a triangular arrangement and the 5-coordinate environment of each Cu atom is formed by two μ-4-NO_2-pyrazolate bridges, two μ_3-Cl caps, and one terminal Cl ligand. While the Cu-N bond lengths of each $[Cu_3(\mu_3\text{-}Cl)_2(\mu\text{-}4\text{-}NO_2\text{-}pz)_3Cl_3]^{2-}$ unit are the same within experimental error, each polymorph exhibits different ligand (chloride and pyrazolate) positions in relation to the Cu_3-plane and different Cu-Cl bond lengths. The geometry of 5-coordinate metal centers can be ideally represented as either square–pyramidal (trans angles $\alpha = \beta = 180°$) or trigonal bipyramidal ($\alpha = 120°$; $\beta = 180°$) where β is the largest angle and α is the second largest. For complexes that deviate from ideal geometries, the τ parameter, defined as follows, provides a quantitative description of the coordination geometry [39]:

$$\tau = \frac{(\beta - \alpha)}{60} \quad (1)$$

Therefore, $\tau = 0$ for a perfectly square pyramidal complex, and $\tau = 1$ for a perfectly trigonal bipyramidal complex. The tau parameters calculated for each one of the three Cu-centers of the polymorphs, along with the total deviation of the three Cu-centers of each polymorph, Σ_τ, are listed in Table 3.

Table 3. Selected bond angles (°) and calculated τ values for **1a–1c**.

		1a		**1b**		**1c**	
Cu1	∠1	Cl1-Cu1-Cl5	119.56	Cl1-Cu1-Cl5	132.23	Cl1-Cu1-Cl5	163.74
	∠2	Cl1-Cu1-Cl4	160.39	Cl1-Cu1-Cl4	147.74	Cl1-Cu1-Cl4	116.35
	∠3	Cl4-Cu1-Cl5	79.99	Cl4-Cu1-Cl5	80.03	Cl4-Cu1-Cl5	79.89
	∠4	N1-Cu1-N6	173.43	N1-Cu1-N6	172.60	N1-Cu1-N6	176.43
	τ	(∠4 − ∠2)/60	0.22	(∠4 − ∠2)/60	0.41	(∠4 − ∠1)/60	0.21
Cu2	∠1	Cl2-Cu2-Cl5	120.58	Cl1-Cu1-Cl5	132.23	Cl2-Cu2-Cl5	133.70
	∠2	Cl2-Cu2-Cl4	160.14	Cl1-Cu1-Cl4	147.74	Cl2-Cu2-Cl4	146.56
	∠3	Cl4-Cu2-Cl5	79.24	Cl4-Cu1-Cl5	80.03	Cl4-Cu2-Cl5	79.73
	∠4	N2-Cu2-N3	173.32	N1-Cu1-N6	172.60	N2-Cu2-N3	172.86
	τ	(∠4 − ∠2)/60	0.22	(∠4 − ∠2)/60	0.54	(∠4 − ∠2)/60	0.44
Cu3	∠1	Cl3-Cu3-Cl5	138.45	Cl1-Cu1-Cl5	132.23	Cl3-Cu3-Cl5	130.39
	∠2	Cl3-Cu3-Cl4	140.69	Cl1-Cu1-Cl4	147.74	Cl3-Cu3-Cl4	148.69
	∠3	Cl4-Cu3-Cl5	80.85	Cl4-Cu1-Cl5	80.03	Cl4-Cu3-Cl5	80.90
	∠4	N4-Cu3-N5	174.94	N1-Cu1-N6	172.60	N4-Cu3-N5	174.36
	τ	(∠4 − ∠2)/60	0.57	(∠4 − ∠1)/60	0.51	(∠4 − ∠2)/60	0.43
Σ_τ		1.01		1.46		1.08	

The tau parameters for complex **1b** are most consistently between square pyramidal and trigonal bipyramidal. The highest tau value of 0.57 is encountered for the Cu3 of **1a**, while the lowest one of 0.21 is for the Cu1 of **1c**; thus, the geometries of these two metal centers better adheres to the trigonal bipyramidal and square pyramidal, respectively. However, the higher total deviation of the three Cu-centers is encountered in **1b**. Figure 2 illustrates the differences in the geometries of these two extremes. For each case, the copper atom and immediate coordination sphere are shown in two different orientations: (1) Cl4 as the axial ligand (square pyramidal) and (2) the nitrogen atoms as the axial ligands (trigonal bipyramidal).

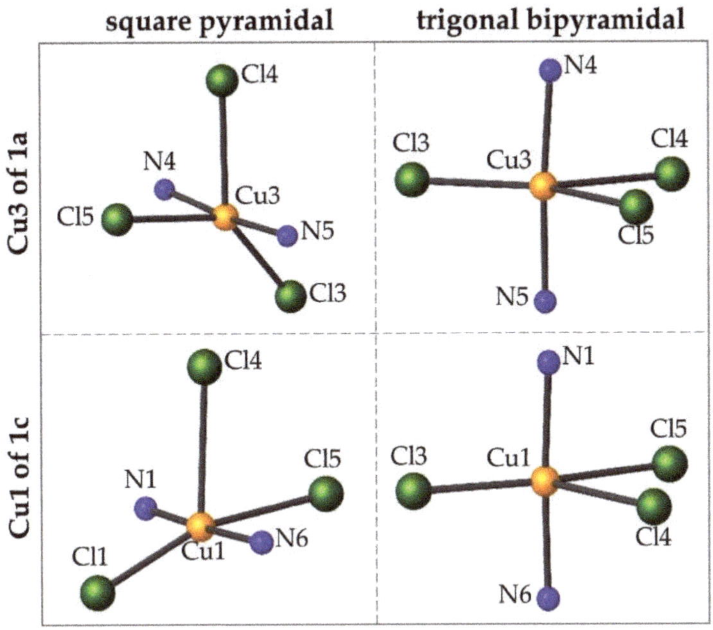

Figure 2. Comparison of the square pyramidal and trigonal bipyramidal geometries for the Cu3 of **1a** (τ = 0.57) and the Cu1 of **1c** (τ = 0.21).

3.1.3. Ligand Position in Relation to the Cu_3 Plane

Another illustrative parameter for describing the three polymorphs is the distance of the three terminal chloride atoms and the three 4-NO_2-pyrazolate ligands from the Cu_3-plane, as reported in Table 4. Here, a centroid was calculated for each pyrazolate ring: pz1 refers to the centroid of the plane that contains N1, N2, C1, C2, and C3; pz2 refers to the centroid of the plane that contains N3, N4, C4, C5, and C6; and pz3 refers to the centroid of the plane that contains N5, N6, C7, C8, and C9. A positive or negative sign is attributed to the ligand being either above or below the Cu-plane, respectively, with each molecule oriented so that Cl5 is above the plane and Cl4 is below the plane.

Table 4. Distances (d, Å) of the terminal chloride ligands and pyrazolates from the Cu_3-plane for **1a–c**.

	Cl1	Cl2	Cl3	$\lvert\Sigma_{d_{Cl}}\rvert$	$\Sigma_{\lvert d_{cl}\rvert}$	pz1	pz2	pz3	$\lvert\Sigma_{d_{pz}}\rvert$	$\Sigma_{\lvert d_{pz}\rvert}$
1a	0.799	0.758	−0.017	1.540	1.574	−0.278	−0.133	−0.075	0.486	0.486
1b	−0.275	−0.116	0.150	0.241	0.541	0.427	−0.145	0.013	0.295	0.585
1c	0.912	−0.248	−0.486	0.178	1.646	0.276	0.018	−0.302	0.008	0.596

Due to the bridging nature of the pyrazolates, the distance between the terminal Cl ligands and the Cu_3-plane is larger than that of the pyrazolate rings (as expected) except for **1b**—in which $\lvert\Sigma_{d_{pz}}\rvert$ and $\Sigma_{\lvert d_{pz}\rvert}$ are larger by 0.054 and 0.044, respectively. **1c** has the largest values for both $\Sigma_{\lvert d_{cl}\rvert}$ and $\Sigma_{\lvert d_{pz}\rvert}$, therefore, exhibiting the most deviation either above or below the Cu_3-plane. However, when the sum is calculated, allowing for the total effect to be taken into consideration—$\lvert\Sigma_{d_{cl}}\rvert$ and $\lvert\Sigma_{d_{pz}}\rvert$—,**1c** has the smallest values of the three polymorphs.

3.1.4. Crystal Packing

As the name suggests, the three polymorphs give rise to vastly different crystal packing motifs. The types of interactions that dominate each polymorph are discussed more in depth in Section 3.3 (vide infra). For simplicity, the crystal packing of the complex is shown separately from that of the cations. The crystal packing viewed parallel to the Cu_3 plane for each molecule is shown in Figure 3. The complexes of **1a** are arranged in pairs that exhibit repulsions between a terminal chloride (Cl2) and the nitro group of a pyrazolate. The Cl2 . . . N7 distance is 3.660 Å. These pairs of cations are eclipsed and extend down the a-axis. The complexes of **1b** are arranged in a manner that allows for the nitro groups of pyrazolates to be stacked with a N7 . . . N7 distance of 3.246 Å. Two other pyrazolates have reciprocal C-H . . . O interactions with a C6 . . . O3 distance of 3.271 Å. These two types of pyrazolate interactions alternate parallel to the b-axis. The complexes of **1c** are arranged in pairs with one nitro group (O1-N7-O2) having interactions with a terminal and capping chloride. The N7 . . . Cl1 distance is 3.767 Å with the chloride ligand bending away from the nitro group. The O1 . . . Cl5 distance is 3.432 Å. Unlike the pairs in **1a**, the pairs in **1c** do not extend into the plane; instead, they rotate ~90° down the plane shown in Figure 3.

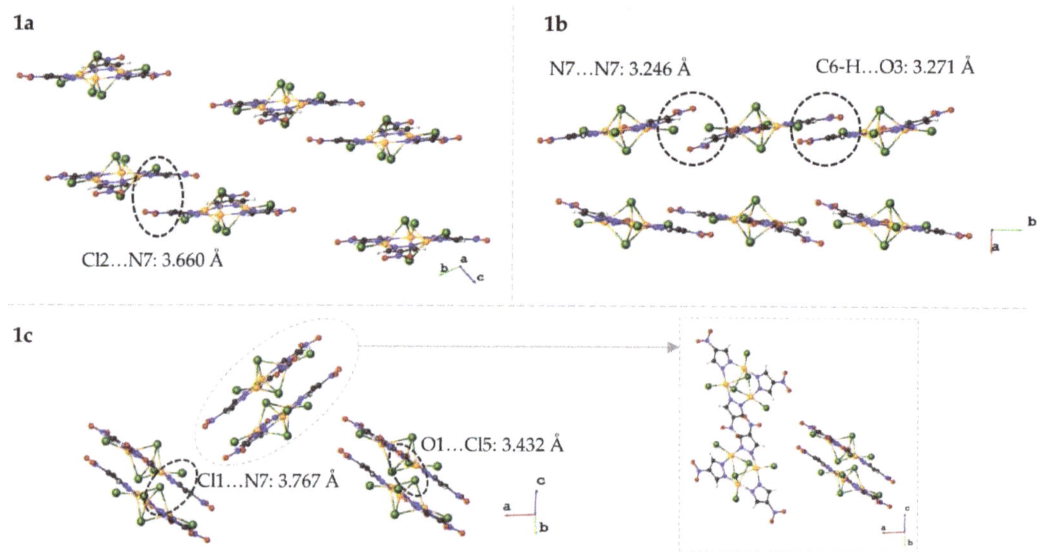

Figure 3. Crystal packing of the dianionic complex viewed approximately parallel to the Cu_3 plane for **1a** (top left), **1b** (top right), and **1c** (bottom). The inset of **1c** shows the rotation of the pair directly behind and in front of the specified pair. Significant interactions are circled and labeled with the identity and distance of the interaction.

The crystal packing of the tetrabutylammonium cations for each molecule are shown in Figure 4. In each polymorph, layers of chains are formed. In **1a**, the chains are approximately linear, running alternatingly parallel to the b-axis and diagonally in the bc plane. The dianionic copper complexes occupy the spaces formed between the diagonal cation chains. The cations in **1b** form sinusoidal waves that propagate parallel to the c-axis; the adjacent layers of cation waves are out of phase. The complex anions occupy the channels created by layers of waves. The cations in **1c** form corrugated layers parallel to the a-axis. Between consecutive layers are pairs of crystallographically non-independent cations. Four complex molecules surround each pair of cations between the layers.

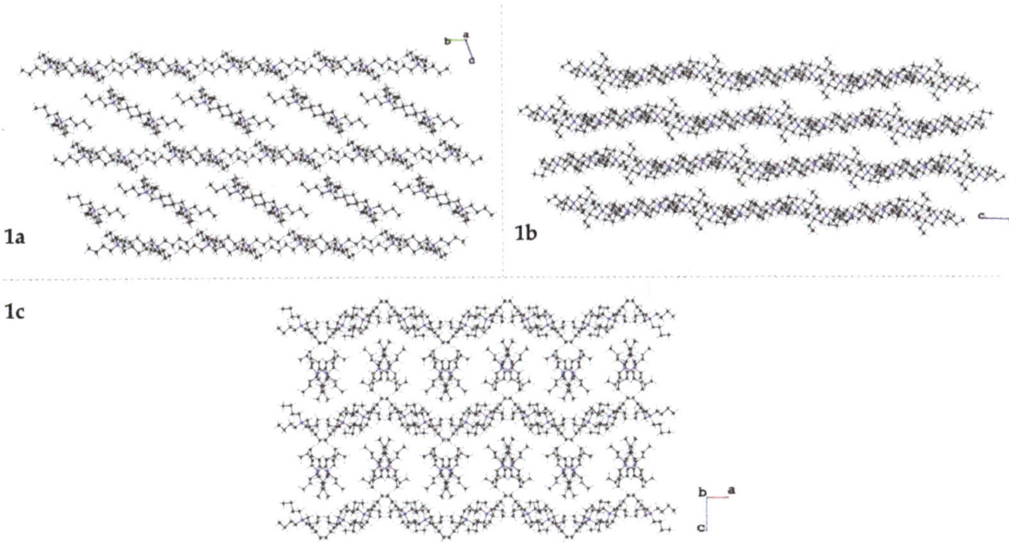

Figure 4. Crystal packing of the two crystallographically independent tetrabutylammonium cations for **1a** (top left), **1b** (top right), and **1c** (bottom).

3.2. DFT Calculations

The potential energies of the three polymorphs were calculated in the gas phase at the B3LYP/6-31G* level of theory. Since calculations of vibrational frequencies were not feasible for molecules of this size, zero-point vibrational energy corrections were not considered. However, this is not expected to affect the computed relative energies significantly. According to the calculations, all three structures correspond to the local minima at the potential energy surface, with **1a** having the lowest energy, and **1b** and **1c** lying respectively 6.9 and 7.8 kcal/mol higher; these results should be interpreted qualitatively, because their differences are within the range of B3LYP accuracy. The main structural difference in the gas phase is the position of the tetrabutylammonium cations with respect to the central $Cu_3(\mu_3\text{-}Cl)_2(\mu\text{-}4\text{-}NO_2\text{-}pz)_3Cl_3$ anion, which is diagonal in **1a**, horizontal in **1b**, and vertical in **1c**.

3.3. Hirshfeld Surface Analysis

Figures 5–7 show the front and back of the d_{norm} Hirshfeld surface for **1a, 1b**, and **1c**, respectively. The "front" face of each molecule is the one that contains μ_3-Cl4 while the "back" face is the one that contains μ_3-Cl5. Although **1a** is the densest polymorph, the Hirshfeld surface of **1b** has the largest number of significant interactions. In total, the Cu complexes of **1a**, **1b**, and **1c** are in contact with five, seven, and four different tetrabutylammonium cations, respectively. In addition, the surface of 1b has significant interactions with one dianionic complex, while **1a** and **1c** do not exhibit complex–complex interactions. The types of complex–cation interactions that dominate each polymorph differ as follows. **1a** exhibits mainly C-H . . . Cl interactions and has two strong interactions with a capping Cl (C27-H . . . Cl5 (3.689(5) Å) and C30-H . . . Cl5 of 3.651(4) Å). **1b** exhibits the strongest interactions involving the nitro group of the pyrazolate ligands (C24-H . . . O4 (3.356(9) Å), C26-H . . . O6 (3.375(8) Å), C10-H . . . O2 (3.404(8) Å, C38-H . . . O3 (3.520(6) Å), and C30-H . . . O3 (3.570(6) Å). **1b** also has an interaction between the 4-NO_2-pyrazolates of two separate Cu complexes: O3 . . . H-C6′ and the reciprocal O3′ . . . H-C6 of 3.271(7) Å. **1c** also has its strongest interaction with the nitro group of a pyrazolate (C21-H . . . O2 (3.36(2) Å) and is the only polymorph to have an interaction with the ring of the 4-NO_2-

pyrazolate (C28-H ... C7/C8 (3.72(3) Å). A more in-depth description of the Hirshfeld surface interactions can be found in the supplemental information.

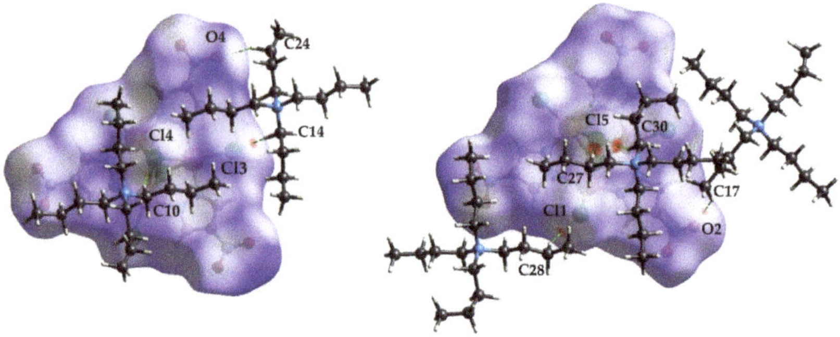

Figure 5. Hirshfeld surface of the front (**left**) and back (**right**) of the $[Cu_3(\mu_3\text{-}Cl)_2(\mu\text{-}4\text{-}NO_2\text{-}pz)_3Cl_3]^{2-}$ unit of **1a** showing the intermolecular interactions with the tetrabutylammonium cations.

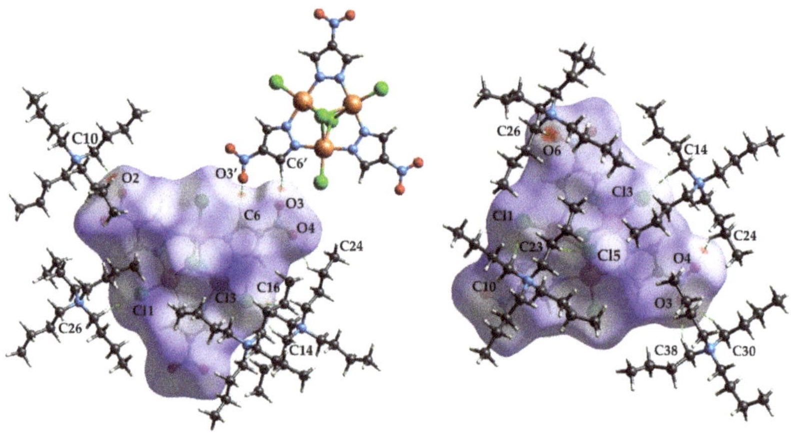

Figure 6. Hirshfeld surface of the front (**left**) and back (**right**) of the $[Cu_3(\mu_3\text{-}Cl)_2(\mu\text{-}4\text{-}NO_2\text{-}pz)_3Cl_3]^{2-}$ unit of **1b** showing the intermolecular interactions with the tetrabutylammonium cations and other Cu_3 dianionic complexes.

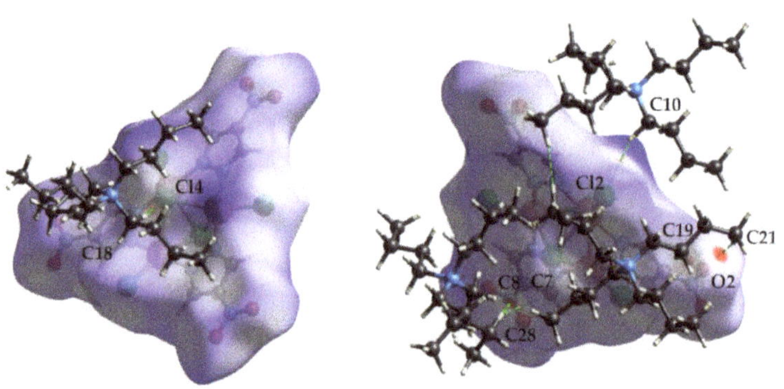

Figure 7. Hirshfeld surface of the front (**left**) and back (**right**) of the $[Cu_3(\mu_3\text{-}Cl)_2(\mu\text{-}4\text{-}NO_2\text{-}pz)_3Cl_3]^{2-}$ unit of **1c** showing the intermolecular interactions with the tetrabutylammonium cations.

From the d_{norm} Hirshfeld surfaces, 2D fingerprint plots were assembled as shown in Figure 8. All three polymorphs show the most interactions (dark blue = small number of interactions, while green = moderate number of interactions) around $d_i \approx 2$ Å, $d_e \approx 1.4$ Å. The main difference between the shape of each plot is the number of interactions at high d_i/d_e values. **1a** is the densest polymorph, and thus its surface has fewer long-range interactions (those above $d_i \approx d_e \approx 2.6$ Å). The other main difference is the number of spikes of each fingerprint plot. **1a** has two main spikes. The first is located roughly at $d_i \approx 1.4$ Å, $d_e \approx 1.1$ Å and continues diagonally upwards. This depicts the O . . . H interactions. The second spike is located roughly at $d_i \approx 1.6$ Å, $d_e \approx 1.0$ Å and continues diagonally upwards. This spike is attributed to the Cl . . . H interactions. **1b** has the highest number of spikes (four) of the polymorphs: (1) $d_i \approx 1.1$ Å, $d_e \approx 1.4$ Å; (2) $d_i \approx d_e \approx 1.2$ Å; (3) $d_i \approx 1.3$ Å, $d_e \approx 1.0$ Å; and (4) $d_i \approx 1.7$ Å, $d_e \approx 1.1$ Å. The O . . . H interactions account for the first and third spikes. The second spike is attributed to the H . . . H interactions, and the fourth spike is attributed to the Cl . . . H interactions. The Cl . . . H interactions of **1b** are not as strong as those of **1a**; however, the O . . . H interactions of **1b** are stronger than those of **1a**. The plot of **1c** shows three spikes: (1) $d_i \approx d_e \approx 1.1$ Å; (2) $d_i \approx 1.4$ Å, $d_e \approx 1.0$ Å; and (3) $d_i \approx 1.7$ Å, $d_e \approx 1.1$ Å. These are attributed to the H . . . H, O . . . H, and Cl . . . H interactions, respectively. Here, the H . . . H interactions are slightly stronger than in **1b**, while the strength of the O . . . H interactions is similar to **1a**, and the strength of the Cl . . . H interactions is similar to **1b**. The breakdown of the 2D fingerprint plots by type of atomistic contribution is shown in Figures S1–S3 of the supplementary information.

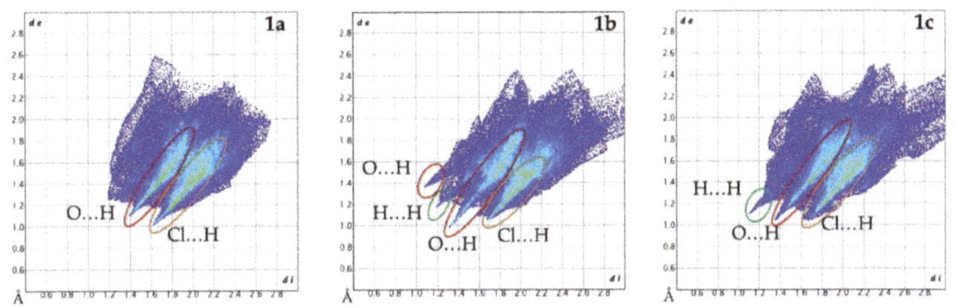

Figure 8. 2D fingerprint plots (d_i/d_e) of **1a** (left), **1b** (middle), and **1c** (right) showing the types of shortest interactions for each Hirshfeld surface.

For all three polymorphs, approximately 93% of all interactions of the complex surface consist of five types: O . . . H, Cl . . . H, H . . . H, C . . . H, and N . . . H. There are small differences in the percentages of these interactions, as shown in Figure 9. The remaining interactions account for ≤2% of the interactions of each polymorph, and the same type of minor interactions do not occur for each polymorph. For example, **1b** has no Cu . . . O interactions or Cl . . . N interactions and is the only polymorph that contains N . . . N interactions. Only **1a** had Cl . . . C interactions. **1c** had no O . . . C interactions.

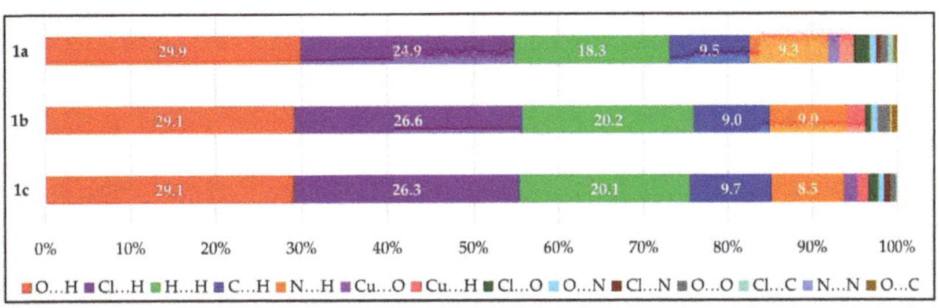

Figure 9. Bar graph, compiling the Hirshfeld surface interactions for **1a–c** analyzed by percent contributions of individual atomic contacts.

4. Conclusions

By varying the crystallization solvent and temperature of $(Bu_4N)_2[Cu_3(\mu_3\text{-}Cl)_2(\mu\text{-}4\text{-}NO_2\text{-}pz)_3Cl_3]$, three true polymorphs of this complex, **1a–c**, have been recognized so far. Aside from the numerous polymorphs of $[M(acac)_3]$ complexes (where M = Fe [40], Cr [41], or Mn [42] and acac = acetylacetonate), we are aware of only a few other examples of metal complexes known in three polymorphic structures under ambient conditions [7,43]. The X-ray crystal structures of the three polymorphic complexes revealed insignificant variations in their bond lengths and angles in contrast to their clearly differentiated 3D packing and, therefore, differences in intermolecular interactions. The DFT calculations showed that, in vacuum, the triclinic **1a** was the ground state polymorph. However, in the solid state, the sum of the various intermolecular interactions (not quantified here) was significant enough to offset the polymorphic instabilities of 6.9 kcal/mol and 7.8 kcal/mol and allow phase transitions to the monoclinic and orthorhombic forms, **1b** and **1c**, respectively. The intermolecular interactions were qualitatively probed by Hirshfeld surface analysis. Interestingly, **1a** was also the densest, therefore the one with the higher number of contacts involving the μ_3-Cl, terminal Cl, and NO_2 groups. However, **1b** had the highest number of significant interactions with the Hirshfeld surface, whereas **1c** had only two strong interactions—involving a terminal Cl and a NO_2 group. It was not possible to assess here the role of the crystallization solvent polarity and boiling point.

Supplementary Materials: The following supporting information can be downloaded at: https://www.mdpi.com/article/10.3390/cryst12111611/s1, Figures S1–S3: Breakdown of 2D fingerprint plot by type of individual atomistic contribution for complexes **1a,b**, respectively. Additional Hirshfeld surface analysis of Figures 5–7 in the main text.

Author Contributions: K.L.R. data analysis and manuscript preparation; L.M. data analysis; G.M. research and data analysis; A.M.M. computational DFT data; R.G.R., supervision and manuscript preparation. All authors have read and agreed to the published version of the manuscript.

Funding: K.L.R. was supported by the U.S. Nuclear Regulatory Commission (NRC) fellowship grant No. 31310018M0012 awarded to FIU.

Data Availability Statement: Not applicable.

Acknowledgments: The authors would like to acknowledge Indranil Chakraborty for his crystallographic expertise.

Conflicts of Interest: The authors declare no conflict of interest.

References

1. Bernstein, J. *Polymorphism in Molecular Crystals*; International Union of Crystallography Monographs on Crystallography; Oxford University Press: Oxford, UK, 2007. [CrossRef]
2. Stahly, G.P. Diversity in Single- and Multiple-Component Crystals. The Search for and Prevalence of Polymorphs and Cocrystals. *Cryst. Grow. Des.* **2007**, *7*, 1007–1026. [CrossRef]
3. Hilfiker, R. *Polymorphism: In the Pharmaceutical Industry*; Wiley-VCH: Weinheim, Germany, 2006.
4. Saha, B.K.; Nath, N.K.; Thakuria, R. Polymorphs with Remarkably Distinct Physical and/or Chemical Properties. *Chem. Rec.* **2002**, e202200173. [CrossRef] [PubMed]
5. Benito, Q.; Le Goff, X.F.; Maron, S.; Fargues, A.; Garcia, A.; Martineau, C.; Taulelle, F.; Kahlal, S.; Gacoin, T.; Boilot, J.-P.; et al. Polymorphic Copper Iodide Clusters: Insights into the Mechanochromic Luminescence Properties. *J. Am. Chem. Soc.* **2014**, *136*, 11311–11320. [CrossRef]
6. Dave, S.; Sahu, R.; Tripathy, B.C. *Electrochemical Applications of Metal-Organic Frameworks*; Elsevier: Amsterdam, The Netherlands, 2022; pp. 17–35.
7. Chai, W.; Hong, M.; Song, L.; Jia, G.; Shi, H.; Guo, J.; Shu, K.; Guo, B.; Zhang, Y.; You, W.; et al. Three Reversible Polymorphic Copper(I) Complexes Triggered by Ligand Conformation: Insights into Polymorphic Crystal Habit and Luminescent Properties. *Inorg. Chem.* **2015**, *54*, 4200–4207. [CrossRef]
8. Muthukumar, P.; Pannipara, M.; Al-Sehemi, A.G.; Moon, D.; Anthony, S.P. Polymorphs of a Copper Coordination Compound: Interlinking Active Sites Enhance the Electrocatalytic Activity of the Coordination Polymer Compared to the Coordination Complex. *CrystEngComm* **2020**, *22*, 425–429. [CrossRef]

9. Wöhlert, S.; Runčevski, T.; Dinnebier, R.E.; Ebbinghaus, S.G.; Näther, C. Synthesis, Structures, Polymorphism, and Magnetic Properties of Transition Metal Thiocyanato Coordination Compounds. *Cryst. Grow. Des.* **2014**, *14*, 1902–1913. [CrossRef]
10. Fromm, K.M.; Doimeadios, J.L.S.; Robin, A.Y. Concomitant Crystallization of Two Polymorphs—A Ring and a Helix: Concentration Effect on Supramolecular Isomerism. *Chem. Commun.* **2005**, *36*, 4548–4550. [CrossRef]
11. White-Morris, R.L.; Olmstead, M.M.; Attar, S.; Balch, A.L. Intermolecular Interactions in Polymorphs of Trinuclear Gold(I) Complexes: Insight into the Solvoluminescence of $AuI_3(MeNCOMe)_3$. *Inorg. Chem.* **2005**, *44*, 5021–5029. [CrossRef]
12. Brog, J.-P.; Chanez, C.-L.; Crochet, A.; Fromm, K.M. Polymorphism, What It Is and How to Identify It: A Systematic Review. *RSC Adv.* **2013**, *3*, 16905. [CrossRef]
13. Angaridis, P.A.; Baran, P.; Boča, R.; Cervantes-Lee, F.; Haase, W.; Mezei, G.; Raptis, R.G.; Werner, R. Synthesis and Structural Characterization of Trinuclear Cu^{II} –Pyrazolato Complexes Containing μ_3 -OH, μ_3 -O, and μ_3 -Cl Ligands. Magnetic Susceptibility Study of $[PPN]_2[(\mu_3\text{-}O)Cu_3(\mu\text{-}pz)_3Cl_3]$. *Inorg. Chem.* **2002**, *41*, 2219–2228. [CrossRef]
14. Boudalis, A.K.; Rogez, G.; Heinrich, B.; Raptis, R.G.; Turek, P. Towards Ionic Liquids with Tailored Magnetic Properties: $Bmim^+$ Salts of Ferro- and Antiferromagnetic $Cu^{II}{}_3$ Triangles. *Dalton Trans.* **2017**, *46*, 12263–12273. [CrossRef]
15. Kreiger, D.I.; Mathivathanan, L.; Raptis, R.G. Coordination Polymers Based on Pyrazole-4-Carboxaldehyde-Containing Cu_3N_6 Metallacycles as Building Units. *CrystEngComm* **2019**, *21*, 3047–3055. [CrossRef]
16. Mathivathanan, L.; Boudalis, A.K.; Turek, P.; Pissas, M.; Sanakis, Y.; Raptis, R.G. Interactions between H-Bonded $[Cu^{II}{}_3(\mu_3\text{-}OH)]$ Triangles; a Combined Magnetic Susceptibility and EPR Study. *Phys. Chem. Chem. Phys.* **2018**, *20*, 17234–17244. [CrossRef] [PubMed]
17. Mathivathanan, L.; Rogez, G.; Ben Amor, N.; Robert, V.; Raptis, R.G.; Boudalis, A.K. Origin of Ferromagnetism and Magnetic Anisotropy in a Family of Copper(II) Triangles. *Chem. Eur. J.* **2020**, *26*, 12769–12784. [CrossRef] [PubMed]
18. Rivera-Carrillo, M.; Chakraborty, I.; Mezei, G.; Webster, R.D.; Raptis, R.G. Tuning of the $[Cu_3(\mu\text{-}O)]^{4+/5+}$ Redox Couple: Spectroscopic Evidence of Charge Delocalization in the Mixed-Valent $[Cu_3(\mu\text{-}O)]^{5+}$ Species. *Inorg. Chem.* **2008**, *47*, 7644–7650. [CrossRef] [PubMed]
19. Mezei, G.; Raptis, R.G. Effect of Pyrazole-Substitution on the Structure and Nuclearity of Cu(II)-Pyrazolato Complexes. *Inorg. Chim. Acta* **2004**, *357*, 3279–3288. [CrossRef]
20. Mezei, G.; Rivera-Carrillo, M.; Raptis, R.G. Effect of Copper-Substitution on the Structure and Nuclearity of Cu(II)-Pyrazolates: From Trinuclear to Tetra-, Hexa- and Polynuclear Complexes. *Inorg. Chim. Acta* **2004**, *357*, 3721–3732. [CrossRef]
21. Boča, R.; Dlháň, L.; Mezei, G.; Ortiz-Pérez, T.; Raptis, R.G.; Telser, J. Triangular, Ferromagnetically-Coupled $Cu^{II}{}_3$–Pyrazolato Complexes as Possible Models of Particulate Methane Monooxygenase (PMMO). *Inorg. Chem.* **2003**, *42*, 5801–5803. [CrossRef]
22. Mezei, G.; Raptis, R.G.; Telser, J. Trinuclear, Antiferromagnetically Coupled Cu^{II} Complex with an EPR Spectrum of Mononuclear Cu^{II}: Effect of Alcoholic Solvents. *Inorg. Chem.* **2006**, *45*, 8841–8843. [CrossRef]
23. Maresca, K.P.; Rose, D.J.; Zubieta, J. Synthesis and Characterization of a Binuclear Rhenium Nitropyrazole Complex $[Re_2O_3Cl_2(PPh_3)_2(C_3H_2N_3O_2)_2]$. *Inorg. Chim. Acta* **1997**, *260*, 83–88. [CrossRef]
24. Perrin, D.D.; Armarego, W.L.F.; Perrin, D.R. *Purification of Laboratory Chemicals*; Pergamon Press: Oxford, UK, 1980.
25. *Data Collection: SMART-NT Software Reference Manual*; Version 5.0; Bruker AXS, Inc.: Madison, WI, USA, 1998.
26. *Data Reduction: SAINT-NT Software Reference Manual*; Version 4.0; Bruker AXS, Inc.: Madison, WI, USA, 1996.
27. Sheldrick, G.M. *SHELXTL-NT*; Version 5.1; Bruker AXS, Inc.: Madison, WI, USA, 1999.
28. Becke, A.D. Density-functional Thermochemistry. III. The Role of Exact Exchange. *J. Chem. Phys.* **1993**, *98*, 5648–5652. [CrossRef]
29. Lee, C.; Yang, W.; Parr, R.G. Development of the Colle-Salvetti Correlation-Energy Formula into a Functional of the Electron Density. *Phys. Rev. B* **1988**, *37*, 785–789. [CrossRef] [PubMed]
30. Hariharan, P.C.; Pople, J.A. The Influence of Polarization Functions on Molecular Orbital Hydrogenation Energies. *Theoret. Chim. Acta* **1973**, *28*, 213–222. [CrossRef]
31. Frisch, M.J.; Trucks, G.W.; Schlegel, H.B.; Scuseria, G.E.; Robb, M.A.; Cheeseman, J.R.; Scalmani, G.; Barone, V.; Mennucci, B.; Petersson, G.A.; et al. *Gaussian 09*; Revision D.01; Gaussian, Inc.: Wallingford, CT, USA, 2009.
32. Dolomanov, O.V.; Bourhis, L.J.; Gildea, R.J.; Howard, J.A.K.; Puschmann, H. OLEX2: A Complete Structure Solution, Refinement and Analysis Program. *J. Appl. Cryst.* **2009**, *42*, 339–341. [CrossRef]
33. Macrae, C.F.; Sovago, I.; Cottrell, S.J.; Galek, P.T.A.; McCabe, P.; Pidcock, E.; Platings, M.; Shields, G.P.; Stevens, J.S.; Towler, M.; et al. Mercury 4.0: From Visualization to Analysis, Design and Prediction. *J. Appl. Cryst.* **2020**, *53*, 226–235. [CrossRef] [PubMed]
34. Spackman, P.R.; Turner, M.J.; McKinnon, J.J.; Wolff, S.K.; Grimwood, D.J.; Jayatilaka, D.; Spackman, M.A. CrystalExplorer: A Program for Hirshfeld Surface Analysis, Visualization and Quantitative Analysis of Molecular Crystals. *J. Appl. Cryst.* **2021**, *54*, 1006–1011. [CrossRef]
35. Spackman, M.A.; Jayatilaka, D. Hirshfeld Surface Analysis. *CrystEngComm* **2009**, *11*, 19–32. [CrossRef]
36. Sundareswaran, S.; Karuppannan, S. Hirshfeld Surface Analysis of Stable and Metastable Polymorphs of Vanillin. *Crys. Res. Technol.* **2020**, *55*, 2000083. [CrossRef]
37. McKinnon, J.J.; Fabbiani, F.P.A.; Spackman, M.A. Comparison of Polymorphic Molecular Crystal Structures through Hirshfeld Surface Analysis. *Cryst. Grow. Des.* **2007**, *7*, 755–769. [CrossRef]
38. Spackman, M.A.; McKinnon, J.J. Fingerprinting Intermolecular Interactions in Molecular Crystals. *CrystEngComm* **2002**, *4*, 378–392. [CrossRef]

39. Addison, A.W.; Rao, T.N.; Reedijk, J.; van Rijn, J.; Verschoor, G.C. Synthesis, Structure, and Spectroscopic Properties of Copper(II) Compounds Containing Nitrogen–Sulphur Donor Ligands; the Crystal and Molecular Structure of Aqua[1,7-bis(N-methylbenzimidazol-2′-yl)-2,6-dithiaheptane]copper(II) perchlorate. *J. Chem. Soc. Dalton Trans.* **1984**, *7*, 1349–1356. [CrossRef]
40. Baker, T.M.; Howard, K.M.; Brennessel, W.W.; Neidig, M.L. Crystal Structure of a Third Polymorph of Tris (acetyl acetonato-κ^2O,O′)iron(III). *Acta Cryst.* **2015**, *E71*, m228–m229. [CrossRef]
41. Morosin, B. The Crystal Structure of Trisacetylacetonatochromium(III). *Acta Cryst.* **1965**, *19*, 131–137. [CrossRef]
42. Geremia, S.; Demitri, N. Crystallographic Study of Manganese(III) Acetylacetonate: An Advanced Undergraduate Project with Unexpected Challenges. *J. Chem. Educ.* **2005**, *82*, 460. [CrossRef]
43. Vrdoljak, V.; Prugovečki, B.; Matković-Čalogović, D.; Hrenar, T.; Dreos, R.; Siega, P. Three Polymorphic Forms of a Monomeric Mo(VI) Complex: Building Blocks for Two Metal–Organic Supramolecular Isomers. Intermolecular Interactions and Ligand Substituent Effects. *Cryst. Grow. Des.* **2013**, *13*, 3773–3784. [CrossRef]

crystals

MDPI

Article

Supramolecular Assemblies in Pyridine- and Pyrazole-Based Coordination Compounds of Co(II) and Ni(II): Characterization, Hirshfeld Analysis and Theoretical Studies

Trishnajyoti Baishya [1], Rosa M. Gomila [2], Miquel Barcelo-Oliver [2], Diego M. Gil [3], Manjit K. Bhattacharyya [1,*] and Antonio Frontera [2,*]

[1] Department of Chemistry, Cotton University, Assam 781001, India
[2] Department de Química, Universitat de les IllesBalears, Crta de Valldemossa km 7.7, 07122 Palma de Mallorca, Spain
[3] INBIOFAL (CONICET—UNT), Instituto de Química Orgánica, Facultad de Bioquímica, Química y Farmacia, Universidad Nacional de Tucumán, Ayacucho 471, San Miguel de Tucumán T4000CAN, Argentina
* Correspondence: manjit.bhattacharyya@cottonuniversity.ac.in (M.K.B.); toni.frontera@uib.es (A.F.)

Abstract: Two new Ni(II) and Co(II) coordination compounds, viz., $[Ni(H_2O)_5(DMAP)](IPhth)\cdot 2H_2O$ (**1**) and $[Co(Hdmpz)_4(H_2O)_2]Cl_2$ (**2**) (where DMAP = 4-dimethylaminopyridine, IPhth = Isophthalate, Hdmpz = 3,5-dimethylpyrazole),were synthesized and characterized using elemental analysis, TGA, spectroscopic (FTIR and electronic) and single-crystal X-ray diffraction techniques. Compound **1** crystallizes as a co-crystal hydrate of Ni(II), whereas compound **2** is a mononuclear compound of Co(II). The crystal structure analysis of compound **1** reveals the presence of various non-covalent interactions such as anion–π, π–π, C–H···π, C–H···C, etc., which stabilize the layered assembly of the compound. In compound **2**, enclathration of counter chloride ions within the supramolecular trimeric host cavity plays a crucial role in the stabilization of the compound. The non-covalent interactions observed in the crystal structures were further studied theoretically, focusing on the cooperative π-stacking interactions between the *DMAP* and *IPhth* counter-ions in **1**. To identify the non-covalent interactions of the compounds, Hirshfeld surfaces and their associated two-dimensional fingerprint regions were analyzed. Theoretical calculations confirm that H-bonding interactions combined with the π-stacking contacts are crucial synthons for the solid-state stability of compound **1**.

Keywords: co-crystal hydrate; supramolecular assemblies; enclathration; cooperative π-stacking; DFT calculations

Citation: Baishya, T.; Gomila, R.M.; Barcelo-Oliver, M.; Gil, D.M.; Bhattacharyya, M.K.; Frontera, A. Supramolecular Assemblies in Pyridine- and Pyrazole-Based Coordination Compounds of Co(II) and Ni(II): Characterization, Hirshfeld Analysis and Theoretical Studies. *Crystals* **2023**, *13*, 203. https://doi.org/10.3390/cryst13020203

Academic Editor: Alexander Y. Nazarenko

Received: 10 December 2022
Revised: 19 January 2023
Accepted: 20 January 2023
Published: 23 January 2023

1. Introduction

Transition-metal-based coordination compounds involving organic ligands have attracted researchers because of their wide potential applications in catalysis, non-linear optics, semiconductor devices and in biology [1–4]. The intriguing self-assemblies of metal–organic compounds have also gained remarkable interest because of their interesting structural topologies and network architectures [5,6]. Metal–organic co-crystals, involving metal centers having a minimum of two components with lattice water molecules, have also received emphasis owing to their wide practical utilities in the pharmaceutical industry, electronic devices as well as in synthetic chemistry [7,8].

Over the last few years, various N-donors, especially with pyridine-based moieties, have been extensively studied to construct supramolecular assemblies of desired architectures and potential applications [9,10]. Similarly, organic moieties containing pyrazole rings have also been exploited to explore self-assembled architectures with fascinating properties [11,12]. Metal–organic compounds of pyrazole- and pyridine-based ligands have been reported to display interesting biological applications [13,14]. The multiple coordination modes of the carboxylate-based aromatic ligands such as terminal, bridging, chelating, syn-anti, anti-anti, etc., have resulted in the formation of metal–organic compounds with fascinating structural topologies [15–20]. Non-covalent interactions are the backbone of supramolecular chemistry which play interesting roles in various fields such as synthesis, catalysis, crystal engineering, design of pharmaceutical agents, molecular biology, etc. [21,22]. Non-covalent contacts involving aromatic rings play crucial roles in the solid-state stability of metal–organic compounds [23] and also in guiding structural topologies of novel functional materials [24,25]. Counter anions present in the crystal lattice of metal–organic compounds not only neutralize the positive charge of complex cationic moieties but are also involved in non-covalent interactions, adding new dimensions in crystal engineering [26,27].

In the present study; we aim to explore the synthesis, detailed structural investigations and supramolecular assemblies of two new Ni(II) and Co(II) compounds involving pyridine and pyrazole ligands. The compounds were characterized using FT-IR, electronic spectroscopic techniques, TGA and elemental analysis. The presence of various non-covalent contactsin the crystal structure of the compounds, such as anion–π, π–π, C–H$\cdots\pi$, C–H$\cdots$C, etc., stabilize the layered assemblies. In compound **2**, enclathration of counter chloride ions within the supramolecular trimeric host cavity plays an important role in the stabilization of the compound. The non-covalent interactions in the crystal structures have been studied using Hirshfeld surface analysis and density functional theory (DFT) calculations, focusing on the π-stacking interactions between the *DMAP* and *IPhth* counter-ions that are further assisted by strong H-bonds involving the water molecules. The supramolecular ternary assembly was further characterized using molecular electrostatic potential (MEP) surface analysis and combined quantum theory of atoms-in-molecules (QTAIM) and non-covalent interaction (NCI) plot computational tools.

2. Materials and Methods

The chemicals used in the present study, viz., nickel(II) chloride hexahydrate, cobalt(II) chloride hexahydrate, isophthalic acid, 4-dimethylaminopyridine and 3,5-dimethyl pyrazole, were purchased from commercial sources and used without further purifications. The elemental analyses of the compounds were carried out using Perkin Elmer 2400 series II CHNS/O analyzer. The FT-IR spectra of the compounds were recorded in the frequency region 4000–500 cm^{-1} using Bruker Alpha (II) Infrared spectrophotometer. The electronic spectra of the compounds were recorded using Shimadzu UV-2600 spectrophotometer. For UV-Vis-NIR spectra, $BaSO_4$ powder was used as reference (100% reflectance). Sherwood Mark 1 magnetic susceptibility balance was used to calculate the room-temperature magnetic moments of the compounds. Thermogravimetric studies were carried out under the flow of N_2 gas using Mettler Toledo TGA/DSC1 STAR^e system at the heating rate of 10 °C min^{-1}.

2.1. Syntheses

2.1.1. Synthesis of $[Ni(H_2O)_5(DMAP)](IPhth)\cdot 2H_2O$ (1)

In 10 mL of de-ionized water, disodium salt of isophthalic acid (0.210 g, 1 mmol) was dissolved in a round-bottom flask. $NiCl_2\cdot 6H_2O$ (0.237 g, 1 mmol) was then slowly added and mechanically stirred for an hour at room temperature. After an hour, *DMAP* (0.122 g, 1 mmol) was slowly added to the reaction mixture and the mixture was kept stirring for another two hours (Scheme 1). Then, the reaction mixture was kept in cooling conditions in a refrigerator (below 4 °C), which yielded pale blue block-shaped single crystals after a few days. Yield: 0.415 g (88.11%). Anal. calcd. for $C_{15}H_{28}N_2NiO_{11}$: C, 38.24%; H, 5.99%; N, 5.95%. Found: C, 38.13%; H, 5.92%; N, 5.87%. IR (KBr pellet, cm^{-1}): 3454(br), 2817(w), 2729(w), 2124(br), 1620(s), 1586(sh), 1483(w), 1431(w), 1387(s), 1342(sh), 1224(w), 795(w), 730(w), 685(w) (s, strong; m, medium; w, weak; br, broad; sh, shoulder).

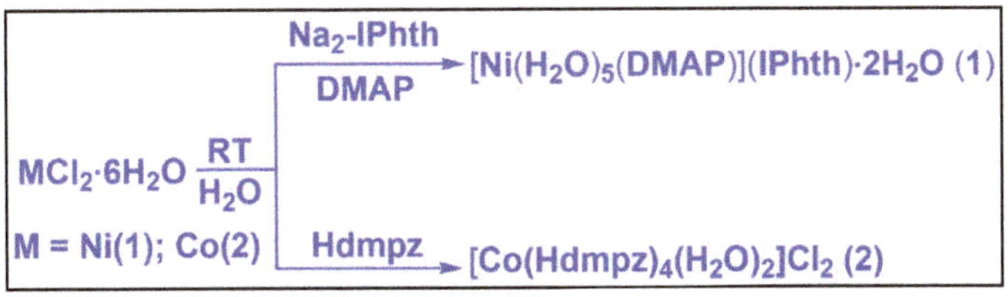

Scheme 1. Syntheses of the compounds **1** and **2**.

2.1.2. Synthesis of $[Co(Hdmpz)_4(H_2O)_2]Cl_2$ (2)

$CoCl_2{\cdot}6H_2O$ (0.237 g, 1mmol) and *Hdmpz* (0.304 g, 4 mmol) were dissolved in de-ionized water (10 mL) in a round-bottom flask and the solution was allowed to stir mechanically for about two hours (Scheme 1). Then, the resulting solution was kept unperturbed in cooling conditions (24 °C), from which red block-shaped crystals were obtained after several days. Yield: 0.480 g (87.27%). Anal. calcd. for $C_{20}H_{36}Cl_2CoN_8O_2$: C, 43.64%; H, 6.59%; N, 20.36%. Found: C, 43.55%; H, 6.51%; N, 20.27%. IR (KBr pellet, cm^{-1}): 3457 (br), 3139 (sh), 3038 (w), 2980 (w), 2920 (m), 2869 (w), 2780 (w), 2714 (w), 2596 (m), 2463 (w), 2294 (m), 2139 (w), 1615 (s), 1571 (s), 1475 (s), 1418 (s), 1285 (s), 1247 (w), 1160 (s), 1040 (s), 988 (w), 782 (s), 701 (m), 612 (m), 517 (m) (s, strong; m, medium; w, weak; br, broad; sh, shoulder).

2.2. Crystallographic Data Collection and Refinement

Single crystals of the compounds **1** and **2** were selected, covered with Parabar 10320 (formally known as Paratone N) and mounted on a cryoloop on a BRUKER D8 Venture diffractometer, with a Photon III 14 detector, using an Incoatec high brilliance ImS DIAMOND Cu tube (λ = 1.54178 Å) equipped with an Incoatec Helios MX multilayer optics. The crystals were kept at 151 K (compound **1**) or 100 K (compound **2**) during data collection. Data reduction and cell refinements were performed using the Bruker APEX4 program [28]. Scaling and absorption corrections were carried out using SADABS [29]. Crystal structures were solved by direct method and refined by full-matrix least-squares technique with SHELXL-2018/3 [30] using WinGX [31] platform. All the non-hydrogen atoms were refined anisotropically. The hydrogen atoms were placed at their calculated positions and refined in the isotropic approximation, except for those in compound 2 attached to water O-atoms, which were located using a Fourier difference map and refined isotropically.

Diamond 3.2 software was used to draw all the structural diagrams [32]. Table 1 contains the crystallographic data and the structure refinement table for the compounds.

Table 1. Crystallographic data and structure refinement details for the compounds **1** and **2**.

Parameters	1	2
Formula	$C_{15}H_{28}N_2NiO_{11}$	$C_{20}H_{36}Cl_2CoN_8O_2$
Formula weight	471.10	550.40
Temp, (K)	151	100.0
Crystal system	Triclinic	Monoclinic
Space group	$P\overline{1}$	$C2/c$
a, (Å)	7.073(4)	10.4681(7)
b, (Å)	11.591(6)	14.1306(10)
c, (Å)	12.804(7)	18.4697(13)
α, (°)	83.465(19)	90
β, (°)	84.37(2)	92.765(3)
γ, (°)	78.857(18)	90
V, (Å^3)	1020.0(9)	2728.9(3)
Z	2	4
Absorption coefficient (mm^{-1})	1.925	6.991
F(0 0 0)	496.0	1156.0
$\rho_{calc}g/cm^3$	1.524	1.340

Table 1. *Cont.*

Parameters	1	2
index ranges	$-8 \leq h \leq 8$ $-13 \leq k \leq 14$, $-14 \leq l \leq 15$	$-12 \leq h \leq 12$, $-16 \leq k \leq 16$, $-22 \leq l \leq 22$
Crystal size, (mm^3)	0.48 × 0.28 × 0.25	0.21 × 0.18 × 0.07
2θ range, (°)	9.956 to 138.182	10.524 to 136.818
Independent reflections	3697 [R_{int} = 0.0669, R_σ = 0.1028]	2504 [R_{int} = 0.0518, R_σ = 0.0227]
Reflections collected	16,188	36,237
Refinement method	Full-matrix least-squares on F^2	Full-matrix least-squares on F^2
Data/restraints/parameters	3697/1/278	2504/0/162
Goodness-of-fit on F^2	1.093	1.147
Final R indices (I > 2σ(I)) (all data)	R_1 = 0.0414, wR_2 = 0.1041 R_1 = 0.0803, wR_2 = 0.1072	R_1 = 0.0759, wR_2 = 0.1804 R_1 = 0.0772, wR_2 = 0.1811
Largest hole and peak (e·Å^{-3})	0.46/−0.41	1.34/−1.46

2.3. *Computational Methods*

2.3.1. Theoretical Study

The energies and wavefunction calculations were performed at the RI-BP86-D3/def2-TZVP [33,34] level of theory and with the Turbomole 7.2 [35] program. Only the position of the H-atoms was optimized, keeping the non-H-atoms frozen. For the octahedral Ni(II) complex, the high spin configuration was considered (triplet state). The quantum theory of atoms-in-molecules (QTAIM) and non-covalent interaction (NCI) plot [36] computational tools were used to characterize the NCIs. The QTAIM analysis [37] was performed using the MULTIWFN program [38] and represented using VMD software [39].

2.3.2. Hirshfeld Analysis

Hirshfeld surface analysis was used for studying intermolecular interactions [40–42] by means of the CrystalExplorer 21 software [43]. The normalized contact distance (d_{norm}) was measured using td_e (the distance from the point to the nearest nucleus external to the surface), d_i (the distance to the nearest nucleus internal to the surface) and the van der Waals (vdW) radii of the atoms involved in the intermolecular contacts. The d_{norm}values allow identifying the different regions participating in the intermolecular interactions. Graphical plots of the Hirshfeld surfaces mapped with d_{norm} function show a color code of red (shorter contacts), white (contacts around the sum of vdW radii of atoms) and blue (longer contacts). The Hirshfeld surfaces were mapped over a d_{norm} range of −0.075 a.u. (red) to +0.75 a.u. (blue), *shape index* of −1.0 a.u. (concave) to +1.0 a.u. (convex) and curvedness of −4.00 a.u. (flat) to +0.4 a.u. (singular). The 2D fingerprints were plotted using the translated 0.6–2.8 Å range, including reciprocal contacts.

3. Results and Discussion

3.1. *Syntheses and General Aspects*

[Ni(H_2O)$_5$(DMAP)](IPhth)·2H_2O (**1**) was prepared by the reaction between one equivalent of $NiCl_2$·6H_2O, one equivalent of *DMAP* and one equivalent of disodium salt of isophthalic acid at room temperature in de-ionized water medium. Similarly, [Co(Hdmpz)$_4$(H_2O)$_2$]Cl_2 (**2**) was synthesized by reacting one equivalent of $CoCl_2$·6H_2O with four equivalents of *Hdmpz* under similar reaction conditions. Both compounds **1** and **2** are soluble in water as well as in common organic solvents. Compounds **1** and **2** exhibit room-temperature (298 K) μ_{eff} values of 2.86 and 3.85 BM, respectively, suggesting the presence of two and three unpaired electrons in the Ni(II) and Co(II) centers of the distorted octahedral coordination spheres of **1** and **2**, respectively [44,45].

3.2. Crystal Structure Analysis

Figure 1 depicts the molecular structure of compound **1**. Table 2 contains the details of the selected bond lengths and bond angles around the Ni(II) center of compound **1**. Compound **1** crystallizes in the triclinic $P\overline{1}$ space group. The Ni(II) center of compound **1** is coordinated to five coordinated aqua ligands and one *DMAP* moiety. The dipositive charge of the complex cationic moiety of compound 1 is neutralized by the presence of one uncoordinated *Iphth* moiety in the crystal lattice. In addition, the asymmetric unit of the compound comprises two uncoordinated water molecules. The coordination geometry around the Ni(II) center is distorted octahedron, where the axial sites are occupied by one coordinated water molecule (O1W) and N1 from the *DMAP* moiety; whereas the equatorial sites are occupied by the remaining four coordinated water molecules (O5W, O2W, O3W and O4W). The equatorial atoms, viz., O5W, O2W, O3W and O4W, are distorted from the equatorial plane with the mean r.m.s. deviation of 0.0304 Å. The average Ni–O and Ni–N bond lengths are well consistent with similar Ni(II) compounds [46,47].

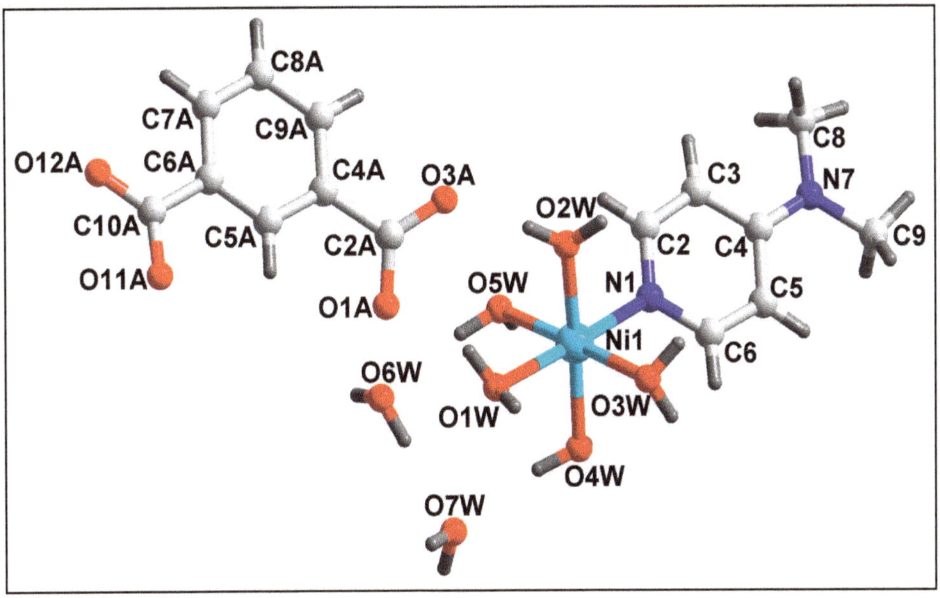

Figure 1. Molecular structure of $[Ni(H_2O)_5(DMAP)](IPhth)\cdot 2H_2O$ (**1**). Aromatic hydrogen atoms are omitted for clarity.

Figure 2 represents the supramolecular trimer formed in the crystal structure of compound **1**, which is stabilized by O–H···O, C–H···O hydrogen bonding, non-covalent C–H···π, anion-π and π-π interactions. As shown in Figure 2, the O–H···O hydrogen bonding interactions are observed between the coordinated water molecules (O2W and O3W) and carboxyl O atoms (O11A and O12A) of uncoordinated *IPhth* moieties having O2W–H2WB···O12A and O3W–H3WA···O11A distances of 1.91 and 1.90 Å, respectively (Table 3). The carboxyl O atom (O1A) is involved in C–H···O hydrogen bonding interactions with the –CH moiety of coordinated *DMAP* having a C9–H9C···O1Adistance of 2.80 Å. The C–H···π interaction [47] is observed between the –CH moiety of coordinated *DMAP* and the aromatic ring of the uncoordinated *IPhth* moiety having C9···Cg and H9A···Cg distances of 3.913(3) and 3.566 Å, respectively (Cg is the ring centroid defined by the atoms C4A-C8A). The corresponding C9–H9A···Cg angle is found to be 108°. The existence of this interaction was further confirmed by combined QTAIM/NCI plot computational tools (vide infra).

Table 2. Selected bond lengths (Å) and bond angles (°) of Ni(II) and Co(II) centers in **1** and **2**, respectively.

Compound 1			
Ni1–O1W	2.0912(2)	O2W–Ni1–O1W	89.52(7)
Ni1–O2W	2.0297(2)	O2W–Ni1–O3W	89.15(7)
Ni1–O3W	2.1004(2)	O2W–Ni1–O4W	174.49(7)
Ni1–O4W	2.0778(2)	O2W–Ni1–O5W	91.25(7)
Ni1–O5W	2.1043(2)	O2W–Ni1–N1	90.137)
Ni1–N1	2.059(2)	O3W–Ni1–O1W	88.67(7)
O4W–Ni1–O1W	85.31(6)	O4W–Ni1–O3W	88.85(7)
O4W–Ni1–O5W	90.59(7)	O5W–Ni1–O1W	89.58(7)
O5W–Ni1–O3W	178.20(6)	N1–Ni1–O1W	178.25(6)
N1–Ni1–O3W	93.03(7)	N1–Ni1–O4W	95.10(7)
N1–Ni1–O5W	88.72(7)		
Compound 2			
Co1–O1W	2.076(5)	O1W–Co1–N1	91.03(1)
Co1–O2W	2.062(6)	O1W–Co1–N1#1	91.03(1)
Co1–N1#1	2.110(5)	O1W–Co1–N6	89.57(1)
Co1–N1	2.110(5)	O1W–Co1–N6#1	89.57(1)
Co1–N6	2.105(5)	O2W–Co1–O1W	180.0
Co1–N6#1	2.105(5)	O2W–Co1–N1#1	88.97(1)
O2W–Co1–N1	88.97(1)	O2W–Co1–N6#1	90.43(1)
O2W–Co1–N6	90.43(1)	N1#1–Co1–N1	177.9(3)
N6–Co1–N1#1	88.32(2)	N6#1–Co1–N1	88.32(2)
N6#1–Co1–N1#1	91.70(2)	N6–Co1–N1	91.70(2)
N6–Co1–N6#1	179.1(3)		

#1 1 − X, +Y, 3/2 − Z.

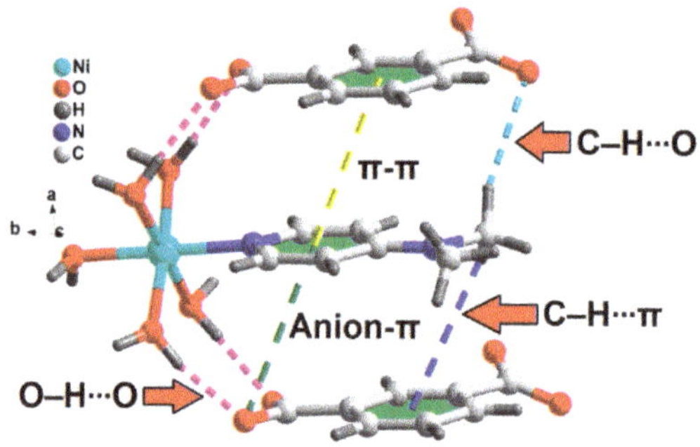

Figure 2. Formation of a supramolecular trimer in compound **1** assisted by O–H···O, C–H···O hydrogen bonding, non-covalent C–H···π, anion-π and π-π interactions.

Moreover, a closer look reveals the presence of a strong anion-π interaction, which is observed between the aromatic ring of the coordinated *DMAP* moiety and the carboxyl O (O12A) atom of the uncoordinated *IPhth* moiety. The distance between the centroid of the aromatic ring and the O12A atom is found to be 3.99(2)Å. The angle between O12A, the ring centroid (Cg) and the C2 atom (defining the aromatic plane) is found to be 95.64°, which suggests the strong nature of the anion-π interaction [48]. Moreover, aromatic π-stacking interactions are also observed between the phenyl and pyridyl rings of uncoordinated *IPhth* and coordinated *DMAP* moieties, respectively, having the centroid(C4A, C5A, C6A, C7A, C8A, C9A)–centroid(C2, C3, C4, C5, C6, N1) distance of 3.857(1) Å. The corresponding slipped angle, the angle between the ring normal and the vector between two ring centroids, is found to be 20.6(2)° [49]. These non-covalent interactions have been further studied theoretically (vide infra). These supramolecular trimers propagate along the crystallographic *a* axis to form the 1D supramolecular chain of the compound (Figure 3).

Table 3. Selected hydrogen bond distances (Å) and angles (deg.) for compounds **1** and **2**.

D–H···A	*d*(D–H)	*d*(D···A)	*d*(H···A)	<(DHA)
		Compound **1**		
O2W–H2WB···O12A#1	0.87	2.741(2)	1.91	160.3
O3W–H3WA···O11A#1	0.87	2.767(2)	1.90	172.7
O5W–H5WB···O12A#2	0.87	2.821(2)	1.95	177.1
O4W–H4WA···O11A#2	0.87	2.762(2)	1.91	165.9
O5W–H5WA···O6W	0.87	2.776(3)	1.92	168.4
O7W–H7WA···O6W#3	0.87	2.824(3)	1.96	174.4
O4W–H4WB···O7W	0.87	2.837(2)	1.98	168.8
		Compound **2**		
O1W–H1W···Cl1#4	1.03(7)	3.039(3)	2.14(7)	167(6)
O2W–H2W···Cl1	0.91(7)	3.030(3)	2.14(7)	169(7)

#1 + X, −1 + Y, +Z, #2 1 + X, −1 + Y, +Z, #3 1 − X, 1 − Y, 1 − Z #4 1/2 − X, −1/2 + Y, 3/2 − Z.

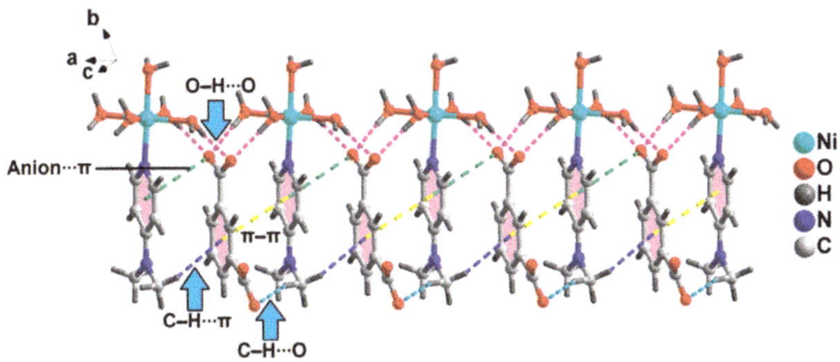

Figure 3. The1D supramolecular chain of compound **1** along the crystallographic *a* axis.

The adjacent 1D supramolecular chains of compound **1** are interconnected via O–H···O and C–H···O hydrogen bonding interactions to form the layered architecture along the crystallographic *ab* plane (Figure S1). The lattice water molecule O6W is involved in O–H···O hydrogen bonding interactions with a coordinated water molecule (O5W) and a carboxyl O atom (O1A) having O5W–H5WA···O6W and O6W–H6WB···O1A distances of 1.92 and 1.96 Å, respectively. Carboxyl O atoms (O1A and O3A) are also involved in O–H···O hydrogen bonding interactions with coordinated water molecules (O2W and O1W) having O2W–H2WA···O1A and O1W–H1WA···O3A distances of 1.80 and 1.95 Å, respectively.

A detailed structural investigation reveals the formation of the supramolecular layered assembly of **1** along the crystallographic *bc* plane assisted by the lattice water molecules. As shown in Figure 4, water molecules (O6W and O7W) are involved in O–H···O hydrogen bonding contacts with two coordinated aqua molecules (O5W and O4W) from two neighboring dimers.

The O5W–H5WA···O6W and O4W–H4WB···O7W distances are found to be 1.91 and 1.97 Å, respectively. Moreover, the –CH moiety of the pyridyl ring of *DMAP* is involved in C–H···O hydrogen bonding interactions with the coordinated aqua molecule (O4W) having a C6–H6···O4W distance of 2.54 Å. Figure 5 depicts the molecular structure of compound 2. Table 2 contains the details of the selected bond lengths and bond angles around the Co(II) center. Compound **2** crystallizes in the monoclinic *C*2/*c* space group. The Co(II) center of compound **2** is hexa-coordinated with four coordinated *Hdmpz* moieties and two coordinated water molecules. The presence of two uncoordinated chloride ions compensates for the dipositive charge of the cationic complex moiety. The X-ray crystallographic analysis reveals that the mononuclear Co(II) complex moiety possesses a two-fold axis of symmetry which passes through the coordinated water molecules (O1W and O2W) and the Co(II) center of the compound. The coordination geometry around the Co(II) center is distorted octahedron, where the axial sites are occupied by the two coordinated water molecules (O1W and O2W) and the equatorial sites are occupied by N1, N1′, N6 and N6′ atoms of the four coordinated *Hdmpz* moieties. The four equatorial atoms, viz., N1, N1′, N6 and N6′, are distorted from the mean equatorial plane with the mean r.m.s. deviation of 0.0046 Å. The average Co–O and Co–N bond lengths are almost consistent with the previously reported Co(II) complexes [50,51].

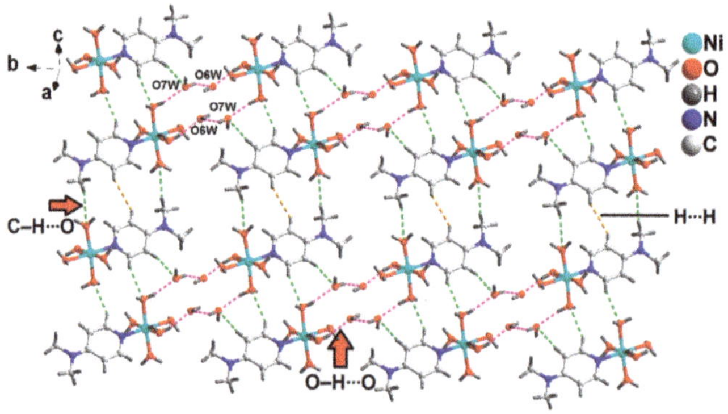

Figure 4. Layered assembly of compound **1** along the crystallographic *bc* plane assisted by lattice water molecules.

As shown in Figure S2, the adjacent monomeric units of compound **2** are involved in the formation of a supramolecular 1D chain stabilized by non-covalent C-H···C interactions between the neighboring monomeric unit having C9–H9···C4 and C4–H4···C9 bond distances of 2.95 and 3.56 Å, respectively, [C(sp^2)–H···C(sp^2), C9···C4 = 3.75(2) Å]. In addition, C-H···C interactions are also observed between the –CH moieties of methyl groups of *Hdmpz* of one monomeric unit and a C atom of *Hdmpz* of a neighboring monomeric unit having C101–H10B···C51 and C51–H51B···C101 distances of 3.11 and 3.19 Å, respectively, [C(sp^3)–H···C(sp^3), C101···C51 = 3.89(1) Å].

The adjacent 1D supramolecular chains of compound **2** are interconnected via non-covalent C–H···C andC–H···π interactions to form the layered architecture along the crystallographic *ac* plane (Figure 6). The –CH moieties of methyl groups of *Hdmpz* are involved in C–H···C interactions with one C atom of methyl groups having C81–H81C···C31 and C31–H31C···C81 distances of 3.68 and 3.76 Å, respectively, [C(sp^3)–H···C(sp^3), C81···C31 = 3.94 Å]. Moreover, C-H···π interactions are also observed involving the H9 atom of the pyrazole moiety and the N1/N2/C3-C5 centroid having H9···centroid separation of 3.417 Å. The corresponding C–H···π angle is found to be 151.9(3)°.

The interesting aspect of the crystal structure of compound **2** is the formation of a trimeric supramolecular host cavity assisted by a number of non-covalent C–H···C interactions. The methyl groups of the *Hdmpz* are involved in various non-covalent C–H···C interactions: [C101–H10A···C81 = 3.29 Å, C(sp^3)–H···C(sp^3), C101···C81 = 3.68 Å; C51–H51B···C101 = 3.19 Å, C(sp^3)–H···C(sp^3), C51···C101 = 3.89 Å; C101–H10B···C51 = 3.11 Å, C(sp^3)–H···C(sp^3), C101···C51 = 3.89 Å; C51–H51B···C101 = 3.19 Å, C(sp^3)–HB···C(sp^3), C51···C101 = 3.89 Å; C51–H51B···C51 = 3.69 Å, C(sp^3)–H···C(sp^3), C51···C51 = 3.92 Å]. In addition, the –CH moieties of pyrazole ring of *Hdmpz* are also involved in C–H···C interactions with the C4 atom of the pyrazole ring of *Hdmpz* of the neighboring monomeric unit having a C9–H9···C4 distance of 2.95 Å, [C(sp^2)–H···C(sp^2), C9···C4 = 2.95 Å]. However, the most fascinating aspect of the crystal structure of compound **2** is the enclathration of the counter chloride ion (Cl1) within the supramolecular trimeric host cavity. Cl1 is interconnected to three monomeric units via C–H···Cl, N–H···Cl and O–H···Cl hydrogen bonding interactions (Figure 7).

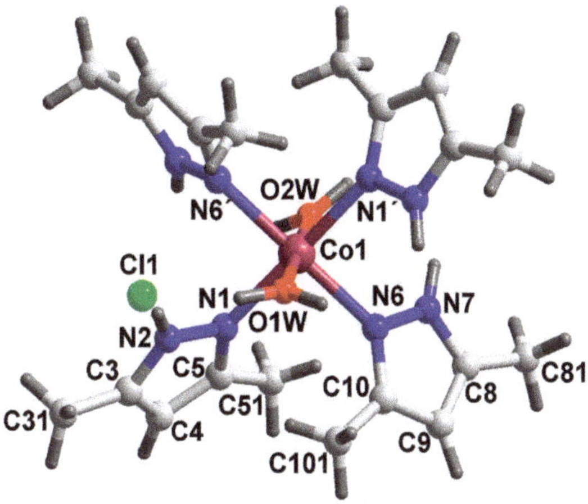

Figure 5. Molecular structure of $[Co(Hdmpz)_4(H_2O)_2]Cl_2$ (**2**). Symmetry codes for N1′ and N6′: 1 − X, +Y, 3/2 − Z.

The counter chloride ion (Cl1) is involved in C–H···Cl hydrogen bonding interactions with the –CH moieties of methyl groups of coordinated *Hdmpz*(C51–H51A···Cl1 = 2.81 Å; C31–H31A···Cl1 = 2.99 Å; C101–H10A···Cl1 = 2.74 Å). N–H···Cl hydrogen bonding interactions are found between the –NH moieties of the pyrazole ring of the coordinated *Hdmpz* and Cl1 atom of the lattice (N7–H7···Cl1 = 2.37 Å, N2–H2···Cl1 = 2.38 Å). Moreover, O–H···Cl interactions between the –OH moieties of coordinated water molecules and the lattice Cl1 (O1W–H1W···Cl1 = 2.05 Å; O2W–H2W···Cl1 = 2.08 Å) are also found to be involved in the enclathration of the counter chloride ion (Cl1) within the supramolecular trimeric host cavity. These enclathrated chloride ions within the supramolecular host cavities stabilize the layered assembly of the compound along the crystallographic *bc* plane (Figure 8).

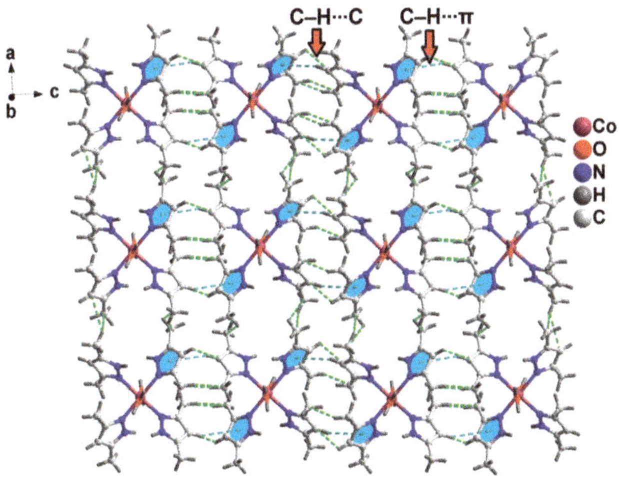

Figure 6. Layered architecture of compound **2** along the crystallographic *ac* plane assisted by non-covalent C–H···C, C–H···π interactions.

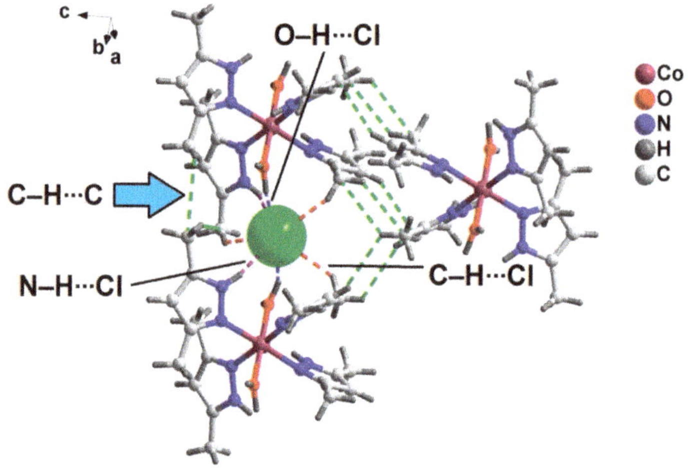

Figure 7. Enclathration of single guest chlorine atom within the supramolecular trimeric host cavity of compound **2**.

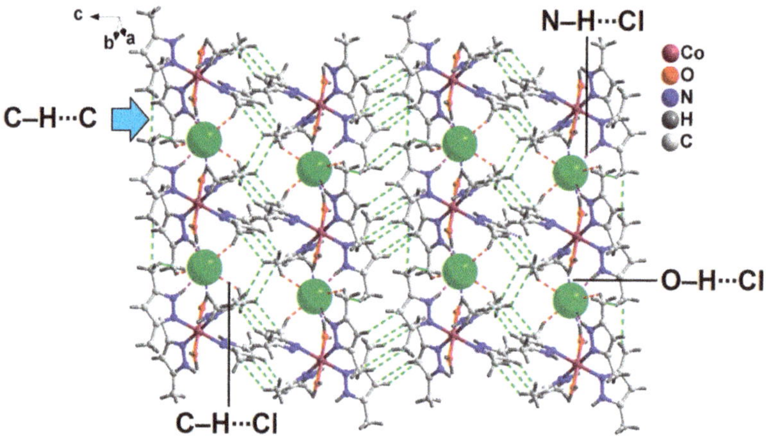

Figure 8. Layered assembly of compound **2** along the crystallographic *bc* plane assisted by the enclathrated chlorine atoms.

3.3. Spectral Studies

3.3.1. FT-IR Spectroscopy

The FT-IR spectra of compounds **1** and **2** have been recorded in the region 4000–500 cm^{-1} (Figure S3). The broad absorption bands at around 3454 cm^{-1}in the FT-IR spectra of the compounds can be correlated to the O-H stretching vibrations of the coordinated and/or lattice water molecules [52]. The FT-IR spectra of **1** and **2** also exhibit bands corresponding to ρ_r (H_2O) (712 cm^{-1}) and ρ_w (H_2O) (635 cm^{-1}), which indicate the presence of coordinated water molecules [53]. The bands at around 1620 and 1342 cm^{-1} in **1** can be assigned to the asymmetric ν_{as}(COO) and symmetric ν_s(COO) stretching vibrations of the carboxylate moiety of the lattice *IPhth* molecule [54]. The presence of the coordinated *DMAP* moiety in compound **1** produces a weak-intensity absorption band in the region 1230–1218 cm^{-1} [55]. For compound 1, the ring stretching vibrations of DMAP have been shifted to 1473, 1593 cm^{-1}, thereby supporting the coordination of *DMAP* through the nitrogen atom of the pyridine ring [56]. Pyridine ring vibrations of coordinated *DMAP* are observed at 1586 and 1387 cm^{-1} [57], while the ring wagging vibrations of *DMAP* are also observed at 685 and 730 cm^{-1} [58]. In the FT-IR spectrum of compound **2**, the band at around 3139 cm^{-1} can be assigned to the ν(N–H) vibrations of the coordinated *Hdmpz* moieties [59], while the ν(C–H) vibrations of the methyl groups of coordinated *DMAP* and *Hdmpz* in compounds **1** and **2** are observed in the region of 2970–2770 cm^{-1} [60]. The absorption bands for C–N, N–N and C=N stretching vibrations of Hdmpz rings are obtained at 1418, 1285 and 1160 cm^{-1} in **2** [61].

3.3.2. Electronic Spectroscopy

The electronic spectra of compounds **1** and **2** in solid as well as in aqueous phases are discussed in detail (see ESI, Figures S4 and S5). The electronic spectra of compounds **1** and **2** corroborate the presence of distorted octahedral Ni(II) and Co(II) centers in the compounds, respectively [62,63]. The peaks for $\pi \rightarrow \pi^*$ transitions of the aromatic ligands are obtained at the expected positions [64].

3.4. Thermogravimetric Analysis

The thermogravimetric curves of compounds **1** and **2** were obtained in the temperature range 25–800 °C under the N_2 atmosphere at the heating rate of 10 °C/min (Figure S6). For compound **1**, the thermal decomposition of 3.85% (calcd. = 3.82%) in the temperature range 30–80 °C can be attributed to the loss of one uncoordinated water molecule [65]. In the temperature range 81–150 °C, the weight loss of 4.14% (calcd. = 3.82%) is due to the decomposition of the other uncoordinated water molecule. The coordinated water molecules decomposed in the temperature range 151–220 °C with the observed weight loss of 17.35% (calcd. = 19.1%) [65]. In the temperature range 221–290 °C, the weight loss of 23.98% (calcd. = 25.93%) is due to the decomposition of the *DMAP* moiety [66]. The *IPhth* moiety is decomposed in the temperature range 291–638 °C with the observed weight loss of 33.79% (calcd. = 35.26%) [67]. For compound **2**, the weight loss of 6.07% (calcd. = 6.54%) in the temperature range 50–210 °C can be assigned to the loss of coordinated water molecules [65]. The observed weight loss of 12.09% (calcd. = 12.88%) in the temperature range 211–250 °C can be assigned to the loss of Cl moieties [68]. Refat et al. have reported similar behavior of Cl moieties in Schiff base compounds of Cu(II), Co(II) and Ni(II) [69]. Three *Hdmpz* moieties undergo thermal decomposition in the temperature range 251–750 °C with the observed weight loss of 54.12% (calcd. = 52.32%) [70].

3.5. Computational Studies

3.5.1. Hirshfeld Study

Hirshfeld surface (HS) analysis was performed to investigate the nature of the intermolecular interactions and their quantitative contributions to the crystal packing of compounds **1** and **2**. Figure 9 shows the HSs mapped over the d_{norm} function for both compounds. The red spots on the surfaces (highlighted as dashed arrows) represent distances shorter than the sum of the vdW radii of atoms and the blue region corresponds to distances longer than the sum of vdW radii. Figure 10 shows the full two-dimensional fingerprint (FP) plots of **1** and **2**.

Compound **1**: The large red spots labeled 3 in the d_{norm} map (Figure 9) are attributed to intermolecular O4W-H4WB···O7W involving the H4WB of the coordinated water molecule and the O7W of the non-coordinated water molecule as the acceptor. The large red regions labeled 1 and 2 in Figure 10 are mainly associated to O1W-H1WB···O3A and O4W-H4WA···O1A hydrogen bonds, respectively. The O12A and O11A of the uncoordinated *IPhth* anion are involved in two hydrogen bonding interactions with two coordinated water molecules, O5W-H5WA···O12A and O2W-H2WA···O11A. These H-bonds are visible on the HS mapped over the d_{norm} function as bright red regions labeled 4 and 5. The presence of O3W-H3WA···O1A hydrogen bonds between one coordinated water molecule and the O1A of the uncoordinated *IPhth* anion is evidenced by the visible large red area labeled 6 in the d_{norm} surface. The red spots labeled 7 on the HS are attributed to weak

C9-H9C···O11A involving the H9C of one methyl group of the coordinated *DMAP* ligand and the O11A of the uncoordinated *IPhth* anion as the acceptor. The proportion of H···O/O···H interactions comprise 37.0% of the total Hirshfeld surface area, and these contacts are identified as a pair of symmetrical spikes at $(d_e + d_i) \approx 1.70$ Å in the FP plots (Figure 10), which is consistent with the intermolecular distances reported in Table 3.

The presence of H···C/C···H contacts (6.30%) is identified by the presence of pronounced wings on both sides of the FP plots (Figure 10).

As was mentioned in the description of the crystal structure, the crystal packing of **1** is also stabilized by π···π stacking interactions involving the centroids of the C4A-C9A and the N1/C2-C6 rings, with an inter-centroid distance of 3.859 Å. In order to identify the π···π stacking interactions, the HSs were mapped with shape index and curvedness properties (Figure 11). The presence of red and blue triangles adjacent to each other (highlighted with a dashed circle) on the HS mapped over the shape index is indicative of π···π stacking interactions (Figure 11a). These contacts are also visible as relatively large and green flat regions indicated as a dashed arrow on the curvedness surface, as shown in Figure 11b. Finally, the C···C contacts involved in the π···π stacking interactions appear as a distinct pale blue to green area (highlighted as a red circle) in the FP plot (Figure 10) at $(d_e + d_i) \approx 3.80$ Å, with a contribution of 3.2% to the total Hirshfeld surface area.

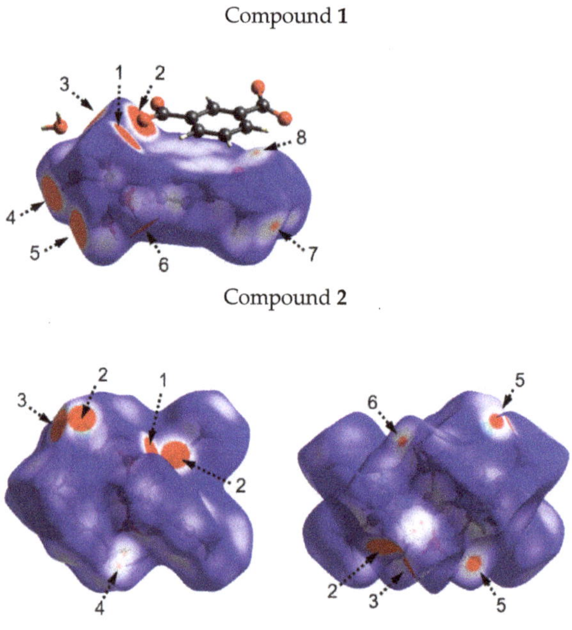

Figure 9. View of the Hirshfeld surfaces of **1** (top) and **2** (bottom) mapped over d_{norm} property. For **2**, the second molecule is rotated by 180° around the horizontal axis of the plot. The labels are discussed in the text.

Compound **2**: The red regions labeled 1 and 3 in the d_{norm} surface (Figure 9) are attributed to O2W-H2W···Cl1 and O1W-H1W···Cl1 involving the H-atoms of the coordinated water molecules and the chlorine anion as the acceptor. The two large red spots labeled 2 on the HS are mainly associated to N7-H7···Cl1 hydrogen bonds involving the N-H moiety of the pyrazole ring as the donor. The red regions labeled 5 and 6 in the d_{norm} map are attributed to C10-H10A···Cl1 and C51-H51A···Cl1 hydrogen bonds, involving the H-atoms of the methyl group and the Cl1 anion as the acceptor. These H···Cl/Cl···H contacts are visible as sharp spikes (labeled 4) in the FP plots (Figure 10) at around $(d_e + d_i) \approx 2.00$ Å, with a contribution of 14.3% to the Hirshfeld surface area. These contacts are visible in the d_{norm} surface as small red spots labeled 4 and comprise 76% of the total HS area.

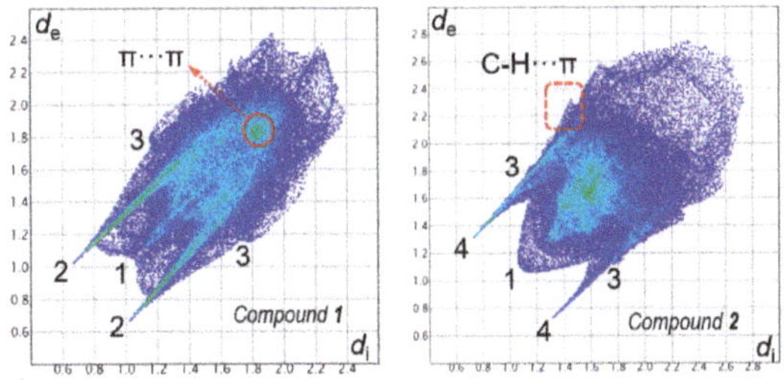

Figure 10. Full two-dimensional fingerprint plots for compounds **1** (left) and **2** (right) showing the spikes corresponding to the main intermolecular interactions. The percentage relative contributions of the intermolecular contacts to the total HS area are discussed in the main text. The FP plots show (1) H···O/O···H, (2) H···C/C···H and (3) H···Cl/Cl···H contacts.

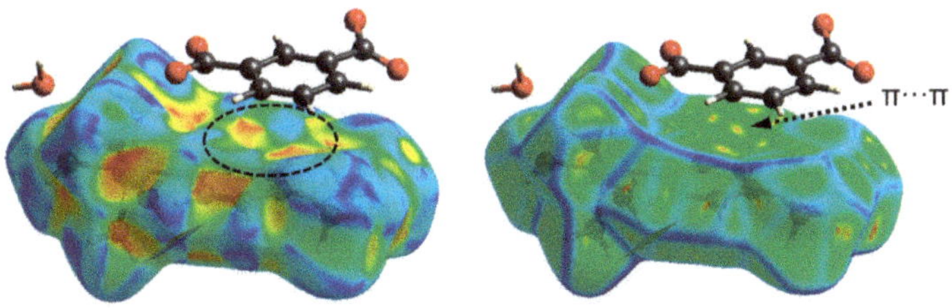

(**a**) (**b**)

Figure 11. Hirshfeld surfaces of **1** mapped over shape index (**a**) and curvedness (**b**) properties.

The crystal packing of **2** shows C-H···π interactions involving the H9 atom of the pyrazole moiety and the N1/N2/C3-C5centroid [d(H4···Cg) = 3.417 Å]. These interactions are visible as a pair of wings in the top left and bottom right region in the FP plot (see Figure 10), which comprise 9.10% of the total Hirshfeld surface area. A view of the shape index confirms the existence of C-H···π interactions showing a large red depression above the pyrazole ring and the blue region surrounding the C9-H9 donor, with both regions highlighted as dashed arrows.

3.5.2. Theoretical Study

The theoretical study is focused on the π-stacked assemblies observed in compound **1** as a combination of anion-π and C-H···π interactions. Moreover, in these assemblies, the *IPhth* anions are H-bonded to the Ni(II) coordinated water molecules that werealso evaluated using the QTAIM analysis.

First, we computed the molecular electrostatic potential (MEP) surface to investigate the most electron-rich and -poor regions of compound **1**. We used the [Ni(H_2O)$_5$(DMAP)](IPhth) model (Figure 12) in order to use a charge-neutral model for the MEP calculations. It can be observed that the MEP values at the coordinated water molecules are very large (+84.7 kcal/mol) and correspond to the MEP maxima. The MEP is also positive over the ring center (+10.7 kcal/mol) and at the H-atoms of the dimethylamino groups (+13.8 kcal/mol). The MEP values are large and negative at the carboxylate groups (−62.7 and −55.2 kcal/mol) of *IPhth* and over the center of the aromatic ring (−37.7 kcal/mol, value not shown in Figure 12).

Figure 13 shows the combined QTAIM/NCI plot of the π-stacked trimer and the binding energy computed from the two possible dimers. The presence of two large RDG isosurfaces that embrace the whole space between the aromatic rings can be observed, thus revealing a strong complementarity between the coordinated *DMAP* and the *IPhth* moieties. Moreover, a multitude of bond-critical points (CP, represented as small red spheres) and bond paths (represented as orange lines) further connect the aromatic rings and the substituents. In fact, the π-stacking assembly can be sub-divided into three interactions (anion–π, π–π and C–H···π), as indicated in Figure 13.

Moreover, an additional C–H···O H-bond is also observed. This intricate combination of interactions explains the large binding energies (−84.1 and −88.7 kcal/mol). Moreover, the combined QTAIM/NCI plot analysis discloses the existence of four OH···O H-bonds, which are characterized by a bond CP, bond path and small blue RDG isosurfaces interconnecting the H and O-atoms. The large interaction energies shown in Figure 10 are partially due to the ion-pair nature of the assembly. We also estimated the contribution of the H-bonds free from the pure Coulombic attraction by using the value of the Lagrangian Kinetic energy density (Gr) measured at the bond CP and the equation proposed by Veneret al. ($E_{ass} = 0.429 \times G_r$) [71] that was specifically developed for H-bonds in X-ray structures. The values are given in (Figure 13) close to the bond CPs and using a red font. The H-bonds are strong, with association energies ranging from −5.3kcal/mol to −6.6 kcal/mol, in line with the MEP surface analysis and the dark blue color of the RDG isosurfaces. The total contribution of the H-bonding interactions is −25.1 kcal/mol.

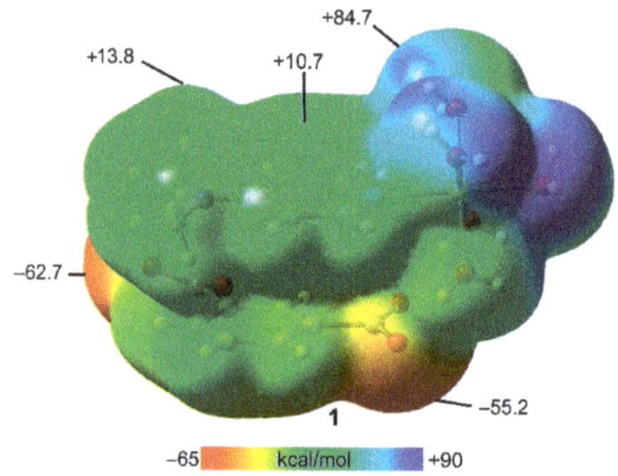

Figure 12. MEP surface of $[Ni(H_2O)_5(DMAP)](IPhth)$ at the RI-BP86-D3/def2-TZVP level of theory (isosurface 0.001 a.u.). Energies in kcal/mol.

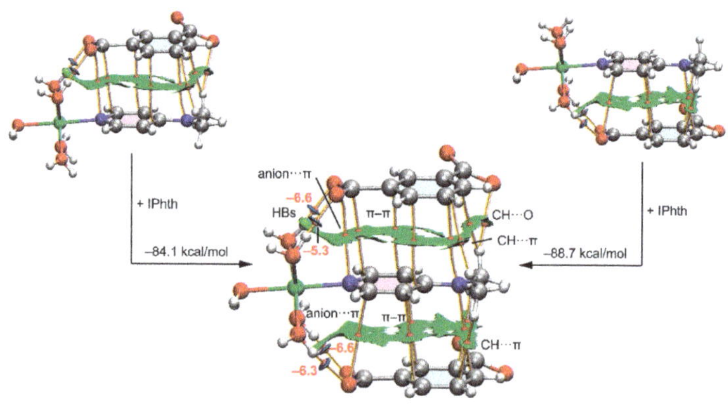

Figure 13. Combined QTAIM analysis (bond CPs in red, bond paths in orange) and NCI surfaces of two dimers of compound **1** (top) and the trimer (bottom). The gradient cut-off is $\rho = 0.04$ a.u., isosurfaces = 0.5, and the color scale is -0.04 a.u. $< (\mathrm{sign}\lambda_2)\rho <$ 0.04 a.u. Only intermolecular interactions are shown. The energies of the H-bonds are given in red adjacent to the bond CPs in kcal/mol. The formation energy of the trimer from the two dimers is also indicated.

4. Conclusions

Two new pyridine- and pyrazole-based coordination compounds of Ni(II) and Co(II) were prepared and characterized using single-crystal X-ray diffraction, electronic, FT-IR spectroscopic techniques, elemental analysis and TGA. Several non-covalent contacts including anion–π, π–π, C–H$\cdots\pi$, C–H$\cdots$C, H$\cdots$H along with H-bonding interactions stabilize the crystal structures. Self-assembled enclathration of counter chloride ions in compound **2** provides additional reinforcement to the crystal structure. All the non-covalent contacts were identified using Hirshfeld surface analysis. The studies reveal that both compounds **1** and **2** were stabilized by unconventional dihydrogen bonding contacts. Theoretical calculations on the supramolecular trimer observed in the crystal structure of **1** reveal that H-bonding interactions combined with the π-stacking contacts play an important role in the solid-state stability of compound **1**. The presence of two large RDG isosurfaces in the NCI plot analysis that embraces the whole space between the aromatic rings reveals the strong complementarity between the coordinated *DMAP* and lattice *IPhth* moieties.

Supplementary Materials: The following supporting information can be downloaded at: https://www.mdpi.com/article/10.3390/cryst13020203/s1, IR and UV-vis spectra of complexes **1** and **2** and additional description of the packing of **1** and **2**, thermogravimetric curves and additional references.

Author Contributions: Conceptualization, M.K.B. and A.F.; methodology, A.F. and D.M.G.; software, A.F. and D.M.G.; formal analysis, A.F. and D.M.G.; investigation, T.B. and R.M.G.; data curation, M.B.-O.; writing—original draft preparation, T.B., M.K.B. and A.F.; writing—review and editing, M.K.B. and A.F.; supervision, M.K.B. and A.F. All authors have read and agreed to the published version of the manuscript.

Funding: Financial supports from ASTEC, DST, Govt. of Assam (grant number: ASTEC/S&T/192(177)/2020–2021/43) and the Gobierno de España, Ministerio de Ciencia e Innovacion (project No. PID2020-115637GB-I00 FEDER funds) are gratefully acknowledged. The authors thank IIT-Guwahati for the TG data.

Data Availability Statement: Not applicable.

Conflicts of Interest: The authors declare no conflict of interest.

References

1. Desiraju, G.R. Crystal engineering: A holistic view. *Angew. Chem. Int. Ed.* **2007**, *46*, 8342–8356. [CrossRef]
2. Gu, J.; Wen, M.; Cai, Y.; Shi, Z.; Nesterov, D.S.; Kirillova, M.V.; Kirillov, A.M. Cobalt(II) coordination polymers assembled from unexplored pyridine-carboxylic acids: Structural diversity and catalytic oxidation of alcohols. *Inorg. Chem.* **2019**, *58*, 5875–5885. [CrossRef]
3. Tang, L.; Tang, L.; Wang, D.; Deng, H.; Chen, K. Metal and ligand effects on the stability and electronic properties of crystalline two-dimensional metal-benzenehexathiolate coordination compounds. *J. Phys. Condens. Matter* **2018**, *30*, 465301–465311. [CrossRef] [PubMed]
4. Abdel-Rahman, L.H.; Abdelhamid, A.A.; Abu-Dief, A.M.; Shehata, M.R.; Bakheeta, A.R. Facile synthesis, X-Ray structure of new multi-substituted aryl imidazole ligand, biological screening and DNA binding of its Cr(III), Fe(III) and Cu(II) coordination compounds as potential antibiotic and anticancer drugs. *J. Mol. Struct.* **2020**, *1200*, 127034–127040. [CrossRef]
5. Adam, R.; Mon, M.; Greco, R.; Kalinke, L.H.A.; Vidal-Moya, A.; Fernandez, A.; Winpenny, R.E.P.; Domenech-Carbo, A.; Leyva-Perez, A.; Armentano, D.; et al. Self-assembly of catalytically active supramolecular coordination compounds within metal-organic frameworks. *J. Am. Chem. Soc.* **2019**, *141*, 10350–10360. [CrossRef]
6. Gu, J.; Cai, Y.; Liang, X.; Wu, J.; Shi, Z.; Kirillov, A.M. Bringing 5-(3,4-dicarboxylphenyl)picolinic acid to crystal engineering research: Hydrothermal assembly, structural features and photocatalytic activity of Mn, Ni, Cu, and Zn coordination polymers. *CrystEngComm* **2018**, *20*, 906–916. [CrossRef]
7. Stahly, G.P. A survey of co-crystals reported prior to 2000. *Cryst. Growth Des.* **2009**, *9*, 4212–4229. [CrossRef]
8. Abidi, S.S.A.; Garg, U.; Azim, Y.; Alam, M.; Gupta, A.K.; Pradeep, C.P.; Azum, N.; Asiri, A.M. Spectroscopic, structural, DFT and molecular docking studies on novel cocrystal salt hydrate of chromotropic acid and its antibiofilm activity. *Arab. J. Sci. Eng.* **2021**, *46*, 353–364. [CrossRef]
9. Ilmi, R.; Al-busaidi, I.J.; Haque, A.; Khan, M.S. Recent progress in coordination chemistry, photo-physical properties, and applications of pyridine-based Cu(I) complexes. *J. Coord. Chem.* **2018**, *21*, 3045–3076. [CrossRef]

10. Shi, L.; Ding, P.; Wang, Y.; Zhang, Y.; Ossipov, D.; Hilborn, J. Self-healing polymeric hydrogel formed by metal-ligand coordination assembly: Design, fabrication, and biomedical applications. *Macromol. Rapid Commun.* **2019**, *40*, 1800837–1800852. [CrossRef]
11. Mogensen, S.B.; Taylor, M.K.; Lee, J. Homocoupling reactions of azoles and their applications in coordination chemistry. *Molecules* **2020**, *25*, 5950. [CrossRef] [PubMed]
12. Silva, V.L.M.; Silva, A.M.S. Recent advances in the synthesis, functionalization and applications of pyrazole-type compounds. *Molecules* **2021**, *26*, 4989. [CrossRef] [PubMed]
13. Abdel-Rahman, L.H.; Abu-Dief, A.M.; Atlam, F.M.; Abdel-Mawgoud, A.A.; Alothman, A.A.; Alsalm, A.M. Chemical, physical, and biological properties of Pd(II), V(IV)O, and Ag(I) complexes of N3 tridentate pyridine-based Schiff base ligand. *J. Coord. Chem.* **2020**, *73*, 3150–3173. [CrossRef]
14. Sarma, P.; Sharma, P.; Gomila, R.M.; Frontera, A.; Barcelo-Oliver, M.; Verma, A.K.; Baruwa, B.; Bhattacharyya, M.K. Charge assisted hydrogen bonded assemblies and unconventional O···O dichalcogen bonding interactions in pyrazole-based isostructural Ni(II) and Mn(II) compounds involving anthraquinonedisulfonate: Antiproliferative evaluation and theoretical studies. *J. Mol. Struct.* **2022**, *1250*, 131883–131897. [CrossRef]
15. Danilescu, O.; Bulhac, I.; Shova, S.; Novitchi, G.; Bourosh, P. Coordination compounds of copper(II) with schiff bases based on aromatic carbonyl compounds and hydrazides of carboxylic acids: Synthesis, structures, and properties. *Russ. J. Coord. Chem.* **2020**, *46*, 838–849. [CrossRef]
16. Gu, J.; Wan, S.; Kirillova, M.V.; Kirillov, A.M. H-Bonded and metal(II)-organic architectures assembled from an unexplored aromatic tricarboxylic acid: Structural variety and functional properties. *Dalton Trans.* **2020**, *49*, 7197–7209. [CrossRef]
17. García-Valdivia, A.A.; Jannus, F.; García-García, A.; Choquesillo-Lazarte, D.; Fernández, B.; Medina-O'Donnell, M.; Lupiáñez, J.A.; Cepeda, J.; Reyes-Zurita, F.J.; Rodríguez-Diéguez, A. Anti-cancer and anti-inflammatory activities of a new family of coordination compounds based on divalent transition metal ions and indazole-3-carboxylic acid. *J. Inorg. Biochem.* **2021**, *215*, 111308–111320. [CrossRef]
18. Shamim, S.; Gul, S.; Rauf, A.; Rashid, U.; Khan, A.; Amin, R.; Akhtar, F. Gemifloxacin-transition metal complexes as therapeutic candidates: Antimicrobial, antifungal, anti-enzymatic, and docking studies of newly synthesized complexes. *Heliyon* **2022**, *8*, 10378–10386. [CrossRef]
19. Scheiner, S. *Non-Covalent Forces*; Springer: Dordrecht, The Netherlands, 2015.
20. Maharramov, A.M.; Mahmudov, M.T.; Kopylovich, M.N.; Pombeiro, A.J.L. *Non-Covalent Interactions in the Synthesis and Design of New Compounds*; John Wiley & Sons, Inc.: Hoboken, NJ, USA, 2016.
21. Desiraju, G.R. Chemistry beyond the molecule. *Nature* **2001**, *412*, 397–400. [CrossRef]
22. Crabtree, R.H. Hypervalency, secondary bonding and hydrogen bonding: Siblings under the skin. *Chem. Soc. Rev.* **2017**, *46*, 1720–1729. [CrossRef] [PubMed]
23. Mundlapati, V.R.; Sahoo, D.K.; Bhaumik, S.; Jena, S.; Chandrakar, A.; Biswal, H.S. Non-covalent carbon-bonding interactions in proteins. *Angew. Chem. Int. Ed.* **2018**, *57*, 16496–16500. [CrossRef] [PubMed]
24. Steed, J.W.; Atwood, J.L. *Supramolecular Chemistry*, 2nd ed.; John Wiley & Sons Ltd.: Chichester, UK, 2009.
25. Groom, C.R.; Bruno, I.J.; Lightfoot, M.P.; Ward, S.C. The Cambridge structural database. *Acta Cryst.* *B* **2016**, *72*, 171–179. [CrossRef] [PubMed]
26. Nath, H.; Sharma, P.; Frontera, A.; Barcelo-Oliver, M.; Verma, A.K.; Das, J.; Bhattacharyya, M.K. Phenanthroline-based Ni(II) coordination compounds involving unconventional discrete fumarate-water-nitrate clusters and energetically significant cooperative ternary π-stacked assemblies: Antiproliferative evaluation and theoretical studies. *J. Mol. Struct.* **2022**, *1248*, 131424–131436. [CrossRef]
27. Sharma, P.; Dutta, D.; Gomila, R.M.; Frontera, A.; Barcelo-Oliver, M.; Verma, A.K.; Bhattacharyya, M.K. Benzoato bridged dinuclear Mn(II) and Cu(II) compounds involving guest chlorobenzoates and dimeric paddle wheel supramolecular assemblies: Antiproliferative evaluation and theoretical studies. *Polyhedron* **2021**, *208*, 115409–115425. [CrossRef]
28. Bruker. *APEX3 User Manual*; Bruker AXS Inc.: Madison, WI, USA, 2015.
29. Bruker. *SADABS, V2.05*; Bruker AXS Inc.: Madison, WI, USA, 1999.
30. Sheldrick, G.M. Crystal structure refinement with SHELXL. *Acta Crystallogr. Sect. A Found. Crystallogr.* **2008**, *64*, 112–122. [CrossRef] [PubMed]
31. Farrugia, L.J. WinGX and ORTEP for Windows: An update. *J. Appl. Crystallogr.* **1999**, *32*, 837–838. [CrossRef]
32. Brandenburg, K. *Diamond 3.1f*; Crystal Impact GbR: Bonn, Germany, 2008.
33. Grimme, S.; Antony, J.; Ehrlich, S.; Krieg, H. A consistent and accurate ab initio parametrization of density functional dispersion correction (DFT-D) for the 94 elements H-Pu. *J. Chem. Phys.* **2010**, *132*, 154104–154119. [CrossRef] [PubMed]
34. Weigend, F. Accurate Coulomb-fitting basis sets for H to Rn. *Phys. Chem. Chem. Phys.* **2006**, *8*, 1057–1065. [CrossRef]
35. Ahlrichs, R.; Bar, M.; Hacer, M.; Horn, H.; Kömel, C. Electronic structure calculations on workstation computers: The program system turbomole. *Chem. Phys. Lett.* **1989**, *162*, 165–169. [CrossRef]
36. Contreras-Garcia, J.; Johnson, E.R.; Keinan, S.; Chaudret, R.; Piquemal, J.P.; Beratan, D.N.; Yang, W. NCIPLOT: A program for plotting non-covalent interaction regions. *J. Chem. Theory Comput.* **2011**, *7*, 625–632. [CrossRef]
37. Bader, R.F.W. Atoms in molecules. *Chem. Rev.* **1991**, *91*, 893–928. [CrossRef]
38. Lu, T.; Chen, F. Multiwfn: A multifunctional wavefunction analyzer. *J. Comput. Chem.* **2012**, *33*, 580–592. [CrossRef]
39. Humphrey, J.W.; Dalke, A.; Schulten, K. VMD: Visual molecular dynamics. *J. Mol. Graph.* **1996**, *14*, 33–38. [CrossRef]

40. McKinnon, J.J.; Spackman, M.A.; Mitchel, A.S. Novel tools for visualizing and exploring intermolecular interactions in molecular crystals. *Acta Crystallogr.* **2004**, *B60*, 627–668. [CrossRef]
41. Spackman, M.A.; Jayatilaka, D. Hirshfeld surface analysis. *CrystEngComm* **2009**, *11*, 19–32. [CrossRef]
42. Spackman, M.A. Molecular electric moments from X-ray diffraction data. *Chem. Rev.* **1992**, *92*, 1769–1797. [CrossRef]
43. Spackman, P.R.; Turner, M.J.; McKinnon, J.J.; Wolff, S.K.; Grimwood, D.J.; Jayatilaka, D.; Spackman, M.A. CrystalExplorer: A program for Hirshfeld surface analysis, visualization and quantitative analysis of molecular crystals. *J. Appl. Cryst.* **2021**, *54*, 1006–1011. [CrossRef]
44. Bora, S.J.; Das, B.K. Synthesis, structure and properties of a fumarate bridged Ni(II) coordination polymer. *J. Mol. Struct.* **2011**, *999*, 83–88. [CrossRef]
45. Gogoi, A.; Islam, S.M.N.; Frontera, A.; Bhattacharyya, M.K. Supramolecular association in Cu(II) and Co(II) coordination complexes of 3,5-dimethylpyrazole: Experimental and theoretical studies. *Inorg. Chim. Acta* **2019**, *484*, 133–141. [CrossRef]
46. Bhattacharyya, M.K.; Devi, P.G.; Dasgupta, D.; Bora, S.J.; Das, B.K. Solid and solution structures and DNA binding properties of $[M^{II}(4\text{-CNpy})_2(SO_4)(H_2O)_3]\cdot H_2O$ for M = Cu, Co, Ni. *Polyhedron* **2012**, *35*, 62–68. [CrossRef]
47. Sharma, P.; Sarma, P.; Frontera, A.; Barceló-Oliver, M.; Verma, A.K.; Sarma, B.; Barthakur, T.; Bhattacharyya, M.K. Energetically significant cooperative π-stacked ternary assemblies in Ni(II) phenanthroline compounds involving discrete water clusters: Anticancer activities and theoretical studies. *J. Mol. Struct.* **2021**, *1229*, 129486–129499. [CrossRef]
48. Das, A.; Choudhury, S.R.; Estarellas, C.; Dey, B.; Frontera, A.; Hemming, J.; Helliwell, H.; Gamez, P.; Mukhopadhyay, S. Supramolecular assemblies involving anion-π and lone pair-π interactions: Experimental observation and theoretical analysis. *CrystEngComm* **2011**, *13*, 4519–4527. [CrossRef]
49. Janiak, C. A critical account on π-π stacking in metal complexes with aromatic nitrogen-containing ligands. *J. Chem. Soc. Dalton Trans.* **2000**, *21*, 3885–3896. [CrossRef]
50. Yakovleva, M.A.; Andreeva, A.A.; Nefedov, S.E. Synthesis and structure of the pyrazolate-bridged cobalt(II) benzoate $Co_2(\mu\text{-dmpz})_2(Hdmpz)_2(OOCPh)_2$. *Russ. J. Inorg. Chem.* **2010**, *55*, 9–21. [CrossRef]
51. Dutta, D.; Sharma, P.; Gomila, R.M.; Frontera, A.; Barcelo-Oliver, M.; Verma, A.K.; Baishya, T.; Bhattacharyya, M.K. Supramolecular assemblies involving unconventional non-covalent contacts in pyrazole-based coordination compounds of Co(II) and Cu(II) pyridinedicarboxylates: Antiproliferative evaluation and theoretical studies. *Polyhedron* **2022**, *224*, 116025–116238. [CrossRef]
52. Sharma, R.P.; Saini, A.; Kumar, J.; Kumar, S.; Venugopalan, P.; Ferretti, V. Coordination complexes of copper(II) with herbicide-trichlorophenoxyacetate: Syntheses, characterization, single crystal X-ray structure and packing analyses of monomeric [Cu(γ-pic)$_3$(2,4,5-trichlorophenoxyacetate)]$\cdot H_2O$, [trans-Cu(en)$_2$(2,4,5-trichlorophenoxyacetate)$_2$]$\cdot 2H_2O$ and dimeric [Cu$_2$(H$_2$tea)$_2$(2,4,5-trichlorophenoxyacetate)$_2$]$\cdot 2H_2O$. *Inorg. Chim. Acta* **2017**, *457*, 59–68.
53. Manna, S.C.; Mistri, S.; Jana, A.D. A rare supramolecular assembly involving ion pairs of coordination complexes with a host–guest relationship: Synthesis, crystal structure, photoluminescence and thermal study. *CrystEngComm* **2012**, *14*, 7415–7422. [CrossRef]
54. Murinzi, T.W.; Hosten, E.; Watkins, G.M. Synthesis and characterization of a cobalt-2,6-pyridinedicarboxylate MOF with potential application in electrochemical sensing. *Polyhedron* **2017**, *137*, 188–196. [CrossRef]
55. Dunstan, P.O.; Khan, A.M. Thermochemical study of 4-(dimethylamino)pyridine complexes of some bivalent metal bromides. *J. Chem. Thermodyn.* **2013**, *66*, 44–49. [CrossRef]
56. Heine, M.; Fink, L.; Schmidt, M.U. 4-Cyanopyridine complexes $[MX_2(4\text{-CNpy})_x]_n$ (with X = Cl, Br and x = 1, 2): Crystal structures, thermal properties and a comparison with $[MX_2(3\text{-CNpy})_x]_n$ complexes. *CrystEngComm* **2020**, *22*, 2067–2082. [CrossRef]
57. Fuhrmann, H.; Brenner, D.; Arndt, P. Octahedral group 4 metal complexes that contain amine, amido, and aminopyridinato ligands: Synthesis, structure, and application in α-olefin oligo-and polymerization. *Inorg. Chem.* **1996**, *35*, 6742–6745. [CrossRef] [PubMed]
58. Dutta, D.; Islam, S.M.N.; Saha, U.; Chetry, S.; Guha, A.K.; Bhattacharyya, M.K. Structural topology of weak non-covalent interactions in a layered supramolecular coordination solid of zinc involving 3-aminopyridine and benzoate: Experimental and theoretical studies. *J. Chem. Crystallogr.* **2018**, *48*, 156–163. [CrossRef]
59. Li, J.; Xing, Y.H.; Zhao, H.Y.; Li, Z.P.; Wang, Z.P.; Zeng, X.Q.; Ge, M.F.; Niu, S.Y. Constructions of a set of hydrogen-bonded supramolecules from reactions of transition metals with 3,5-dimethylpyrazole and different dicarboxylate ligands. *Inorg. Chim. Acta* **2009**, *362*, 2788–2795. [CrossRef]
60. Titi, A.; Shiga, T.; Oshio, H.; Touzani, R.; Hammouti, B.; Mouslim, F.; Warad, I. Synthesis of novel $Cl_2Co_4L_6$ cluster using 1-hydroxymethyl-3,5-dimethylpyrazole (LH) ligand: Crystal structure, spectral, thermal, Hirschfeld surface analysis and catalytic oxidation evaluation. *J. Mol. Struct.* **2020**, *1199*, 126995–127006. [CrossRef]
61. Direm, A.; Tursun, M.; Parlak, C.; Cherif, N.B. Trans-dichlorotetrakis (1H-pyrazole-κN2) copper(II): Synthesis, crystal structure, hydrogen bonding graph-sets, vibrational and DFT studies. *J. Mol. Struct.* **2015**, *1093*, 208–218. [CrossRef]
62. Mautner, F.A.; Scherzer, M.; Berger, C.; Fischer, C.R.; Vicente, R.; Massoud, S.S. Synthesis and characterization of five new thiocyanato-and cyanato-metal(II) complexes with 4-azidopyridine as co-ligand. *Polyhedron* **2015**, *85*, 20–26. [CrossRef]
63. Bhattacharyya, M.K.; Dutta, D.; Islam, S.M.N.; Frontera, A.; Sharma, P.; Verma, A.K.; Das, A. Energetically significant antiparallel π-stacking contacts in Co(II), Ni(II) and Cu(II) coordination solids of pyridine-2,6-dicarboxylates: Antiproliferative evaluation and theoretical studies. *Inorg. Chim. Acta* **2020**, *501*, 119233–119245. [CrossRef]

64. Basumatary, D.; Lal, R.A.; Kumar, A. Synthesis, and characterization of low- and high-spin manganese(II) complexes of polyfunctionaladipoyldihydrazone: Effect of coordination of N-donor ligands on stereo-redox chemistry. *J. Mol. Struct.* **2015**, *1092*, 122–129. [CrossRef]
65. Mandal, T.; Dey, A.; Pathak, S.; Islam, M.; Konar, S.; Ortega-Castro, J.; Seth, S.K.; Ray, S.; Frontera, A.; Mukhopadhyay, S. Structures, photoresponse properties and DNA binding abilities of 4-(4-pyridinyl)-2-pyridone salts. *RSCAdv.* **2019**, *9*, 9663–9677. [CrossRef]
66. Jian, F.; Zheng, J.; Zhao, P.; Li, Y.F. Synthesis, characterization and density functional calculations on dichloro-bis(4-dimethylaminopyridine) cobalt(II) complex [Co(II)($C_7H_{10}N_2$)$_2$$Cl_2$]. *J. Coord. Chem.* **2008**, *61*, 705–714. [CrossRef]
67. Du, M.; Zhang, Z.; You, Y.; Zhao, X. R-Isophthalate (R = –H, –NO_2 and –COOH) as modular building blocks for mixed-ligand coordination polymers incorporated with a versatile connector 4-amino-3,5-bis(3-pyridyl)-1,2,4-triazole. *CrystEngComm* **2008**, *10*, 306–321. [CrossRef]
68. Yang, Q.; Chen, S.; Gao, S. Two Mn(II) chloride complexes containing guest molecules. *J. Therm. Anal. Calorim.* **2007**, *89*, 567–571. [CrossRef]
69. Refat, M.S.; El-deen, I.M.; Anwer, Z.M.; El-ghol, S. Bivalent transition metal complexes of coumarin-3-yl thiosemicarbazone derivatives: Spectroscopic, antibacterial activity and thermogravimetric studies. *J. Mol. Struct.* **2009**, *920*, 149–162. [CrossRef]
70. Gogoi, A.; Das, A.; Frontera, A.; Verma, A.K.; Bhattacharyya, M.K. Energetically significant unconventional π-π contacts involving fumarate in a novel coordination polymer of Zn(II): *In-vitro* anticancer evaluation and theoretical studies. *Inorg. Chim. Acta* **2019**, *493*, 1–13. [CrossRef]
71. Vener, M.V.; Egorova, A.N.; Churakov, A.V.; Tsirelson, V.G. Intermolecular hydrogen bond energies in crystals evaluated using electron density properties: DFT computations with periodic boundary conditions. *J. Comput. Chem.* **2012**, *33*, 2303–2309. [CrossRef]

crystals

MDPI

Article

Unprecedented Coordination Compounds with 4,4′-Diaminodiphenylethane as a Supramolecular Agent and Ditopic Ligand: Synthesis, Crystal Structures and Hirshfeld Surface Analysis

Nicoleta Craciun [1,2], Diana Chisca [1,2], Elena Melnic [1] and Marina S. Fonari [1,*]

[1] Institute of Applied Physics USM, Academiei Str. 5, MD-2028 Chisinau, Moldova
[2] Faculty of Biology and ChemistryIon Creangă State Pedagogical University, Ion Creangă Str. 1, MD-2069 Chisinau, Moldova
* Correspondence: marina.fonari@ifa.md or fonari.xray@gmail.com

Abstract: In this pioneering research, mononuclear coordination complexes and coordination polymers were obtained using the conformationally flexible ditopic ligand 4,4′-diaminodiphenylethane and different metal salts (nitrates, sulfates, tetrafluoroborates and perchlorates). Seven new products, including the mononuclear complexes $[Cd(2,2'\text{-bpy})_3](ClO_4)_2](dadpe)(4,4'\text{-bpy})$ (**1**), $[Ni(dadpe)_2(H_2O)_4](SO_4)\cdot H_2O$ (**2**), one-dimensional coordination polymers $\{[Zn(NO_3)(dadpe)(dmf)_2](NO_3)\}_n$ (**3**), $\{[Cd(2,2'\text{-bpy})_2(dadpe)](ClO_4)_2\}_n$ (**4**), and two-dimensional coordination polymers, $\{[Cd(4,4'\text{-bpy})_2(H_2O)_2](ClO_4)_2(dadpe)(EtOH)_2\}_n$ (**5**), $\{[Co(4,4'\text{-bpy})_2(H_2O)_2](BF_4)_2(dadpe)(EtOH)_2\}_n$ (**6**) and $\{[Cd(adi)(dadpe)](H_2adi)\}_n$ (**7**), (dadpe=4,4′-diaminodiphenylethane, 2,2′-bpy=2,2′-bipyridine, 4,4′-bpy=4,4′-bipyridine, H_2adi=adipic acid) were produced. The synthesized compounds were characterized by FTIR and single-crystal X-ray diffraction analyses. The dadpe was recorded as a neutral guest in the crystals of mononuclear complex **1** and in coordination polymers **5** and **6**. In compound **2,** two dadpe ligands coordinate in a monodentate mode and occupy two trans-positions in the $[Ni(H_2O)_4(dadpe)_2]^{2+}$ octahedral complex cation. Coordination polymers **3** and **4** represent single chains originating from dadpe as a bidentate linker in both. The H-donor's possibilities of amino groups were utilized in the interconnection of coordination chains into H-bonded networks via $NH(NH_2)\cdots O$ hydrogen bonds. The isostructural coordination polymers **5** and **6** comprise similar cationic square grids $[M(4,4'\text{-bpy})_2(H_2O)_2]^{2+}$ [M=Cd (**5**), M=Co (**6**)], with **sql** topology balanced by the charge-compensated anions, while dadpe and EtOH as neutral guests are situated in the interlayer space. The neutral 2D coordination network in **7** with the **sql** topology originates from both adi and dadpe linkers as bidentate-bridging ligands, and the neutral H_2adi is entrapped as a guest in crystal lattice. The impact of different types of intermolecular interactions was evaluated by Hirshfeld surface analysis.

Keywords: transition metals; synthesis; crystal structure; non-covalent interactions; Hirshfeld surface

Citation: Craciun, N.; Chisca, D.; Melnic, E.; Fonari, M.S. Unprecedented Coordination Compounds with 4,4′-Diaminodiphenylethane as a Supramolecular Agent and Ditopic Ligand: Synthesis, Crystal Structures and Hirshfeld Surface Analysis. *Crystals* **2023**, *13*, 289. https://doi.org/10.3390/cryst13020289

Academic Editor: Waldemar Maniukiewicz

Received: 24 January 2023
Revised: 3 February 2023
Accepted: 5 February 2023
Published: 8 February 2023

1. Introduction

For decades, 4,4′-bipyridine (4,4′-bpy), its longer homologues and modified derivatives were widely used as bidentate-bridging ligands in the synthesis of coordination polymers, giving rise either to cationic polymeric arrays when combined with easy-leaving anions [1–4], or to neutral coordination networks when used in a partnership with strongly coordinated anions such as polycarboxylates [5–8]. Although ditopic amino-ligands are weaker bases than the five- and six-membered heterocyclic N-bases, the presence of terminal amino groups gives them the advantage of simultaneous coordination with the metal and participation in different hydrogen bonds and weak interactions reinforcing the crystal lattice [9–11]. Moreover, the NH_2-group in organic luminescent materials is regarded as an electron-donating group suitable for binding the electron-accepting metal ions. Compounds containing such amino ligands show luminescence through intraligand p–p* transitions [12].

So far, a significant number of coordination compounds, including silver and cadmium coordination polymers, were documented [13] for 4,4′-diaminodiphenylmethane [10–12,14]. Silver and cadmium coordination polymers were also obtained with the more extended ditopic ligand, 4,4′(1,4-phenylenediisopropylidene)bis(aniline) [15,16]. The 3,3′-diaminobiphenylsulfone was reported to trap heavy metals (Cu(II), Hg(II)) in the form of coordination compounds [17,18]. On the other hand, no coordination compounds with 4,4′-diaminodiphenylethane (dadpe, also known under the names 4,4′-diaminobibenzyl, 4,4′-ethylenedianiline), the closest homologue of 4,4′-diaminodiphenylmethane, were yet reported. The survey of CSD (Version 5.43, November 2022 updates) only disclosed the dadpe in the form of hydrate [19], and as a guest in the inclusion compound with beta-cyclodextrin (β-CD) [20]. In the crystal of the hydrate, a dadpe molecule was registered in an extended trans-conformation. Alternatively, the same molecule exhibited one extended and two bent conformations in the inclusion complex with β-CD [20]. The expected and registered conformational flexibility of dadpe was an additional benefit for its use in the synthesis of flexible coordination networks. The flexible CPs reveal some advantages and attractive applications compared with the rigid frameworks. For example, they show pore opening and significantly increased adsorption capacity [21]. In this pioneering work, we report the synthesis, IR spectroscopic characterization and crystal structures for seven coordination compounds obtained from the different metal salts, $Cd(ClO_4)_2{\cdot}2H_2O$, $Cd(NO_3)_2{\cdot}4H_2O$, $NiSO_4{\cdot}7H_2O$, $Co(BF_4)_2{\cdot}6H_2O$, $Zn(NO_3)_2{\cdot}6H_2O$ and a ditopic dadpe ligand. New coordination compounds include mononuclear complexes $[Cd(2,2'\text{-bpy})_3](ClO_4)_2](\text{dadpe})(4,4'\text{-bpy})$ (**1**), $[Ni(\text{dadpe})_2(H_2O)_4](SO_4){\cdot}2H_2O$ (**2**), one-dimensional (1D) coordination polymers $\{[Zn(NO_3)(\text{dadpe})(\text{dmf})_2](NO_3)\}_n$ (**3**), $\{[Cd(2,2'\text{-bpy})_2(\text{dadpe})](ClO_4)_2\}_n$ (**4**), and two-dimensional (2D) coordination polymers, $\{[Cd(4,4'\text{-bpy})_2(H_2O)_2](ClO_4)_2(\text{dadpe})(EtOH)_2\}_n$ (**5**), $\{[Co(4,4'\text{-bpy})_2(H_2O)_2](BF_4)_2(\text{dadpe})(EtOH)_2\}_n$ (**6**) and $\{[Cd(\text{adi})(\text{dadpe})](H_2\text{adi})\}_n$ (**7**). The distribution of intermolecular interactions in compounds **1**, **2**, **3** and **7**, which reveal different structural functions of dadpe ligand (as a neutral guest, a monodentate terminal ligand, and a bidentate-bridging ligand), was evaluated by Hirshfeld surface analysis.

2. Materials and Methods

2.1. Materials and Measurements

The starting salts, organic ligands and solvents were obtained from commercial sources (Sigma-Aldrich, St. Louis, MO, USA) and were used without further purification. The IR(ATR) spectra were recorded on a FTIR Spectrum-100 Perkin Elmer spectrometer in the range of 4000–650 cm^{-1}. Elemental analysis was performed on a Vario EL III Element Analyzer.

2.2. Synthesis

2.2.1. $[Cd(2,2'\text{-bpy})_3](ClO_4)_2](\text{dadpe})(4,4'\text{-bpy})$ (**1**)

Compound **1** was prepared with the hydrothermal method. Dadpe 0.04 g (0.2 mmol) was dissolved in 7 mL EtOH. In this solution, 0.03 g (0.1 mmol) $Cd(ClO_4)_2{\cdot}2H_2O$, 0.045 g (0.3 mmol) 2,2′-bpy and 0.045 g (0.3 mmol) 4,4′-bpy, were added successfully. The obtained solution was placed in a 20 mL Teflon-lined stainless steel autoclave, which was then sealed and heated to 100 °C for 20 h. The solution was filtered and left for slow evaporation at room temperature. After 72 h, colorless crystals were filtered and dried in the air (yield 57%). Anal. Calc. for $C_{42}H_{36}CdCl_2N_8O_8$ (%): C, 52.32; H, 3.76; N, 11.62. Found: C, 52.98; H, 3.49; N, 11.04. IR-ATR (cm^{-1}): 3439 m, 3355 m, 1611 m, 1515 s, 14,382, 1278 m, 1193 w, 1083 s, 1024 m, 970 w, 819 s, 762 s, 735 w, 648 w, 621 m, 527 s, 488 w, 411 m.

2.2.2. $[Ni(\text{dadpe})_2(H_2O)_4](SO_4){\cdot}H_2O$ (**2**)

Dadpe 0.02 g (0.1 mmol) was dissolved in 5 mL EtOH. After that, 0.02 g (0.1 mmol) of $NiSO_4{\cdot}7H_2O$ was added. The reaction mixture was refluxed with stirring for 20 min at 80 °C. A white precipitate formed and was dissolved with 3 mL of H_2O_{dist}. The solution was filtered and left for slow evaporation at room temperature. After 7 days, the crystals of yellow color were filtered, and dried in the air (yield ~60%). Anal. Calc. for $C_{28}H_{44}N_4NiO_{10}S$ (%): C, 48.92; H, 6.45; N, 8.15. Found: C, 48.38; H, 5.99; N, 8.24. *IR-ATR* (cm^{-1}): 3392 m, 3310 m, 2976 w, 2917 w, 2854 w, 1614 s, 1514 s, 1455 w, 1281 m, 1250 m, 1173 s/h, 1057 s, 939 w, 833 s, 537 s.

2.2.3. {[Zn(NO_3)(dadpe)(dmf)$_2$](NO_3)}$_n$ (**3**)

Dadpe 0.02 g (0.1 mmol) was dissolved in 10 mL mixture CH_3CN:EtOH (5:5). In this solution, 0.018 g (0.1 mmol) $Zn(NO_3)_2{\cdot}6H_2O$ and 3 mL DMF were added. The mixture was positioned in an ultrasonic bath, at ambient temperature and atmospheric pressure for 30 min. The solution was filtered and left for slow evaporation at room temperature. After 150 days, light-yellow crystals were filtered, and dried in the air (yield ~47%). *Anal. Calc.* for $C_{20}H_{30}N_6O_8Zn$ (%): C, 43.85; H, 5.52; N, 15.34. Found: C, 43.35; H, 5.19; N, 14.99. *IR-ATR* (cm^{-1}): 3440 m, 3376 m, 3355 m, 3300 m, 3230 m, 3146 m, 3011 m, 2915 m, 2852 m, 2360 s, 2343 s, 1617 s, 1515 s, 1475 w, 1437 m, 1363 m, 1313 s, 1280 s, 1250 m, 1170 w, 1113 s, 1084 s, 1015 s, 947 w, 913 w, 820 s, 762 m, 735 m, 669 m, 621 m, 527 s, 408 m.

2.2.4. {[Cd(2,2′-bpy)$_2$(dadpe)](ClO_4)$_2$}$_n$ (**4**)

To a cold solution of dadpe 0.04 g (0.2 mmol) in 7 mL EtOH, $Cd(ClO_4)_2{\cdot}2H_2O$ 0.03 g (0.1 mmol) and 2,2′-bpy 0.05 g (0.3 mmol) were added and continuously stirred. The obtained solution was placed in a 20 mL Teflon-lined stainless steel autoclave which was then sealed and heated to 80 °C for 17 h. The solution was filtered and left for slow evaporation at room temperature. After 24 h, crystals of a light green color were filtered and dried in the air (yield ~58%). Anal. Calc. for $C_{34}H_{32}CdCl_2N_6O_8$ (%): C, 48.85; H, 3.86; N, 10.05. Found: C, 48.95; H, 2.99; N, 10.25. IR-ATR (cm^{-1}): 3320 m, 3258 m, 1595 m, 1515 m, 1438 s, 1318 m, 1246 w, 1226 m, 1083 s, 1013 s, 936 m, 823 m, 756 s, 736 n, 618 s, 521 m, 478 w.

2.2.5. {[Cd(4,4′-bpy)$_2$(H_2O)$_2$](ClO_4)$_2$(dadpe)(EtOH)$_2$}$_n$ (**5**)

To a cold solution of dadpe 0.04 g (0.2 mmol) in 7 mL EtOH, $Cd(ClO_4)_2{\cdot}2H_2O$ 0.04 g (0.2 mmol) and 4,4′-bpy 0.03 g (0.1 mmol) were added. The reaction mixture was refluxed with stirring for 20 min at 80 °C. The solution was filtered and left for slow evaporation at room temperature. After 6 days, light brown crystals were filtered and dried in the air (yield ~63%). Anal. Calc. for $C_{38}H_{48}CdCl_2N_6O_{12}$ (%): C, 47.34; H, 5.02; N, 8.72. Found: C, 47.94; H, 4.99; N, 8.94. *IR-ATR* (cm^{-1}): 3440 m, 3355 m, 3113 w, 3069 w, 1595 s, 1518 s, 1477 s, 1437 s, 1321 m, 1280 m, 1249 s, 1177 w, 1163 w, 1076 s, 1013 s, 977 m, 840 m, 762 s, 762 m, 68 m, 619 s, 528 m.

2.2.6. {[Co(4,4′-bpy)$_2$(H_2O)$_2$](BF_4)$_2$(dadpe)(EtOH)$_2$}$_n$ (**6**)

Dadpe 0.04 g (0.2 mmol) was dissolved in 7 mL EtOH. To this solution, 0,034 g (0.1 mmol) $Co(BF_4)_2{\cdot}6H_2O$ and 0.03 g (0.2 mmol) 4,4′-bpy were added successfully. The obtained solution was placed in a 20 mL Teflon-lined stainless steel autoclave which was then sealed and heated to 100 °C for 20 h. The solution was filtered off and left for slow evaporation at a room temperature. After 24 h, light brown crystals were filtered off and dried in the air (yield ~46%). Anal. Calc. for $C_{38}H_{48}B_2CoF_8N_6O_4$ (%): C, 51.55; H, 5.46; N, 9.49. Found: C, 51.99; H, 5.09; N, 8.99. *IR-ATR* (cm^{-1}): 3553 m, 3480 m, 3359 m, 1605 s, 1535 m, 1515 s, 1490 m, 1411 s, 1394 s/h, 1301 w, 1221 s, 1173 w, 1064 s, 1049 s, 1003 s, 972 s/h, 883 m, 850 m, 809 s, 732 m, 681 w, 628 s, 585 m, 541 m, 519 m, 469 m.

2.2.7. {[Cd(adi)(dadpe)](H_2adi)}$_n$ (**7**)

In a first beaker, 0.02 g (0.1 mmol) of dadpe was dissolved in 3 mL MetOH. In another beaker, 0.03 g (0.1mmol) of $Cd(NO_3)_2{\cdot}4H_2O$ and 0.01 g (0.1 mmol) of H_2adi were dissolved in 5 mL EtOH. The solutions were mixed upon stirring. To the white precipitate, 5 mL

H_2O_{dist}. was added. The mixture was positioned in an ultrasonic bath at ambient temperature and atmospheric pressure for 30 min. The solution was filtered off and left for slow evaporation at a room temperature. After 30 days, light yellow crystals were filtered off and dried in the air (yield ~61%). Anal. Calc. for $C_{26}H_{34}CdN_2O_8$ (%): C, 50.78; H, 5.57; N, 4.56. Found: C, 51.06; H, 5.89; N, 4.99. *IR-ATR* (cm^{-1}): 3239 m, 3151 m, 2925 m, 1694 s, 1617 m, 1512 s, 1455 m, 1408 s, 1331 m, 1263 m, 1229 s, 1183 m, 1136 s, 1060 s, 820 s, 656 s, 542 s, 452 s.

2.3. Single Crystal X-ray Analysis

Single crystal X-ray diffraction measurements for **1–7** were carried out on an Xcalibur E diffractometer equipped with a CCD area detector and a graphite monochromator, utilizing MoKα radiation at a room temperature. Final unit cell dimensions were obtained and refined on entire datasets. All calculations to solve the structures and to refine the models proposed were carried out with the programs SHELXS97 and SHELXL2014 [22,23]. The disordering problems were resolved for weakly bound perchlorate anion in **1**, **4**, **5**, tetrafluoroborate anion in **6**, dadpe ligand in **4**, and EtOH solvent molecule in **5** and **6**. Whenever necessary, restraints were imposed on the geometry and displacement parameters of disordered molecules. C-bound H-atoms were positioned geometrically and treated as riding atoms using SHELXL default parameters with $U_{iso}(H) = 1.2U_{eq}(C)$ and $U_{iso}(H) = 1.5U_{eq}(CH_3)$. N(O)-bound H-atoms were found predominantly from the difference Fourier maps and refined at the final stages using geometric constraints to keep the geometry of amino groups and water molecules reliable. The X–ray data and details of the refinement for **1–7** are summarized in Table 1. The principal bond distances and angles in **1–7** have common values and are summarized in Table S1. The geometric parameters of H-bonds are given in Table S2 (see Supporting Information file). The figures were produced using MERCURY [24]. The solvent-accessible areas were evaluated using MERCURY [24] and PLATON [25] facilities. CIF files for **1–7** that contain all crystallographic details may be obtained free of charge from the Cambridge Crystallographic Data Centre (https://www.ccdc.cam.ac.uk/structures/, accessed on 16 January 2023) by quoting the CCDC deposition numbers 2236422-2236428.

Table 1. Crystal data and structure refinement parameters for **1**–**7**.

	1	2	3	4	5	6	7
CCDC deposition number	2236422	2236423	2236424	2236425	2236426	2236427	2236428
Empirical formula	$C_{42}H_{36}CdCl_2N_8O_8$	$C_{28}H_{44}N_4NiO_{10}S$	$C_{20}H_{30}N_6O_8Zn$	$C_{34}H_{32}CdCl_2N_6O_8$	$C_{38}H_{48}CdCl_2N_6O_{12}$	$C_{38}H_{48}B_2CoF_8N_6O_4$	$C_{26}H_{34}CdN_2O_8$
T, K	293(2)	293(2)	293(2)	293(2)	293(2)	293(2)	293(2)
FW (g mol^{-1})	964.09	687.44	547.87	835.95	964.12	885.37	614.95
Crystal system	Triclinic	Monoclinic	Orthorhombic	Monoclinic	Monoclinic	Monoclinic	Monoclinic
Space group	P-1	$P2_1/c$	$Pbca$	$P2_1/c$	$C2/c$	$C2/c$	$P2/c$
a/Å	11.4066(6)	18.195(2)	17.2741(10)	10.5752(5)	19.5958(5)	19.0858(14)	13.0429(4)
b/Å	13.2070(6)	10.053(2)	10.4778(4)	13.6755(6)	11.7421(2)	11.4429(4)	5.1903(2)
c/Å	14.1328(8)	9.1989(7)	28.0153(11)	24.4455(19)	20.1263(6)	22.5537(18)	19.9688(6)
α/deg	96.087(4)	90	90	90	90	90	90
β/deg	104.337(5)	101.660(7)	90	91.488(5)	108.939(3)	121.947(11)	100.079(3)
γ/deg	96.049(4)	90	90	90	90	90	90
V/Å^3	2031.89(19)	1648.0(3)	5070.6(4)	3534.1(4)	4380.3(2)	4179.6(6)	1330.96(8)
Z	2	2	8	4	4	4	2
D_{calcd} Mg/m^3	1.576	1.385	1.435	1.571	1.462	1.407	1.534
μ/mm^{-1}	0.734	0.711	1.023	0.829	0.686	0.493	0.871
F(000)	980	728	2288	1696	1984	1836	632
Reflections collected	11770	5343	12232	13628	7798	7782	4527
Independent reflections	7176 [R(int) = 0.0303]	2906 [R(int) = 0.0242]	4692 [R(int) = 0.0615]	6523 [R(int) = 0.0471]	4070 [R(int) = 0.0204]	4083 [R(int) = 0.0293]	2620 [R(int) = 0.0200]
Data/restraints/parameters	7176/92/572	2906/22/239	4692/19/320	6523/82/524	4070/169/341	4083/114/339	2620/15/193
GOF	1.001	1.000	1.000	1.005	1.092	0.999	1.004
R indices [$I > 2\sigma(I)$], $R1$, $wR2$	0.0464, 0.1036	0.0498, 0.1413	0.0559, 0.0797	0.0600, 0.1230	0.0373, 0.0916	0.0525, 0.1419	0.0293, 0.0720
R indices (all data), $R1$, $wR2$	0.0630, 0.1139	0.0711, 0.1573	0.1162, 0.0938	0.1301, 0.1524	0.0441, 0.0949	0.0734, 0.1545	0.0336, 0.0749

3. Results and Discussion

3.1. FTIR Characterization of 1–7

The FTIR spectra for compounds 1–7 confirm the presence of the ligands used in the syntheses (Figure 1 and Figures S1–S7 in the Supporting Information file). The occurrence of the dadpe ligand is indicated by stretching vibrations of amino groups registered in two spectral ranges: at high frequencies, 3440–3330 cm^{-1} characteristic for hydrogen stretching, and in the range 1277–1250 cm^{-1} which is typical for asymmetric and symmetric stretching vibrations of C-N bond of aromatic amines [19]. This shows great similarity with the aniline stretching C-N mode at 1276 cm^{-1} [26]. At high frequencies, the broad IR bands attributed to the stretching O-H vibrations of coordinated water molecules in **2**, **5**, **6** were also registered. The asymmetric and symmetric methylene C-H stretching vibrations come out as 2976 and 2854 cm^{-1} bands in **2**, and a 2852 cm^{-1} band in **3** in the IR spectra. The stretching vibration of the central C-C bond is mainly localized at 1083–1003 cm^{-1}. Four types of bending modes (scissoring, twisting, wagging, and rocking) are inherent for the terminal -NH_2 and central –CH_2- groups of ethane fragment, but they occur with weak intensities, and for NH_2-group were registered at 1515, and 850–735 cm^{-1}, respectively. It is worth noting that all of these modes are strongly mixed with vibrations of phenyl and pyridyl rings, and the rings' predominantly stretching vibrations are assigned to the range of 1650–1400 cm^{-1}, while the rings' predominantly bending vibrations are assigned to the range of 1400 to 400 cm^{-1}. In **7**, the characteristic asymmetric stretching vibrations of the carboxylic group appear at 1617 cm^{-1}, and the bands in the range of 1450 to 1308 cm^{-1} are attributed to the symmetric stretching mode of carboxylic groups. The strong IR band at 1694 cm^{-1} also indicates the presence of neutral H_2adi in the crystal structure. The IR bands in the region of 1310 to 1000 cm^{-1} are attributed to the –CH– in-plane or out-of-plane bending modes, ring breathing, and ring deformation frequencies of the pyridine ring, respectively.

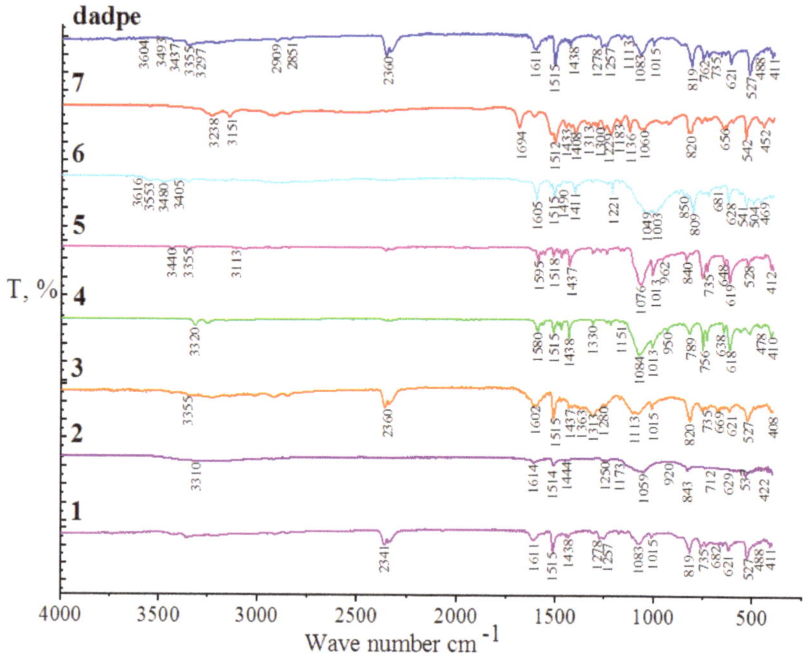

Figure 1. Combined diagram of IR spectra for **1–7** and dadpe ligand.

3.2. X-ray Study

3.2.1. Mononuclear Complexes

Competition of 2,2′-bpy and 4,4′-bpy with dadpe as a weaker N-base resulted in mononuclear complex $[Cd(2,2'\text{-bpy})_3](ClO_4)_2](\text{dadpe})(4,4'\text{-bpy})$ (**1**). The asymmetric part of the triclinic unit cell comprises one complex cation $[Cd(2,2'\text{-bpy})_3]^{2+}$, one non-bound charge-compensated ClO_4^- anion and halves of centrosymmetric 4,4′-bpy and dadpe molecules as the guests (Figure 2a). A Cd atom is situated in general position, and its N_6-octahedral coordination geometry forms three 2,2′-bpy ligands coordinated in bidentate-chelate coordination modes with Cd-N distances varying in the narrow range, 2.324(3)–2.379(3) Å. Geometry of the complex cation is consistent with the reported examples [27,28]. In the crystal, stacking interactions prevail, since each blade of the three-blade propeller-type $[Cd(2,2'\text{-bpy})_3]^{2+}$ cation participates in its own stacking pattern, being centrosymmetric dimer 2,2′-bpy/2,2′-bpy with an interplanar separation of 3.491 Å, and two stacking trimers, namely 2,2′-bpy/4,4′-bpy/2,2′-bpy (with 4,4′-bpy situated at equidistant distances, 3.47–3.49 Å between two 2,2′-bpy), and 2,2′-bpy/dadpe/2,2′-bpy (with 2,2′-bpy-2,2′-bpy separation of 7.427 Å, Figure 2b). The centrosymmetric dadpe molecule takes an extended conformation with phenyl rings situated in parallel planes and a -C-CH_2-CH_2-C- torsion of 180.0°. No meaningful hydrogen bonds with participation of dadpe amino groups were registered in this structure; alternatively, the outer-sphere ClO_4^- anion was involved only in weak CH⋯O interactions (Table S2).

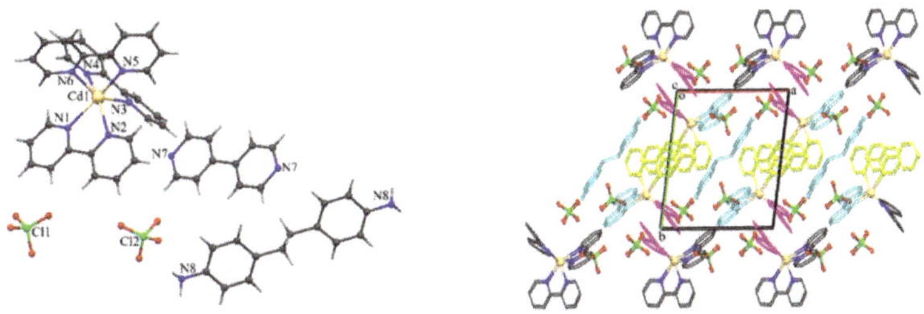

(**a**) (**b**)

Figure 2. Compound **1**. (**a**) View of components in **1** with partial labelling scheme. (**b**) Fragment of crystal packing with stacking patterns colored magenta for 2,2′-bpy/2,2′-bpy dimer; yellow for 2,2′-bpy/4,4′-bpy/2,2′-bpy trimer; and cyan for 2,2′-bpy/dadpe/2,2′-bpy trimer.

Mononuclear complex $[Ni(\text{dadpe})_2(H_2O)_4](SO_4)\cdot 2H_2O$ (**2**) comprises an octahedral complex cation $[Ni(H_2O)_4(\text{dadpe})_2]^{2+}$, an outer-sphere charge-balanced sulfate anion and a solvent water molecule (Figure 3a). A Ni atom, lying on an inversion center, is bound with two dadpe ligands [Ni–N 2.143(3) Å] situated in trans-positions and four water molecules [Ni–O 2.046(2) Å]. Two identical monodentate dadpe ligands have extended conformations with –C-CH_2-CH_2-C- torsion of 178.3(4)°, and provide the longitude of the complex cation of 27.121(6) Å. Components in the crystal are linked in the 3D H-bonded network via a plethora of hydrogen bonds, with participation of both coordinated and non-coordinated amino groups (Figure 3b, Table S2).

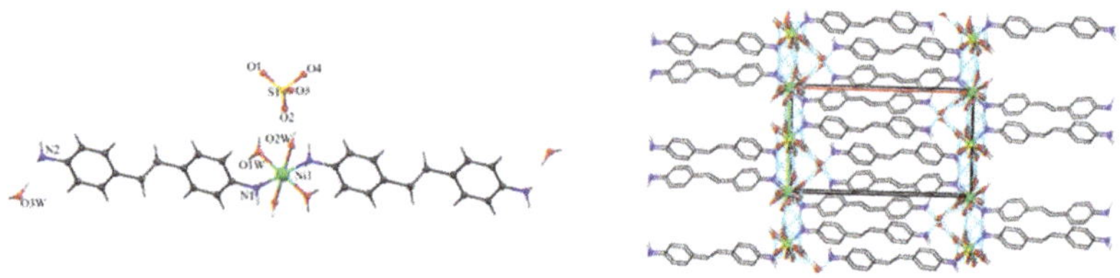

(a) (b)

Figure 3. Compound **2**. (**a**) View of complex **2** with partial labelling scheme. (**b**) Crystal packing with indication of principal hydrogen bonds.

3.2.2. 1D Coordination Polymers

Compound {[Zn(NO_3)(dadpe)(dmf)$_2$](NO_3)}$_n$ (**3**) comprises a linear positively charged coordination polymeric chain [Zn(NO_3)(dadpe)(dmf)$_2$]$_n^+$ running along the coordination c axis and one outer sphere charge-compensated nitrate anion. The dadpe takes a slightly twisted extended trans-conformation, with a –C-CH_2-CH_2-C- torsion of 177.3(4) ° and an interplanar Ph/Ph angle of 6.36 °, and provides Zn···Zn separation of 14.041(8) Å along the coordination chain. The Zn atom is situated in general position and takes a distorted N_2O_4 octahedral coordination geometry originating from one nitrate anion that coordinates in a bidentate-chelate mode, two monodentate trans-situated dmf molecules and two bidentate-bridging dadpe ligands situated in neighboring positions [the bond angle N(5*)-Zn(1)-N(4) = 106.62(12)°, Table S1] and each facing the O-atom of the chelate nitrato-group (Figure 4a). The rather similar coordination environment for Zn(II) was registered in 4,13-bis(2-aminobenzyl)-4,13-diaza-18-crown-6-bis(nitrato-O,O′)-zinc(II) where two nitrate anions are fixed similarly at long Zn-O(NO_3) distances, 2.336 and 2.370 Å [29]. The proximal arrangement of dadpe ligands in **3** is reinforced by their involvement in two NH(NH_2)··· O(NO_3) hydrogen bonds (Table S2), with an outer-sphere nitrate anion situated in a perching position and reproaching these ligands (Figure 4a). The cationic chains situated parallel with the c axis are interconnected along the b-axis by NH(NH_2)··· O(NO_3) hydrogen bonds from the same O-atoms of the outer sphere nitrate anion again (Table S2), thus combining the coordination chains in the H-bonded thick layer situated parallel to the *bc* plane (Figure 4b). All the meaningful hydrogen bonds, with the participation of two NH_2-groups, are realized within this H-bonded thick layer restricted by the coordinated nitrate anions (Figure 4c).

Similar to compound **3**, compound {[Cd(2,2′-bpy)$_2$(dadpe)](ClO_4)$_2$}$_n$ (**4**) crystallizes as a linear positively charged polymeric coordination chain {[Cd(2,2′-bpy)$_2$(dadpe)]$_n^{2+}$ (Figure 5a), but with both ClO_4^- as outer sphere counter-ions. A Cd(II) atom takes a distorted octahedral N_6 coordination geometry originating from two bidentate-chelate 2,2′-bpy ligands and two bidentate-bridging dadpe ligands that are situated in neighboring cis-positions, angle N(NH_2)-Cd-N(NH_2) = 95.9(2)°, with Cd-N bonds ranging 2.345(5)-2.371(5) Å. The dadpe linker has a twisted conformation with an interplanar angle of 73.44° between the phenyl rings, affording Cd···Cd separation of 13.6755(8) Å along the coordination chain running parallel to the b axis. Such a twisted conformation of dadpe makes it possible for the intrachain 2,2′-bpy/Ph to partially overlap with the proximity of the Cd atom. In this structure, in a mode similar to **3**, one of two crystallographically unique ClO_4^- anions chelates the neighboring NH_2-groups via a couple of NH(NH_2)··· O(ClO_4^-) H-bonds, while the second ClO_4^- anion mediates the neighboring chains along the crystallographic *a* axis via two other H-atoms of each amino group, thus combining the coordination chains in porous H-bonded layer parallel to the AB plane (Figure 5b). Similar to **3**, amino groups of bidentate-bridging coordinated dadpe ligands are completely involved in the H-bonds in the crystal. The neighboring coordination chains meet in the crystal by the 2,2′-bpy edges without overlap (Figure 5c).

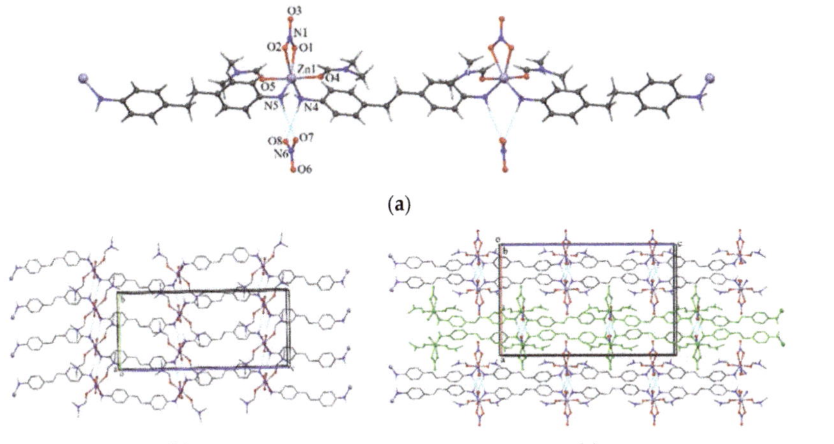

(**a**)

(**b**) (**c**)

Figure 4. Compound **3**. (**a**) Fragment of cationic polymeric chain with outer sphere nitrate anions attached by H-bonds. (**b**) Association of coordination chains in the H-bonded layer, view along the *a*-axis; (**c**) Stacking of the H-bonded layers in the crystal in the ABA manner along the *a* axis.

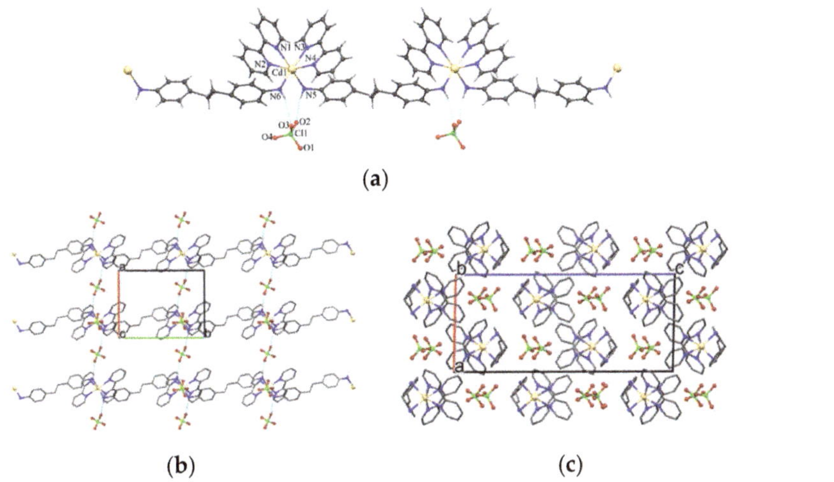

(**a**)

(**b**) (**c**)

Figure 5. Compound **4**. (**a**) Fragment of cationic polymeric chain with outer-sphere perchlorate-anions attached by H-bonds. (**b**) Association of coordination chains in the H-bonded layer, view along the *b*-axis; (**c**) Crystal packing.

3.2.3. 2D Coordination Polymers

Two isostructural compounds {[Cd(4,4′-bpy)$_2$(H_2O)$_2$](ClO_4)$_2$(dadpe)(EtOH)$_2$}$_n$ (**5**) and {[Co(4,4′-bpy)$_2$(H_2O)$_2$](BF_4)$_2$(dadpe)(EtOH)$_2$}$_n$ (**6**) crystallize in the monoclinic C2/c space group (Table 1) and comprise the positively charged square grid frameworks [M(4,4′-bpy)$_2$(H_2O)$_2$]$^{2+}$ [M=Cd (**5**), M=Co (**6**)] with **sql** topology, outer-sphere charge-compensated anions [ClO_4^- (**5**), BF_4^- (**6**)], and the neutral guests, dadpe and EtOH (Figure 6a). Two coordinated 4,4′-bpy ligands obey C_i and C_2 point group symmetries and have planar and twisted geometries. The rectangular meshes in the cationic grids have linear dimensions of 11.742 × 11.786 Å^2 in **5** and 11.443 × 11.514 Å^2 in **6**. The inspection of CSD [13] revealed only few examples with the same 2D grids exclusively for Cu(II) as a central metal [1,30,31]. In those seminal works, the fascinating properties of such networks were registered as dynamic anion exchange, possibility to accommodate tremendous aggregates such as polyoxometalates (POMs) in the voids of 2D networks. The reported polymers **5** and **6** also demonstrated the possibility of impressive loading of these crystals through incorporation of framework-regulated anions and bulky neutral molecules; dadpe and EtOH solvent were held in the interlayer space via an extensive H-bonding network including neutral and anionic guest species and water molecules of coordinated scaffold (Table S2). Similar to compound **1**, in compounds **5** and **6**, the dadpe guest molecules reside on inversion centers and take an ideal trans-conformation. The meaningful solvent accessible volumes (SAV) comprise 2151.7 Å^3 (49.1%) of 4380.3 Å^3 unit cell volume in **5** and 2063.2 Å^3 (49.4%) of 4179.6 Å^3 unit cell volume in **6**, as calculated by PLATON after evacuation of dadpe and EtOH molecules (Figure 6b).

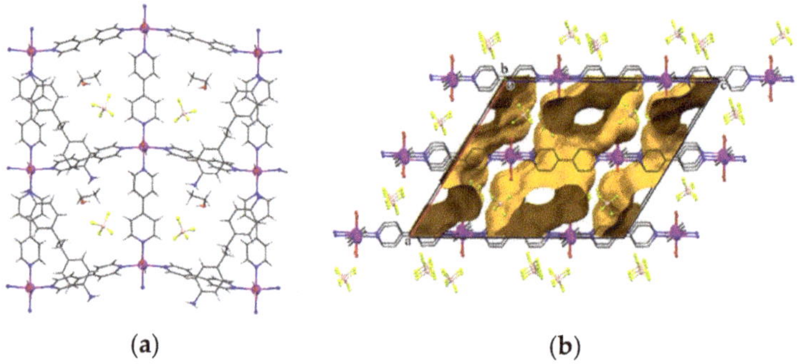

Figure 6. Compound **6**. (**a**) Fragment of cationic 2D grid with outer-sphere tetrafluoroborate anions and dadpe and EtOH guest molecules. (**b**) Crystal packing with voids shown in yellow.

Coordination polymer {[Cd(adi)(dadpe)](H$_2$adi)}$_n$ (**7**) crystallizes in the monoclinic *P*2/*c* space group and represents a 2D coordination network with **sql** topology (Figure 7a). Each Cd(II) atom takes a distorted octahedral N$_2$O$_4$ coordination geometry originating from adipate anions and neutral dadpe-bridging ligands. The same ligands are situated in trans-positions in Cd polyhedron and adi ligand coordinates in a bidentate-chelate bridging mode, while dadpe coordinates in a bidentate-bridging mode. Both coordinated ligands reside on conversion centers, and dadpe undergoes extended trans-conformation identical to that registered in **1**, **5** and **6**, while adi undergoes significantly twisted conformation with resolved disordering of the central methylene fragment. The rhombohedral meshes in the coordination layer have dimensions 10.175 × 15.102 Å^2, calculated as Cd···Cd separations across the adi and dadpe-bridging ligands, respectively. The coordination layers are situated parallel to the *ac* coordination plane (Figure 7b) and accommodate H$_2$adi molecule as a guest held in the crystal via NH···O and OH···O hydrogen bonds. Voids, occupied by H$_2$adi molecules, comprise 328.51 Å^3 or 24.7% of the unit cell volume (Figure 7c).

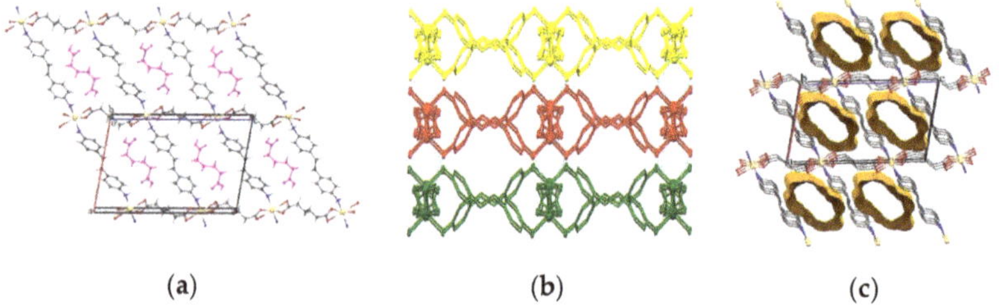

Figure 7. Compound **7**. (**a**) Fragment of 2D grid with H-bonded H$_2$adi shown in magenta color. (**b**) Stacking of coordination layers. (**c**) Crystal packing with voids shown in yellow.

3.3. *Hirshfeld Surface Modeling of Coordination Compounds* **1–3**, **7**

Various studies show that the way coordination compounds pack in the crystals is influenced by intermolecular interactions, and different methods are used to study their types. One of these methods is Hirshfeld surface analysis [32]. The size and shape of Hirshfeld surfaces (HS) help identify intermolecular interactions and classify molecular crystals in terms of packing similarities [33]. The CrystalExplorer program is used to generate Hirshfeld surfaces, and in our case, we used CrystalExplorer 17.5 [34].

We chose to perform the Hirshfeld surface analysis for four coordination compounds which differ in dimensionality, these being discrete mononuclear complexes **1** and **2**, 1D and 2D coordination polymers, **3** and **7**, with dadpe revealing different structural functions, as the neutral guest in **1**, the terminal monodentate ligand with one amino group free from metal coordination in **2**, and a bidentate-bridging ligand in **3** and **7**.

The main contributions to the total HS areas of the molecules **1–3**, **7** are depicted by the decomposed d_{norm} surfaces (Figure 8a–d) and by the full (Figure 9) and decomposed fingerprint plots (Figures S8–S11 in Supporting Information file). The eight principal types of interaction are H···H, H···O/O···H, H···C/C···H, H···N/N···H, C···O/O···C, C···N/N···C, C···C and O···O contact. The predominant interactions in all cases are H···H; they appear in the middle of the scattered points in the fingerprint maps, ranging from 34.4% in **1** to 53% in **2**. With significantly less contribution, the next most important interactions are H···O/O···H, which vary from 22.4% in **2** to 36.6% in **3**. The C···C contacts associated with π-π stacking interactions are present in three compounds **1**, **2** and **3**, but a more obvious contribution is observed in compound **1**, which on the other hand does not reveal any pronounced hydrogen-bonding interactions (Figure 10).

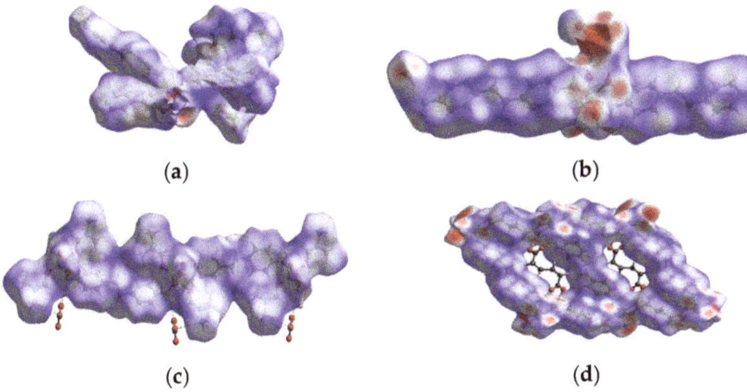

Figure 8. Hirshfeld surfaces mapped with d_{norm} ranging from −0.40 Å (red) to 2.40 Å (blue) for crystal structures **1** (**a**), **2** (**b**), **3** (**c**) and **7** (**d**).

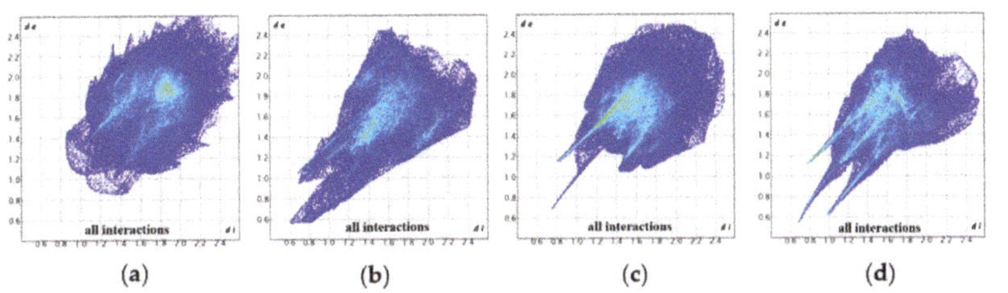

Figure 9. 2D fingerprint plots with di and d_e ranging from 1.0 to 2.4 Å for **1** (**a**), **2** (**b**), **3** (**c**) and **7** (**d**).

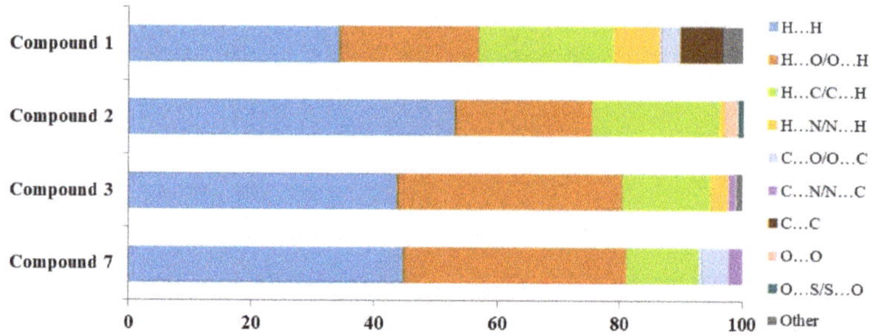

Figure 10. The relative contributions of different types of intermolecular interactions on the basis of Hirshfeld surface analysis of **1–3** and **7**.

4. Conclusions

For the first time, Zn/Ni/Co/Cd coordination compounds have been reported and obtained with participation of ditopic flexible ligand, 4,4′-diaminodiphenylethane (dadpe). The dadpe, as a neutral guest, was held in the crystals either via π-π stacking interactions or via NH···O hydrogen bonds; meanwhile, as a bidentate-bridging ligand, it gave rise to the 1D coordination polymers or 2D polymeric network used in combination with an adipato ligand. Both coordinated and free amino groups of dadpe participate in hydrogen-bonding systems, thus diversifying the supramolecular architectures. Hirshfeld surface analysis, as a useful tool, helped to evaluate the distribution of intermolecular interactions in the systems with different functions of dadpe ligand.

Supplementary Materials: The following supporting information can be downloaded at: https://www.mdpi.com/article/10.3390/cryst13020289/s1, Table S1: Bond lengths and angles for **1–7**, Table S2: Hydrogen bonds for **1–7**; Figures S1–S7: Infrared spectra for compounds **1–7** and starting ligands; Figures S8–S11: 2D fingerprint plots for distribution of intermolecular contacts in **1–3**, **7**.

Author Contributions: N.C.—Investigation, data curation, visualization; D.C.—Investigation, data curation, methodology; E.M.—Investigation, data curation, visualization, original manuscript writing; M.S.F.—Conceptualization, Validation, Writing—original draft preparation. All authors have read and agreed to the published version of the manuscript.

Funding: This research was funded by the projects of Agentia Natională pentru Cercetare si Dezvoltare (ANCD) 20.80009.5007.15 and 20.80009.5007.28.

Data Availability Statement: Any additional details are available from the authors.

Conflicts of Interest: The authors declare no conflict of interest.

References

1. Noro, S.; Kitaura, R.; Kondo, M.; Kitagawa, S.; Ishii, T.; Matsuzaka, H.; Yamashita, M. Framework engineering by anions and porous functionalities of Cu(II)/4,4′-bpy coordination polymers. *J. Am. Chem. Soc.* **2002**, *124*, 2568–2583. [CrossRef] [PubMed]
2. McManus, G.; Perry IV, J.J.; Perry, M.; Wagner, B.D.; Zaworotko, M.J. Exciplex Fluorescence as a Diagnostic Probe of Structure in Coordination Polymers of Zn^{2+} and 4,4′-Bipyridine Containing Intercalated Pyrene and Enclathrated Aromatic Solvent Guests. *J. Am. Chem. Soc.* **2007**, *129*, 9094–9101. [CrossRef] [PubMed]
3. Melnic, E.; Coropceanu, E.B.; Forni, A.; Cariati, E.; Kulikova, O.V.; Siminel, A.V.; Kravtsov, V.C.; Fonari, M. Discrete Complexes and One-Dimensional Coordination Polymers with $[Cu(II)(2,2'\text{-bpy})]^{2+}$ and $[Cu(II)(phen)]^{2+}$ Corner Fragments: Insight into Supramolecular Structure and Optical Properties. *Cryst Growth Des.* **2016**, *16*, 6275–6285. [CrossRef]
4. Croitor, L.; Coropceanu, E.B.; Chisca, D.; Baca, S.G.; van Leusen, J.; Kogerler, P.; Bourosh, P.; Kravtsov, V.C.; Grabco, D.; Pyrtsac, C.; et al. Effects of Anion and Bipyridyl Bridging Ligand Identity on the Co(II) Coordination Networks. *Cryst. Growth Des.* **2014**, *14*, 3015–3025. [CrossRef]
5. Du, M.; Li, C.-P.; Liu, C.-S.; Fang, S.-M. Design and Construction of Coordination Polymers with Mixed-Ligand Synthetic Strategy. *Coord. Chem. Rev.* **2013**, *257*, 1282–1305. [CrossRef]

6. Pal, A.; Chand, S.; Senthilkumar, S.; Neogi, S.; Das, M.C. Structural Variation of Transition Metal Coordination Polymers Based on Bent Carboxylate and Flexible Spacer Ligand: Polymorphism, Gas Adsorption and SC-SC Transmetallation. *CrystEngComm* **2016**, *18*, 4323–4335. [CrossRef]
7. Lozovan, V.; Kravtsov, V.C.; Coropceanu, E.B.; Siminel, N.; Kulikova, O.V.; Costriucova, N.V.; Fonari, M.S. Seven Zn(II) and Cd(II) 1D Coordination Polymers Based on Azine Donor Linkers and Decorated with 2-Thiophenecarboxylate: Syntheses, Structural Parallels, Hirshfeld Surface Analysis, and Spectroscopic and Inclusion Properties. *Polyhedron* **2020**, *188*, 114702. [CrossRef]
8. Czylkowska, A.; Pietrzak, A.; Szczesio, M.; Rogalewicz, B.; Wojciechowski, J. Crystal Structures, Hirshfeld Surfaces, and Thermal Study of Isostructural Polymeric Ladders of La(III) and Sm(III) Coordination Compounds with 4,4′-Bipyridine and Dibromoacetates. *Materials* **2020**, *13*, 4274. [CrossRef]
9. Carlucci, L.; Ciani, G.; Proserpio, D.H.; Porta, F. New metal–organic frameworks and supramolecular arrays assembled with the bent ditopic ligand 4,4-diaminodiphenylmethane. *CrystEngComm* **2006**, *8*, 696–706. [CrossRef]
10. Luo, J.; Hong, M.; Wang, R.; Cao, R.; Shi, Q.; Weng, J. Self-Assembly of Five Cadmium(II) Coordination Polymers from 4,4′-Diaminodiphenylmethane. *Eur. J. Inorg. Chem.* **2003**, *9*, 1778–1784. [CrossRef]
11. Chisca, D.; Croitor, L.; Melnic, E.; Petuhov, O.; Kulikova, O.; Fonari, M.S. Six transition metal–organic materials with the ditopic 4,4′-diaminodiphenylmethane ligand: Synthesis, structure characterization and luminescent properties. *Polyhedron* **2020**, *192*, 114844. [CrossRef]
12. Zhao, D.; Liu, X.-H.; Zhao, Y.; Wang, P.; Liu, Y.; Azam, M.; Al-Resayes, S.I.; Lu, Y.; Sun, W.Y. Luminescent Cd(II)–organic frameworks with chelating NH_2 sites for selective detection of Fe(III) and antibiotics. *J. Mat. Chem. A.* **2017**, *5*, 15797–15807. [CrossRef]
13. Groom, C.R.; Bruno, I.J.; Lightfoot, M.P.; Ward, S.C. The Cambridge Structural Database. *Acta Cryst. B* **2016**, *72*, 171–179. [CrossRef] [PubMed]
14. Kokunov, Y.V.; Kovalev, V.V.; Gorbunova, Y.E. Layered Structure of the Silver Coordination Polymer with Nonrigid Aromatic Diamine $\{Ag[CH_2(C_6H_4NH_2)_2]_2(CH_3C_6H_4NH_2)\}NO_3$. *Russ. J. Inorg. Chem.* **2007**, *52*, 1877–1882. [CrossRef]
15. Kokunov, Y.V.; Gorbunova, Y.E.; Kovalev, V.V. Synthesis and Structure of Silver Coordination Polymer with Extended Ditopic Ligand Containing Terminal Amino Groups $[Ag(C_{24}H_{28}N_2)_{1.5}]NO_3$. *Russ. J. Inorg. Chem.* **2012**, *57*, 953–958. [CrossRef]
16. Kokunov, Y.V.; Gorbunova, Y.E.; Kovalev, V.V.; Kozyukhin, S.A. Synthesis, Crystal Structure, and Luminescence Properties of the Tetranuclear Complex of Cadmium(II) Acetate with 4,4′(1,4-Phenylenediisopropylidene)bisaniline. *Russ. J. Inorg. Chem.* **2012**, *57*, 1553–1558. [CrossRef]
17. Ltaief, H.; Ben Ali, S.; Mahroug, A.; Ferretti, V.; Graça, M.P.F.; Belhouchet, M. A new copper hybrid compound based on 3,3′-diaminodiphenylsulfone as ligand: Growth, crystal structure, spectroscopic analysis, and thermal behavior. *J. Mol. Struct.* **2023**, *1273*, 134334. [CrossRef]
18. Ltaief, H.; Mahroug, A.; Paoli, P.; Rossi, P.; Belhouchet, M. A new hybrid compound based on mercury and 3,3′-diaminobiphenylsulfone studied by a combined experimental and theoretical approach. *J. Mol. Struct.* **2020**, *1220*, 128760. [CrossRef]
19. Smirnov, A.N.; Odintsova, O.V.; Starova, G.L.; Solovyeva, E.V. X-ray and vibrational analysis of amino and chloro bibenzyl 4,4′-derivatives supported by quantum chemical calculations. *J. Mol. Struct.* **2020**, *1202*, 127287. [CrossRef]
20. Giastas, P.; Yannakopoulou, K.; Mavridis, I.M. Molecular structures of the inclusion complexes b-cyclodextrin ± 1,2-bis(4-aminophenyl)ethane and b-cyclodextrin ± 4,4′-diaminobiphenyl; packing of dimeric b-cyclodextrin inclusion complexes. *Acta Cryst.* **2003**, *B59*, 287–299. [CrossRef]
21. Behera, N.; Duan, J.; Jin, W.; Kitagawa, S. The chemistry and applications of flexible porous coordination polymers. *EnergyChem* **2021**, *3*, 100067. [CrossRef]
22. Sheldrick, G.M. A short history of SHELX. *Acta Cryst.* **2008**, *A64*, 112–122. [CrossRef] [PubMed]
23. Sheldrick, G.M. Crystal structure refinement with SHELXL. *Acta Cryst.* **2015**, *C71*, 3–8. [CrossRef]
24. Macrae, C.F.; Bruno, I.J.; Chisholm, J.A.; Edgington, P.R.; McCabe, P.; Pidcock, E.; Rodriguez-Monge, L.; Taylor, R.; Streek, J.; Wood, P.A. Mercury CSD 2.0—New features for the visualization and investigation of crystal structures. *J. Appl. Cryst.* **2008**, *41*, 466–470. [CrossRef]
25. Spek, A.L. Structure validation in chemical crystallography. *Acta Cryst.* **2009**, *D65*, 148–155. [CrossRef]
26. David, A. Thornton, D.A. Metal Complexes of Aniline: Infrared and Raman Spectra. *J. Coord. Chem.* **1991**, *24*, 261–289. [CrossRef]
27. Park, H.M.; Hwang, I.H.; Bae, J.M.; Jo, Y.D.; Kim, C.; Kim, H.-Y.; Kim, Y.; Kim, S.-J. Anion Effects on Crystal Structures of Cd^{II} Complexes Containing 2,2′-Bipyridine: Photoluminescence and Catalytic Reactivity. *Bull. Korean Chem. Soc.* **2012**, *33*, 1517. [CrossRef]
28. Zhang, W.; Jiang, Z.; Lu, L. Tris(2,2′-bipyridine-kappa^{2}N,N′)cadmium(II) bis(perchlorate) hemihydrate. *Acta Crystallogr.* **2009**, *E65*, m7. [CrossRef]
29. Vaiana, L.; Platas-Iglesias, C.; Esteban-Gomez, D.; Avecilla, F.; de Blas, A.; Rodriguez-Blas, T. Receptor versus Counterion: Capability of *N,N*′-Bis(2-aminobenzyl)-diazacrowns for Giving Endo- and/or Exocyclic Coordination of Zn^{II}. *Eur. J. Inorg. Chem.* **2007**, *2007*, 1874–1883. [CrossRef]
30. Kong, X.-J.; Ren, Y.-P.; Zheng, P.-Q.; Long, Y.-X.; Long, L.-S.; Huang, R.-B.; Zheng, L.-S. Construction of Polyoxometalates-Based Coordination Polymers through Direct Incorporation between Polyoxometalates and the Voids in a 2D Network. *Inorg. Chem.* **2006**, *45*, 10702–10711. [CrossRef]

31. Lu, J.Y.; Fernandez, W.A.; Ge, Z.; Abboud, K.A. A novel two-fold interpenetrating 3D $4^2.8^4$ network self-assembled from a new 1D coordination polymer. *New J. Chem.* **2005**, *29*, 434–438. [CrossRef]
32. Hirshfeld, F.L. Bonded-atom fragments for describing molecular charge densities. *Theor. Chim. Acta.* **1977**, *44*, 129–138. [CrossRef]
33. Wolff, S.K.; Grimwood, D.J.; McKinnon, J.J.; Turner, M.J.; Jayatilaka, D.; Spackman, M.A. *Crystal Explorer*; University of Western Australia: Perth, Australia, 2012.
34. Spackman, M.A.; Jayatilaka, D. Hirshfeld Surface Analysis. *CrystEngComm* **2009**, *11*, 19–32. [CrossRef]

crystals

MDPI

Article

New Low-Dimensional Organic–Inorganic Lead Halide Hybrid Systems Directed by Imidazo[1,5-*a*]pyridinium-Based Cation or Imines: Synthesis, Structures, Non-Covalent Interactions and Optical Properties

Olga Yu. Vassilyeva [1,*], Elena A. Buvaylo [1], Oksana V. Nesterova [2], Alexandre N. Sobolev [3] and Dmytro S. Nesterov [2,*]

1 Department of Chemistry, Taras Shevchenko National University of Kyiv, 64/13 Volodymyrska Str., 01601 Kyiv, Ukraine
2 Centro de Química Estrutural, Institute of Molecular Sciences, Instituto Superior Técnico, Universidade de Lisboa, Av. Rovisco Pais, 1049-001 Lisboa, Portugal
3 School of Molecular Sciences, M310, University of Western Australia, Perth, WA 6009, Australia
* Correspondence: vassilyeva@univ.kiev.ua (O.Y.V.); dmytro.nesterov@tecnico.ulisboa.pt (D.S.N.)

Abstract: The organic–inorganic lead halide hybrids comprising semiconducting perovskite components and organic modules have proven to be promising candidates for optoelectronic applications. The modulation of the inorganic components as optical centres by diverse organic cationic templates is under intense investigation. Herein, we successfully prepared new one-dimensional lead halide hybrid perovskites $[L1]_{2n}[Pb_2Cl_6]_{n\infty}\cdot nH_2O$ (**1**) and $[PbBr_2(L2)]_{n\infty}\cdot 0.5nH_2O$ (**2**), and the dimeric complex $[PbBr_2(L3)]_2$ (**3**) in water media. In **1**, 2-(2-hydroxyethyl)-2H-imidazo[1,5-*a*]pyridinium cation $[L1]^+$ resulted from the oxidative condensation–cyclization between formaldehyde, ethanolamine and 2-pyridinecarbaldehyde (2-PCA); the polydentate Schiff base ligands L2 and L3 formed in the in situ condensation of 2-PCA and ethanolamine or ethylenediamine, respectively. The lead chloride hybrid **1** contains the previously unreported type of a $[Pb_2Cl_6]_\infty$ double chain constructed from three-edge- and five-edge-sharing $PbCl_6$ octahedra, and cations forming π-bonded stacks aligned along the inorganic wires. In the crystal of **2**, pairs of the double-side organically decorated $[PbBr_2(L2)]_\infty$ chains built of corner-sharing $PbBr_3N_2O$ octahedra arrange hydrophilic channels to host water molecules. In the solid state, the identically stacked dimers of **3** form columns parallel to the *ab* plane with the Pb_2Br_4 moieties in the column being strictly coplanar. Hirshfeld surface analysis was used to rationalize the packing patterns through hydrogen bonds of O−H···O/Cl and C−H···O/Cl types with the involvement of OH groups of $[L1]^+$, L2 and water molecules in **1** and **2**, as well as C–H···Br hydrogen bonding in **2** and **3**. The QTAIM analysis of non-covalent interactions in **1–3** was performed. According to the analysis of the solid-state UV–visible reflectance spectra by a Tauc plot, the optical band gap values of **1**, **2** and **3** as direct gap semiconductors were estimated to be 3.36, 3.13 and 2.96 eV, respectively.

Keywords: crystal structure; organic–inorganic hybrid; lead(II) perovskite; Schiff-base ligand; Hirshfeld surface analysis; QTAIM analysis; DFT calculations; reflectance spectra; band gap

Citation: Vassilyeva, O.Y.; Buvaylo, E.A.; Nesterova, O.V.; Sobolev, A.N.; Nesterov, D.S. New Low-Dimensional Organic–Inorganic Lead Halide Hybrid Systems Directed by Imidazo[1,5-*a*]pyridinium-Based Cation or Imines: Synthesis, Structures, Non-Covalent Interactions and Optical Properties. *Crystals* **2023**, *13*, 307. https://doi.org/10.3390/cryst13020307

Academic Editors: Helmut Cölfen and Leonid Kustov

Received: 28 January 2023
Revised: 6 February 2023
Accepted: 10 February 2023
Published: 13 February 2023

1. Introduction

The versatile optoelectronic properties of organic–inorganic hybrid metal halides make them an attractive alternative to photovoltaic devices utilizing a conventional crystalline silicon solar cell or emerging dye-sensitized solar cells, organic tandem cells and quantum dot cells [1]. In addition, due to their structural richness, semiconducting, electrical and optical properties, as well as processability in solution using low temperature techniques, perovskite-based hybrids are promising materials for use in other optoelectronic devices such as light-emitting diodes (LEDs), photodetectors, semiconductor optical amplifiers and lasers [2,3].

Considering the solid-state arrangement, lead halide hybrid perovskites are built of $PbHal_6$ octahedra that share corners, edges or faces to form structures with dimensionality varying from zero (0D) to three (3D), templated by organic cations. The dimensionality and geometry of the inorganic framework determine the optical and electrical properties of the material, while the organic cation can alter/tune its optoelectronic characteristics by modifying this framework. The broadband emission of lead halide hybrid perovskites was proposed to originate from the intrinsic excited electron–hole pairs bound to a deformable lattice (self-trapped excitons), rather than from emissive extrinsic dopants or surface defect sites [4]. There is growing evidence that 1D systems—which often combine several modes of connectivity within a single lead(II) halide chain, depending on the number of bridging halides—create the strong quantum confinement to enable easier exciton self-trapping [5,6].

Recently, we proposed an advantageous synthetic procedure for preparing organic–inorganic hybrid halometalate salts with substituted imidazo[1,5-*a*]pyridinium cations [7–12]. Imidazo[1,5-*a*]pyridines are fused nitrogen-containing bicyclic systems of interest in many research areas, e.g., material science and pharmaceuticals [13,14]. They also demonstrate enhanced fluorescence intensity combined with high quantum yield [15,16]. The organic cation formation occurs in the interaction of equimolar amounts of two aldehydes—formaldehyde (FA) and 2-pyridinecarbaldehyde (2-PCA)—with amine in aqueous media [8,11,12]. The oxidative cyclocondensation is catalyzed by acid that is conveniently introduced as the amine adduct; the cation prepared in situ does not require isolation. The reaction of the preformed heterocyclic cation synthesized using methylamine hydrochloride and metal halides yielded hybrid compounds $[L]_n[PbCl_3]_{n\infty}$, $[L]_2[ZnCl_4]$ [8] and $[L]_2[CdCl_4]$ [11], as well as mixed-halide analogues of the latter [12], where L^+ is 2-methyl-imidazo[1,5-*a*]pyridinium cation. The photophysical properties of the organic–inorganic 1D perovskite $[L]_n[PbCl_3]_{n\infty}$ and 0D pseudo-layered hybrid $[L]_2[ZnCl_4]$ were presumed to originate from the synergistic effects of the electronic structure of the cation and the solid-state architectures.

In the present work, the developed approach was extended to replace $CH_3NH_2{\cdot}HCl$ with ethanolamine (Ea) hydrochloride, as summarized in Scheme 1, to obtain lead halide hybrid compounds with another electron-deficient cation, 2-(2-hydroxyethyl)-2H-imidazo[1,5-*a*]pyridinium $[L1]^+$. Given that hydrogen bonds may impact important properties of the material due to their directionality and collective strength, Ea was chosen to examine the effect of introducing hydroxyl (–OH) functionality in the cation onto the resulting hybrid structure. The $[L1]^+$ cation templated the formation of a 1D chloroplumbate(II) wire in $[L1]_{2n}[Pb_2Cl_6]_{n\infty}{\cdot}nH_2O$ (**1**), the hybrid perovskite structure of which was confirmed by X-ray crystal structure analysis. Further replacement of chloride with bromide anion was realized by the addition of the necessary acid component, HBr, directly to the reaction media. For comparison reasons, the study was supplemented with an analogous synthesis involving the 'NH_2' alternative of Ea, ethylenediamine (En), to probe hydrogen bonding with amino group. In contrast, the isolated Pb(II) bromides appeared to bear Schiff-base ligands *N*-(2-pyridylmethylene)ethanolamine, L2, and *N,N′*-bis(2-pyridylmethylene)ethylenediamine, L3, produced in the amine-aldehyde condensation reactions between 2-PCA and Ea or En, respectively. Herein, we report the preparation, crystal structures, spectroscopic characterization and optical study of the new 1D hybrid lead halide perovskites $[L1]_{2n}[Pb_2Cl_6]_{n\infty}{\cdot}nH_2O$ (**1**) and $[PbBr_2(L2)]_{n\infty}{\cdot}0.5nH_2O$ (**2**), and the dimeric Schiff-base complex $[PbBr_2(L3)]_2$ (**3**). The Hirshfeld surface (HS) analysis was used to examine the packing patterns through non-covalent interactions. The quantum theory of atoms in molecules (QTAIM) was employed to perform the topological analysis of the electron density.

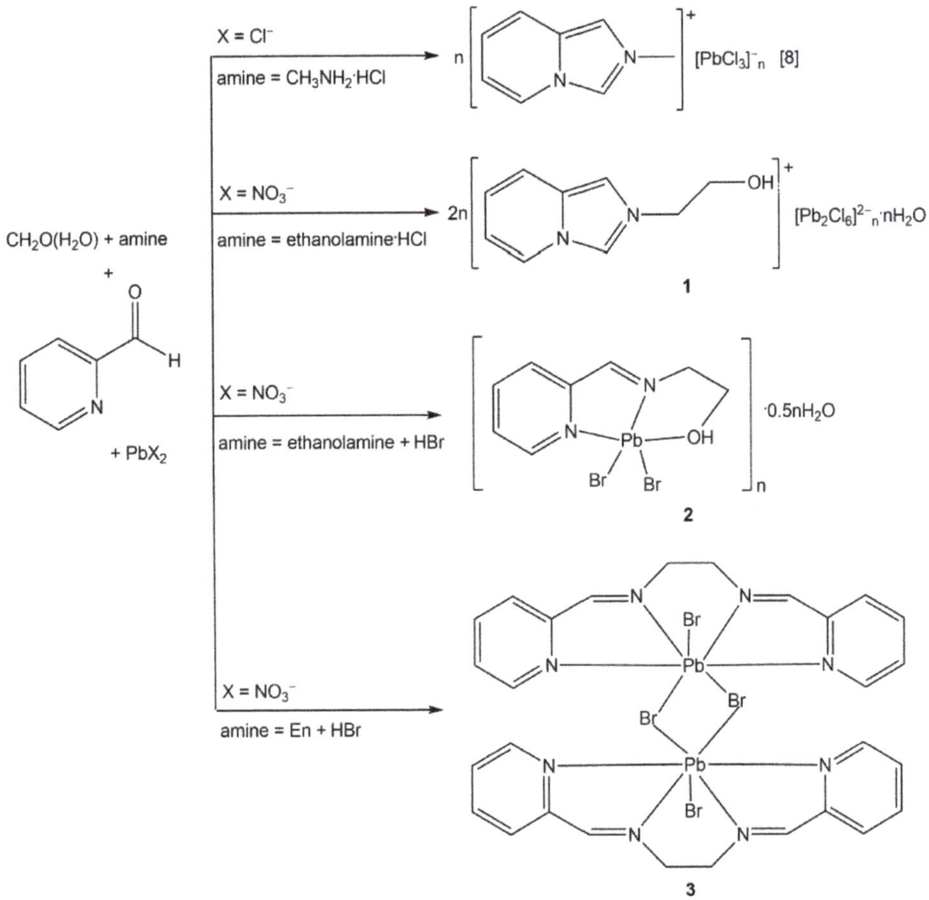

Scheme 1. Reaction schemes and structural formulas of $[L]_n[PbCl_3]_{n\infty}$ [8], $[L1]_{2n}[Pb_2Cl_6]_{n\infty}\cdot nH_2O$ (**1**), $[PbBr_2(L2)]_{n\infty}\cdot 0.5nH_2O$ (**2**) and $[PbBr_2(L3)]_2$ (**3**).

2. Experimental Section

2.1. Materials and General Methods

For the synthesis of lead compounds, 2-PCA (Merck) was used as received; all other chemicals were purchased from local suppliers and used without further purification. All solvents were of AP-grade; all the experiments were carried out in air. Elemental analyses for C, H and N were performed with a Perkin–Elmer 2400 analyzer. The ^{1}H NMR spectra of **1–3** in DMSO-d_6 were measured using a Mercury 400 Varian spectrometer at 400 MHz at r.t. The chemical shifts (δ) values are given in ppm downfield from internal Me_4Si. J values are in hertz. The FT–IR spectra were recorded on a PerkinElmer 1600 FT–IR instrument from KBr pellets in the 400–4000 cm^{-1} region. Optical diffuse reflectance measurements were performed using a Shimadzu UV-2600i spectrophotometer equipped with the 60 mm integrating sphere operating in the 220–1400 nm region. Ground powder samples were placed in a powder sample holder; $BaSO_4$ was used as the reference of 100% reflectance. The reflectance data were converted to absorption according to Kubelka–Munk function $\alpha/S = (1 - R)^2\,(2R)^{-1}$, where R represents the reflectance and α and S are the absorption and scattering coefficients, respectively, from which the band gap values were estimated by a Tauc plot [17,18]. HS and fingerprint plots were generated by *CrystalExplorer* 21.5 program (revision 608bb32) [19].

2.1.1. Synthesis of $[L1]_{2n}[Pb_2Cl_6]_{n\infty}\cdot nH_2O$ (**1**)

FA solution was prepared by dissolving paraform (0.13 g, 4.5 mmol) in 10 mL boiling deionized water in a 50 mL conic flask. After cooling at r.t., solid Ea·HCl (0.39 g, 4 mmol) was introduced into the FA solution, which was stirred vigorously at r.t. for half an hour. Then, the pale-yellow solution was filtered and left open overnight. The next day, 2-PCA (0.38 mL, 4 mmol) was added and the solution was kept stirring for 1 h while getting brownish. A subsequent dropwise addition of $Pb(NO_3)_2$ (0.33 g, 1 mmol) dissolved in DMF (5 mL) induced the precipitation of a small amount of a yellow residue. The mixture was kept stirring for another hour, then filtered and the brown solution was allowed to stand at r.t. Colourless needles of **1** suitable for X-ray crystallography were formed within a few days after addition of ^{i}PrOH (2 mL). The crystals were filtered off, washed with ^{i}PrOH and finally dried in air. Yield based on $Pb(NO_3)_2$: 47%. FT–IR (ν, cm^{-1}): 3400vs, 3116s, 3072, 3060, 2954, 2928, 2880, 1654, 1560, 1544, 1508, 1448, 1384, 1350, 1328, 1258, 1224, 1150s, 1132, 1076, 946, 922, 880, 796s, 750, 666, 636, 428. ^{1}H NMR (400 MHz, DMSO-d_6): δ (ppm) 10.01 (s, 1H, =N–CH=N–), 8.79 (d, J = 7.0 Hz, py), 8.32 (s, 1H, =C=CH−N–), 7.87 (d, J = 9.2 Hz, 1H, py), 7.26 (dd, J = 9.1, 6.7 Hz, 1H, py), 7.16 (dd, J = 9.9, 3.8 Hz, 1H, py), 5.23 (s, 1H, OH), 4.67 (m, 2H, CH_2), 3.92 (m, 2H, CH_2). Analysis calculated for $C_{18}H_{24}Cl_6N_4O_3Pb_2$ (971.49): C, 22.25; H 2.49; N 5.77%. Found: C 22.39; H 2.28; N 5.52%.

2.1.2. Synthesis of $[PbBr_2(L2)]_{n\infty}\cdot 0.5nH_2O$ (**2**)

Compound **2** was synthesized by adopting the procedure similar to that of **1** using Ea (0.24 mL, 4 mmol) and HBr (0.15 mL, 4 mmol) in place of solid Ea·HCl, and dry $Pb(NO_3)_2$ instead of its solution. Orange prisms of **2** suitable for X-ray crystallography were formed within a week after addition of ^{i}PrOH (3 mL) to the orange solution. Yield based on $Pb(NO_3)_2$: 31%. FT–IR (ν, cm^{-1}): 3392br, 3180br, 3086, 3062, 3016, 2930, 2900, 2872, 1654, 1590, 1568, 1440, 1314, 1224, 1154, 1108, 1048vs, 1006, 988, 912, 874, 782s, 634, 570, 510, 408. ^{1}H NMR (400 MHz, DMSO-d_6): δ (ppm) 8.76 (m, 1H, py), 8.50 (s, 1H, −N=CH−py), 8.00 (m, 1H, py), 7.98 (m, 1H, py), 7.62 (m, 1H, py), 4.40 (m, 1H, OH), 3.76 (m, 2H, CH_2), 3.71 (m, 2H, CH_2). Analysis calculated for $C_8H_{11}Br_2N_2O_{1.5}Pb$ (526.20): C, 18.26; H 2.11; N 5.32%. Found: C 18.45; H 2.15; N 5.47%.

2.1.3. Synthesis of $[PbBr_2(L3)]_2$ (**3**)

Compound **3** was synthesized by adopting the procedure similar to that of **2** using En (0.26 mL, 4 mmol) in place of Ea. Yellow plate-like crystals of **3** suitable for X-ray crystallography were formed within a week after addition of ^{i}PrOH (3 mL) to the orange solution. Yield based on $Pb(NO_3)_2$: 38%. FT–IR (ν, cm^{-1}): 3076, 3060, 3008, 2920, 2898, 1662, 1652, 1586s, 1564, 1474, 1434, 1374, 1306s, 1218, 1148, 1104, 1036, 1000s, 986, 942, 782, 748, 626, 506, 406. ^{1}H NMR (400 MHz, DMSO-d_6): δ (ppm) 8.99 (m, 2H, py), 8.86 (s, 2H, −N=CH−py), 8.01 (m, 4H, py), 7.60 (m, 2H, py), 4.17 (s, 4H, $=NCH_2CH_2N=$). Anal. Calcd for $C_{14}H_{14}Br_2N_4Pb$ (605.30): C, 27.78; H 2.33; N 9.26%. Found: C 28.04; H 2.45; N 9.38%.

*2.2. Single Crystal Structure Determination of **1–3***

Crystallographic data for the structures were collected on an Oxford Diffraction Gemini (**1**) and Bruker D8 Quest diffractometers (**2**, **3**) using Mo $K\alpha$ (λ = 0.71073 Å) radiation. Following analytical absorption corrections and solution by direct methods, the structures were refined against F^2 with full-matrix least-squares using the program SHELXL-2019/2 [20]. Anisotropic displacement parameters were employed for the non-hydrogen atoms. The (N22, N23A) cation in **1** was modelled as being disordered over two sets of sites with site occupancies constrained to 0.701(5) and its complement. The hydroxyl and water molecule hydrogen atoms (**1**, **2**) were located from the experimental data and refined with O−H distances restrained to their idealized values. Other hydrogen atoms were added at calculated positions and refined by the use of a riding model with isotropic displacement parameters based on those of the parent atom. Details of the data collection and processing, structure solution and refinement are summarized in Table 1, while the selected bond lengths and angles are presented in Table 2. CCDC 2217143, 2223148 and 2223147 contain the supplementary crystallographic data for this paper. These data can be obtained free of charge from the Cambridge Crystallographic Data Centre via www.ccdc.cam.ac.uk/data-request/cif.

Table 1. Crystallographic parameters and refinement data for **1**–**3**.

Compound	1	2	3
Empirical formula	$C_{18}H_{24}Cl_6N_4O_3Pb_2$	$C_{16}H_{22}Br_4N_4O_3Pb_2$	$C_{28}H_{28}Br_4N_8Pb_2$
Formula weight	971.49	1052.39	1210.60
Temperature/K	100(2)	296(2)	296(2)
Crystal system	Triclinic	Orthorhombic	Triclinic
Space group	$P\overline{1}$ (No. 2)	*Pccn* (No. 56)	$P\overline{1}$ (No. 2)
a/Å	7.5232(3)	9.5345(14)	8.9423(5)
b/Å	11.3164(5)	33.493(5)	9.3263(8)
c/Å	16.5513(6)	7.9626(12)	12.3991(7)
α/°	105.748(3)	90	82.812(3)
β/°	96.367(3)	90	70.156(2)
γ/°	90.885(3)	90	61.408(2)
V/Å^3	1346.32(10)	2542.7(6)	853.18(10)
Z	2	4	1
Calculated density D_c/g cm^{-3}	2.396	2.749	2.356
Absorption coefficient μ/mm^{-1}	13.112	19.534	14.569
F_{000}	900	1896	556
θ range for data collection/°	2.547 to 26.372	2.809 to 28.337	2.975 to 29.994
Completeness to θ_{max}	0.999	0.996	0.997
Reflections collected	11,257	77,380	44,646
Independent reflections	5508 [R_{int} = 0.0330]	3175 [R_{int} = 0.0690]	5257 [R_{int} = 0.0844]
Min. and max. transmission	0.260 and 1.0	0.3359 and 0.7457	0.3640 and 0.7461
Data/restraints/parameters	5508/237/341	3175/3/138	5257/0/190
Reflections ($I > 2\sigma(I)$)	4556	2783	4455
GoF	1.000	1.264	1.052
$R1$, w$R2$ [$I > 2\sigma(I)$]	0.0278, 0.0536	0.0469, 0.1013	0.0446, 0.1158
$R1$, w$R2$ (all data)	0.0382, 0.0581	0.0547, 0.1044	0.0550, 0.1245
Largest diff. peak and hole/e Å^{-3}	1.498 and −1.159	1.420 and −2.146	2.430 and −1.690
CCDC no.	2217143	2223148	2223147

2.3. Theoretical Calculations

The ORCA 5.0.3 package [21,22] was used for all DFT calculations. Range-separated ωB97M-V functional [23] with the ZORA-def2-TZVPP basis sets [24] and SARC/J auxiliary basis set [25,26] were used for all atoms except of Pb, for which the SARC-def2-TZVPP [25] basis set was applied. *AutoAux* keyword [27] was used to generate other auxiliary basis sets in all cases. The zero-order regular approximation (ZORA) [28] was used because of the presence of heavy elements. The crystal field was accounted for by means of the conductor-like polarizable continuum model (C-PCM) [29] with ε = infinity. The SCF optimization convergence criteria were settled with *VeryTightSCF* keywords, and integration grids of high density (*Defgrid3* keyword) were employed. Analysis of bond critical points and non-covalent interactions indexes was performed using the Multiwfn 3.8 program [30]. Visualization of the reduced density gradient [31] isosurfaces was performed by means of VMD 1.9.4a53 program [32].

3. Results and Discussion

3.1. Synthesis and Characterization

The mechanism of the formation of a substituted imidazo[1,5-*a*]pyridinium cation has been suggested by us before [8]. In the present study, the primary addition of Ea·HCl to FA aqueous solution yields 2-(methyleneamino)ethanol (Scheme 2). The subsequent 2-PCA attack on the protonated Schiff base A initiates condensation with the intermediate B. The intramolecular nucleophilic attack onto the aldehyde carbon atom by the amine nitrogen brings the product of cyclization C, which experiences an irrevocable dehydration reaction affording the aromatic heterocyclic cation $[L1]^+$. The dehydration of the five-membered ring results in the formation of two double bonds and the delocalization of the positive charge on N atoms in a resonance structure. The reaction does not require the presence of metal ions.

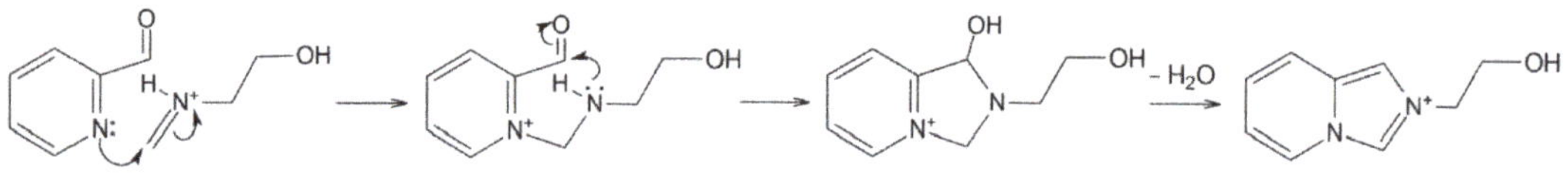

Scheme 2. Proposed mechanism of the formation of $[L1]^+$cation.

The lead chloride hybrid perovskite **1** formed by self-assembly from the in situ prepared monovalent organic cations, Pb^{2+} and chloride ions at an overall $1PbCl_2{:}4CH_2O{:}4Ea{\cdot}HCl{:}4(2\text{-}PCA)$ mole ratio. It is insoluble in alcohols and water and slightly soluble in *N*,*N*-dimethylformamide (DMF) and dimethylsulfoxide (DMSO); it is indefinitely stable in air. Similar reaction procedures that used Ea or En combined with HBr did not afford the respective imidazo[1,5-*a*]pyridinium cations, presumably due to the insufficient acid strength.

The IR spectra of **1**–**3** confirmed the presence of aromatic rings, alkyl and other functional groups (Figures S1–S3). The spectrum of the hybrid compound **1** demonstrates a distinctive pattern, which was suggested to characterize the imidazo[1,5-*a*]pyridinium skeleton [8,11,12]: the sharp very strong peaks attributed to aromatic C–H vibrations (3116–3060 cm^{-1}), medium intensity sharp bands at 1654 and 1544 cm^{-1} associated with heterocyclic ring stretching, an intense absorption at 1150 cm^{-1} and three peaks in the region of out-of-plane C–H bending (796, 750 and 636 cm^{-1}). The very intense band, due to ν(O–H) vibration at 3400 cm^{-1}, dominated the spectrum; another intense absorption was ascribed to C–O stretching (1076 cm^{-1}). In the case of compounds **2** and **3**, the aromatic ν(C–H) bands above 3000 cm^{-1} are much weaker. The spectrum of compound **2** is dominated by a peak at 1048 cm^{-1} due to ν(C–O), while the absorption of moderate intensity due to O–H stretching (3392 cm^{-1}) is broad. ν(C=N) + ν(C=C) stretching frequencies of the metal coordinated Schiff bases are observed at 1654, 1590 cm^{-1} (**2**) and 1652, 1586 cm^{-1} (**3**).

The compounds are easily distinguishable by their ^1H NMR spectra that demonstrate the sets of signals, as well as the correct alkyl/aromatic proton ratios expected for $[L1]^+$ cations and the Schiff base ligands. In the room temperature (r.t.) spectrum of compound **1** in DMSO-d_6, CH protons in the imidazolium ring appear as singlets at δ = 10.01 and 8.32 ppm, the alkyl groups protons are observed as multiplets at 4.67 and 3.92 ppm. The imine protons in **2** and **3** are detected at 8.50 and 8.86 ppm, respectively, while the protons of CH_2 groups appear as two multiplets at 3.76 and 3.71 (**2**) and one multiplet at 4.17 ppm (**3**). The hydroxyl protons in **1** and **2** are observed distinctly at 5.23 and 4.40 ppm, respectively.

3.2. *Structural Description of* **1–3**

The asymmetric unit of the lead chloride hybrid **1** is comprised of two $[L1]^+$ cations, $[Pb_2Cl_6]^{2-}$ anionic fragment and water molecule of crystallization (Figure 1). Two crystallographically non-equivalent cations configurations are very similar; the (N22, N23A) cation is disordered over two sites, with occupancies of 0.701(5) and 0.299(5)-. The pyridinium rings bond distances of their fused heterocyclic cores are uneventful; the N–C bond lengths in the imidazolium entities fall in the range 1.331(7)–1.414(6) Å. The N atoms are planar, showing a sum of three angles of 360°. In the cores, the five- and six-membered rings are nearly coplanar, with the respective dihedral angles of 1.35, 1.73 and 2.98°.

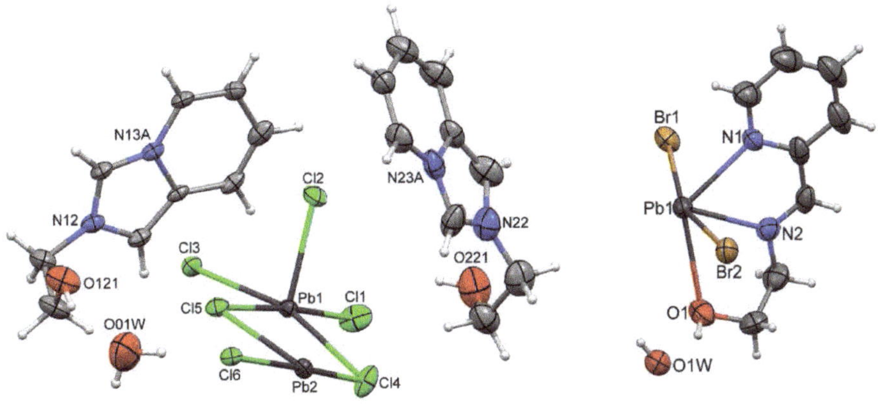

1 2

Figure 1. Asymmetric units and principal atom labelling for $[L1]_{2n}[Pb_2Cl_6]_{n\infty}{\cdot}nH_2O$ (**1**) and $[PbBr_2(L2)]_{n\infty}{\cdot}0.5nH_2O$ (**2**) with the 50% probability ellipsoids. The major component of the disordered (N22, N23A) cation of **1** is shown.

The distorted octahedral environment of Pb1 atom in the anionic motif consists of two terminal, two μ_2-Cl and two μ_3-Cl atoms, while the distorted octahedron around Pb2 is built of two μ_2-Cl and four μ_3-Cl ligands (Figure 2a). The metal–chloride distances vary in the range 2.6509(14)–3.1516(14) Å. The average value of 2.907 Å is reasonable for Pb(II) ions in chloroplumbates(II) with strong ionic Pb–Cl bonds [33,34]; the *cis* and *trans* angles at the metal atom fall in the ranges of 77.50(4)–107.33(5)° and 162.86(5)–177.20(4)°, respectively (Table 2), showing a strong degree of distortion of the metal polyhedra. The $PbCl_6$ octahedra are joined by three-edge- and five-edge-sharing into a 1D twin lead(II) chloride wire of 10.344 Å width, lying in the *a*-axis direction (Figure 2a,b). The bulkier 2-(2-hydroxyethyl)-2H-imidazo[1,5-*a*]pyridinium cation in **1**, if compared to its methyl analogue [8], expectedly templated the formation of a low-dimensional anionic perovskite frame, however, with a different type of a double chain. This set of connectivity modes within a double chain was not observed in the lead(II) chloride hybrid compounds in accordance with our survey of the Cambridge Structural Database [35]. The type of edge-sharing within a double haloplumbate(II) chain shown in the inset of Figure 2b is more frequent and found in, for example, 1D lead chloride perovskites with substituted diammonium and piperidinium dications [6,36].

Table 2. Selected bond lengths (Å) and angles (°) in **1**–**3** [a].

$[L1]_{2n}[Pb_2Cl_6]_{n\infty}{\cdot}nH_2O$ (1)			
Pb1–Cl1	2.7587(16)	Pb2–Cl4	2.7924(15)
Pb1–Cl2	2.6509(14)	Pb2–Cl3 [1]	2.8177(14)
Pb1–Cl3	2.8223(13)	Pb2–Cl6 [2]	2.8730(13)
Pb1–Cl4	3.0750(14)	Pb2–Cl5	2.9355(13)
Pb1–Cl5	3.0472(14)	Pb2–Cl6	2.9613(14)
Pb1–Cl6 [3]	3.1516(14)	Pb2–Cl5 [3]	2.9957(14)
Cl2–Pb1–Cl1	93.17(5)	Cl4–Pb2–Cl3 [1]	94.36(5)
Cl2–Pb1–Cl3	86.32(4)	Cl4–Pb2–Cl6 [2]	96.36(4)
Cl1–Pb1–Cl3	94.13(5)	Cl3 [1]–Pb2–Cl6 [3]	82.39(4)
Cl2–Pb1–Cl5	90.30(4)	Cl4–Pb2–Cl5	87.03(4)
Cl1–Pb1–Cl5	175.07(4)	Cl3 [1]–Pb2–Cl5	94.79(4)
Cl3–Pb1–Cl5	89.58 (4)	Cl6 [2]–Pb2–Cl5	175.73(3)
Cl2–Pb1–Cl4	107.33(5)	Cl4–Pb2–Cl6	177.20(4)

Table 2. *Cont.*

Cl1−Pb1−Cl4	95.37(4)	Cl3 [1]−Pb2−Cl6	87.68(4)
Cl3−Pb1−Cl4	162.86(5)	Cl6 [2]−Pb2−Cl6	85.80(4)
Cl5−Pb1−Cl4	80.25(4)	Cl5−Pb2−Cl6	90.89(4)
Cl2−Pb1−Cl6 [3]	161.97(4)	Cl4−Pb2−Cl5 [3]	92.01(5)
Cl1−Pb1−Cl6 [3]	95.85(5)	Cl3 [1]−Pb2−Cl5 [3]	173.17(4)
Cl3−Pb1−Cl6 [3]	77.50(4)	Cl6 [2]−Pb2−Cl5 [3]	99.36(4)
Cl5−Pb1−Cl6 [3]	81.77(4)	Cl5−Pb2−Cl5 [3]	83.07(4)
Cl4−Pb1−Cl6 [3]	87.35(4)	Cl6−Pb2−Cl5 [3]	85.88(4)
$[PbBr_2(L2)]_{n\infty}\cdot 0.5nH_2O$ (2)			
Pb1−N1	2.584(7)	Pb1−Br2	2.9777(10)
Pb1−N2	2.562(7)	Pb1−Br1 [4]	3.0764(10)
Pb1−O1	2.839(7)	Pb1−Br1	3.0293(11)
N2−Pb1−N1	64.4(3)	O1−Pb1−Br1 [4]	146.36(15)
N2−Pb1−O1	61.8(2)	Br2−Pb1−Br1 [4]	89.35(3)
N1−Pb1−O1	123.3(2)	N2−Pb1−Br1	84.41(17)
N2−Pb1−Br2	85.72(18)	N1−Pb1−Br1	80.21(16)
N1−Pb1−Br2	80.57(16)	O1−Pb1−Br1	110.23(16)
O1−Pb1−Br2	79.24(16)	Br2−Pb1−Br1	160.70(3)
N2−Pb1−Br1 [4]	149.39(18)	Br1−Pb1−Br1 [4]	90.73(3)
N1−Pb1−Br1 [4]	85.02(18)		
$[PbBr_2(L3)]_2$ (3)			
Pb1−N1	2.745(6)	Pb1−Br1	3.0632(9)
Pb1−N2	2.625(5)	Pb1−Br2	2.9599(9)
Pb1−N3	2.621(6)	Pb1−Br1 [3]	3.2525(8)
Pb1−N4	2.808(6)		
N2−Pb1−N1	61.15(17)	N2−Pb1−Br1	84.71(12)
N3−Pb1−N1	123.10(18)	N1−Pb1−Br1	93.02(11)
N3−Pb1−N2	63.60(19)	N4−Pb1−Br1	85.58(12)
N3−Pb1−N4	60.4(2)	Br2−Pb1−Br1	174.31(2)
N2−Pb1−N4	121.93(19)	N3−Pb1−Br1 [3]	149.42(14)
N1−Pb1−N4	176.39(16)	N2−Pb1−Br1 [3]	145.63(13)
N3−Pb1−Br2	81.33(13)	N1−Pb1−Br1 [3]	87.48(13)
N2−Pb1−Br2	89.91(12)	N4−Pb1−Br1 [3]	89.05(15)
N1−Pb1−Br2	85.97(12)	Br2−Pb1−Br1 [3]	102.31(3)
N4−Pb1−Br2	95.75(12)	Br1−Pb1−Br1 [3]	83.22(2)
N3−Pb1−Br1	94.60(13)		

[a] Symmetry codes: [1] $x-1, y, z$; [2] $-x, -y+1, -z+1$; [3] $-x+1, -y+1, -z+1$; [4] $1/2-x, y, 1/2+z$.

In the crystal, non-equivalent $[L1]^+$ cations of **1** form separate stacks parallel the *a*-axis with the fused heterocyclic cores of neighboring entities being strictly coplanar (Figure 2c). Centrosymmetically related trans-oriented $[L1]^+$ cations are stacked with varying levels of offset, showing the ring centroid distances of 3.456, 3.651 Å for (N12, N13A) and 3.839, 5.258 Å for (N22, N23A) cations enabling the possibility of weak π-bonding [37]. Hydrogen bonds of O−H···O/Cl and C−H···O/Cl types with the involvement of the OH group of the organic cation and water molecule (Table 3) strengthen the hybrid salt structure. The introduction of OH group on the cation enabled a stronger interaction between cationic and anionic counterparts in **1** through additional conventional hydrogen bonding, which is absent in the lead chloride hybrid with methyl derivative of $[L1]^+$ cation [8]. The new type of a double chain realized in **1** may also be considered a result of such a modification of the organic cation.

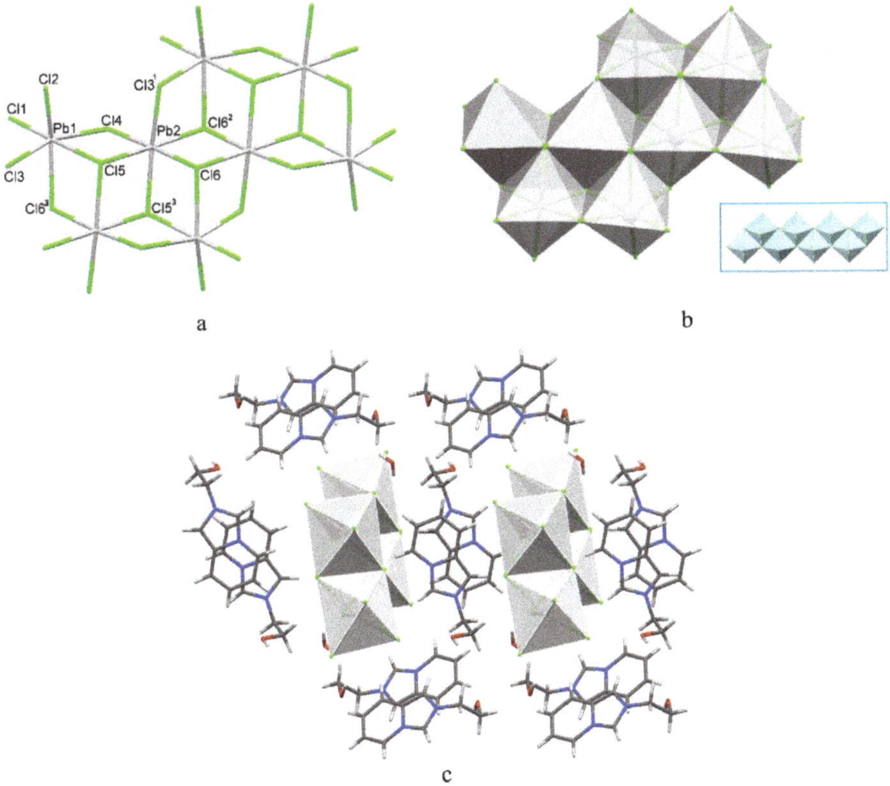

Figure 2. (**a**) Fragment of the 1D twin offset chloroplumbate(II) chain in **1**. The symmetry codes have the same numbering as those in Table 2; (**b**) polyhedral representation of the same fragment; the inset shows a common type of edge-sharing within a double haloplumbate(II)wire; (**c**) fragment of the crystal packing of **1** demonstrating spatial arrangement of the organic and inorganic counterparts (the major component of the disordered (N22, N23A) cation is shown).

Table 3. Geometry of hydrogen bonds for **1**–**3** (Å and °) [a].

D–H···A	d(D–H)	d(H···A)	d(D···A)	∠(DHA)
$[L1]_{2n}[Pb_2Cl_6]_{n\infty}\cdot nH_2O$ (1) [b]				
C11–H11···Cl5	0.95	2.90	3.627(5)	134
C12–H12B···Cl2 [1]	0.99	2.96	3.862(6)	152
C13–H13···Cl2 [1]	0.95	2.71	3.558(6)	148
C14–H14···Cl6 [2]	0.95	2.86	3.506(5)	126
C15–H15···Cl6 [2]	0.95	2.86	3.509(5)	127
C17–H17···Cl3 [3]	0.95	2.96	3.529(5)	120
C17–H17···Cl5	0.95	2.88	3.649(5)	139
C22–H22B···Cl1 [3]	0.99	2.81	3.705(11)	150
C23–H23···Cl1 [3]	0.95	2.84	3.481(10)	126
C24–H24···Cl3 [3]	0.95	2.75	3.597(12)	149
C25–H25···O01W [2]	0.95	2.48	3.325(12)	148
C27–H27···Cl2 [4]	0.95	2.85	3.716(14)	153
C221–H22C···O221 [5]	0.99	2.39	3.105(13)	128
C221–H22D···Cl4	0.99	2.80	3.753(12)	161
O01W–H01A···Cl1 [6]	0.85	2.34	3.179(7)	172

Table 3. *Cont.*

O01W–H01B···Cl4 [7]	0.85	2.37	3.168(6)	155
O121–H121···O01W	0.84	1.94	2.710(8)	151
O221–H221···Cl1	0.84	2.24	2.967(8)	145
$[PbBr_2(L2)]_{n\infty}\cdot 0.5nH_2O$ (2)				
O1–H1···O1W	0.84(2)	2.06(7)	2.829(10)	152(13)
O1W–H1W···O1 [8]	0.84(2)	1.92(3)	2.740(9)	164(4)
C1–H1A···Br1 [9]	0.93	2.90	3.657(13)	139.3
C2–H2A···Br1 [10]	0.93	2.94	3.845(13)	164.3
C4–H4A···Br1 [11]	0.93	2.90	3.631(11)	136.3
C6–H6A···Br2 [12]	0.93	2.89	3.819(9)	172.8
C8–H8A···Br2 [13]	0.97	3.14	3.881(10)	134.4
$[PbBr_2(L3)]_2$ (3)				
C2–H2A···Br1 [14]	0.93	2.88	3.690(8)	147.0
C4–H4B···Br2 [3]	0.93	3.09	3.890(9)	145.2
C6–H6A···Br2 [5]	0.93	3.15	3.977(7)	149.7
C7–H7A···Br2 [5]	0.97	3.16	4.095(8)	161.4
C11–H11A···Br2 [15]	0.93	3.07	3.869(10)	145.1
C13–H13A···Br1 [13]	0.93	2.97	3.684(9)	134.9

[a] Symmetry codes: [1] $-x+1, -y, -z+1$; [2] $-x, -y, -z+1$; [3] $x-1, y, z$; [4] $-x+1, -y, -z$; [5] $-x+1, -y+1, -z$; [6] $-x+1, -y+1, -z+1$; [7] $-x, -y+1, -z+1$; [8] $-x+1/2, y, z+1/2$; [9] $-x+1/2, y, z+1/2$; [10] $-x, -y+1, -z$; [11] $-x-1/2, y, z+1/2$; [12] $-x-1/2, y, z-1/2$; [13] $x, -y+3/2, z-1/2$; [14] $x-1, y+1, z$; [15] $x+1, y-1, z$; [16] $x+1, y, z$. [b] H bonds for the major component of the disordered (N22, N23A) cation are given.

Lead halide hybrid **2**, which crystallizes in the orthorhombic space group *Pccn*, consists of a 1D monobromo-bridged lead(II) polymer and water molecules of crystallization. The asymmetric unit of **2** is shown in Figure 1. In the double-sided, organically decorated chain structure, the nearest six-coordinate environment of each metal atom is formed by one tridentate Schiff base, one terminal and two μ_2-Br ligands (Figure 3a,b). The neutral Schiff base ligand chelates the Pb1 ion by the N1 atom of the pyridyl ring, the imine N2 atom and O1 atom of the ethanol group with an average value of the Pb1−N distances of 2.57 Å, and the Pb−O1 bond [2.839(7) Å] being significantly elongated (Table 2). The metal distances to terminal and bridging bromide atoms are similar, with an average of 3.03 Å falling in the usual range for 1D lead(II) bromide systems [38,39]. The additional Pb···Br2{$1/2-x$, y, $-1/2+z$} contact of 3.378 Å is appreciably larger than the sum of the Shannon ionic radii of the octahedral lead(II) cation and bromide anion [$r(Pb^{2+}) + r(Br^-) = 3.15$ Å] and is barely significant in the construction of the bromoplumbate(II) chain. The bond angles at the Pb1 atom vary from 61.8(2) to 123.3(2)° and from 146.36(15) to 160.70(3)° (Table 2), evidencing the severe distortion of the metal polyhedron. The closest Pb···Pb separation within the chain is 4.4144(6) Å. In the crystal, two bromoplumbate(II) chains related by a crystallographic two-fold axis form a hydrophilic channel to host water molecules held by O-H···O hydrogen bonds (Figure 3c, Table 3). Numerous C-H···Br contacts result in a 3D supramolecular architecture.

Complex **3** belongs to the triclinic space group $P\bar{1}$; the dimeric molecule is located on an inversion centre (Figure 4). The geometry of Pb1 atom is that of an irregular pentagonal bipyramid, the equatorial plane consisting of four nitrogens from the tetradentate chelate L3 and the centrosymmetrically related bromide ligand Br1{$1-x$, $1-y$, $1-z$} of the dimer, with two other bromides—Br1 and Br2—occupying apical positions. The Pb–N bond distances within the range of 2.621(6)–2.804(6) Å (Table 2) agree well with those of two other known Pb(II) complexes with L3 [40,41]. The Pb1–Br1{$1-x, 1-y, 1-z$} bond is elongated at 3.2525(8) Å, in contrast to the average Pb1–Br1/Br2 bond length of 3.014 Å. The *cis* angles at the lead atom vary from 60.4(2) to 95.75(12)°; the *trans* angles fall in the range 121.93(19)−176.39(16)°. In the dimer, the Pb···Pb′ separation is 4.7239(5) Å.

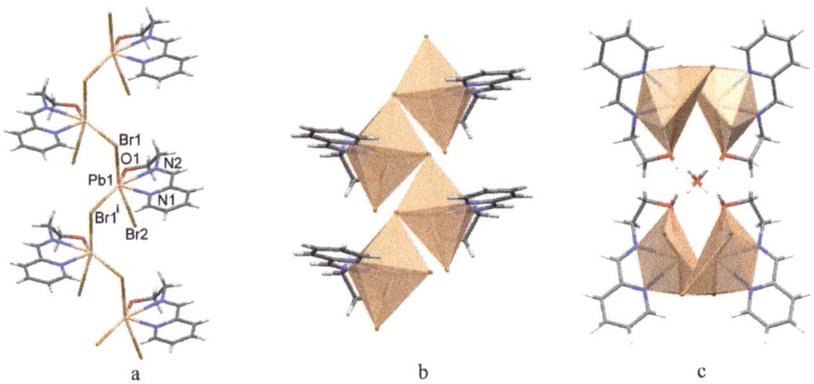

Figure 3. (**a**) Fragment of the double-side organically decorated bromoplumbate(II) chain in **2** [symmetry code: ^{i} $1/2 - x, y, 1/2 + z$]; (**b**) polyhedral representation of the same fragment; (**c**) fragment of the crystal packing viewed down the *c*-axis showing spatial arrangement of two bromoplumbate(II) chains to form a hydrophilic channel filled with water molecules.

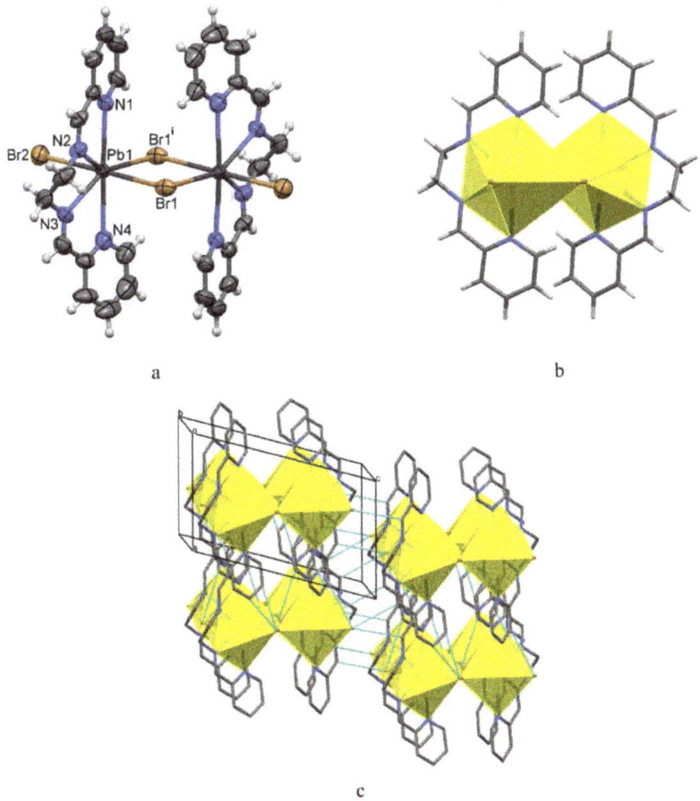

Figure 4. (**a**) Molecular structure and principal atom labelling of $[PbBr_2(L3)]_2$ (**3**), ellipsoids are shown at the 50% probability level [symmetry code: ^{i} $1 - x, 1 - y, 1 - z$]; (**b**) the same molecule in a polyhedral form; (**c**) fragment of the 3D supramolecular network in **3** stabilized with C–H···Br hydrogen bonds (shown in blue, H atoms omitted).

In the solid state, the dimeric molecules of **3** are stacked identically, forming columns parallel to the *ab* plane, with the Pb_2Br_4 moieties in the column being strictly coplanar (Figure 4c). The minimal Pb···Pb distance in the lattice is 8.1348(6) Å; no effective π-overlap is observed. C–H···Br hydrogen bonding (Table 3) consolidates an extended supramolecular 3D network structure (Figure 4c). The contacts strength depends on their direction in the crystal lattice. In the *ab* plane, the interactions are represented by four distinct crystallographically independent C–H···Br hydrogen bonds with *d*(C···Br) < 3.9 Å (Table 3). Every dimeric molecule of **3** forms 16 respective contacts with the four nearest molecules in the *ab* plane. The dimeric molecules along the *c* axis are bridged by weaker C–H···Br hydrogen bonds with *d*(C···Br) > 3.9 Å (Table 3).

3.3. Hirshfeld Surface and QTAIM Analyses

The set of intermolecular interactions in complex **1** can be viewed as a combination of contacts between crystallographically independent (N12, N13A) and (N22, N23A) $[L1]^+$ cations, negatively charged $[Pb_2Cl_6]_n^{2n-}$ chains and water molecules. The HS of both $[L1]^+$ cations surrounding a fragment of the $[Pb_2Cl_6]_n^{2n-}$ chain is shown in Figure 5. Here, and further on, the colours on the surface correspond to shortened contacts (red), van der Waals (vdW) interactions (white) and contacts with longer distances (blue).

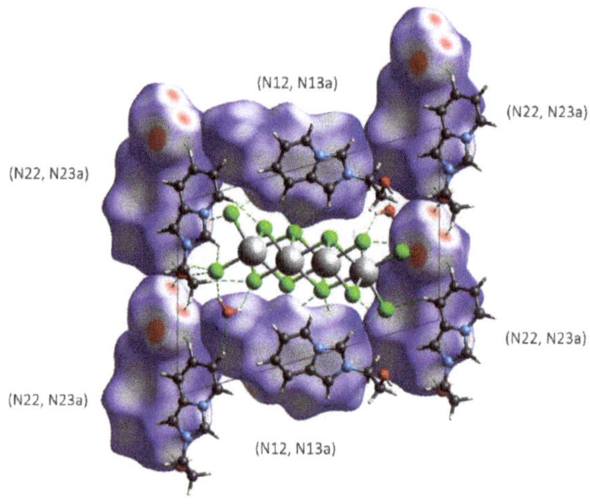

Figure 5. Hirshfeld surface mapped over d_{norm} of (N12, N13a) and (N22, N23A) $[L1]^+$ cations for compound **1** (view along the *a* axis, the major component of the disordered (N22, N23A) cation is shown).

Both $[L1]^+$ cations reveal involvement in several distinct hydrogen bonds (Table 3), which are identifiable in the fingerprint plots (Figure 6). The plots are characterized by broad segments of H···H interactions (Figure 6), while the C···C ones are much less abundant (5.4 and 4.3% for (N12, N13A) and (N22, N23A) independent cations, respectively, Figure S4).

The weak interactions between the stacked cations were evaluated by the analysis of reduced density gradient (RDG) [31] of the electron density calculated at the ωB97M-V/ZORA-def2-TZVPP level using crystallographic atomic coordinates (Figure 7). In line with the HS analysis results, the non-covalent contacts in the tetramer $\{(N12, N13A)\text{-}[L1]_4\cdot 2H_2O\}^{4+}$ are mostly characterized by the broad vdW areas between stacked imidazo[1,5-*a*]pyridine moieties. These findings indicate that $[L1]^+$ cations in their stacks are too offset to engage in appreciable π-bonding. At the same time, the chain of (N12, N13A)-$[L1]^+$ cations is supported by the strong attractive interactions involving water molecules (O121–H121···O01W), for which the electron density at the bond critical point—$\rho_{BCP}(\mathbf{r})$ (3, −1) between the H121 and O01W atoms—is 0.02504 a.u. According to the dependence for charged assemblies reported in [42], the obtained value of $\rho_{BCP}(\mathbf{r})$ corresponds to 39.3 kJ mol^{-1} of binding energy (BE) between the (N12, N13A)-[L1]$^+$ cation and water molecule. However, as the BE vs. $\rho_{BCP}(\mathbf{r})$ dependences [42] were developed for assemblies with a single overall charge (in contrast to 4+ charge of the tetramer), the obtained binding energy value has only indicative significance.

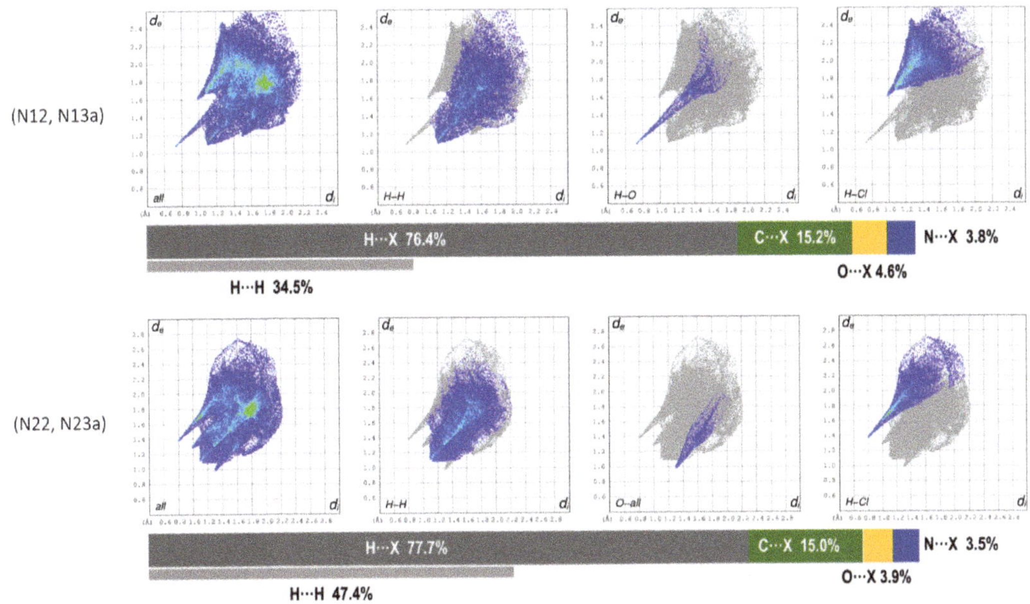

Figure 6. Fingerprint plots of (N12, N13A) and (N22, N23A, major component) [L1]$^+$ cations in the crystal structure of **1** showing the specific $X_i \cdots X_e$ interactions (where X = Cl, O or H) and their contributions (in %).

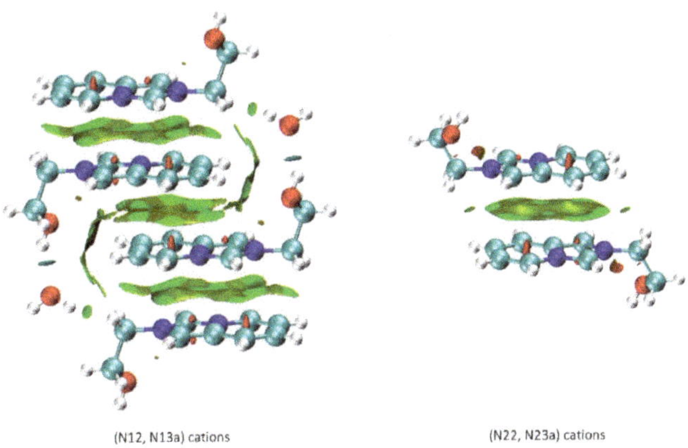

Figure 7. Reduced density gradient surfaces (isovalue of 0.5) showing vdW interactions (green area) between [L1]$^+$ cations and strong hydrogen bonds with participation of water molecules in the crystal structure of **1** (the major component of the disordered (N22, N23A) cation is shown). The colour scheme corresponds to the sign(λ_2)$\rho(\mathbf{r})$ function with the second largest eigenvalue of the Hessian of electron density, λ_2, at respective points: negative (blue, strong attraction), nearly zero (green, vdW interaction), positive (red, strong repulsion).

The monomeric fragment $[PbBr_2(L2)]$ of the 1D chain in complex **2** was used to construct the Hirshfeld surface (Figure 8). The fingerprint plots disclose several distinct non-covalent contacts mostly H···O and H···Br in nature (Figures 8 and S5). The strongest hydrogen bonds between O1 and O1W atoms are reflected by sharp "peaks" having the shortest d_e and d_i distances in the fingerprint plots (Figures 8 and S5). The $\rho_{BCP}(\mathbf{r})$ electron density at BCP between H1 and O1W atoms was estimated as 0.01986 a.u., from which the binding energy of the respective contact can be evaluated as –15.4 kJ mol^{-1} (according to the equation for non-charged assemblies [42]). This energy strength can be classified as "weak-to-medium" [42]. The RGD isosurfaces of the vdW contacts between the aromatic groups in **2** are shown in Figure 9.

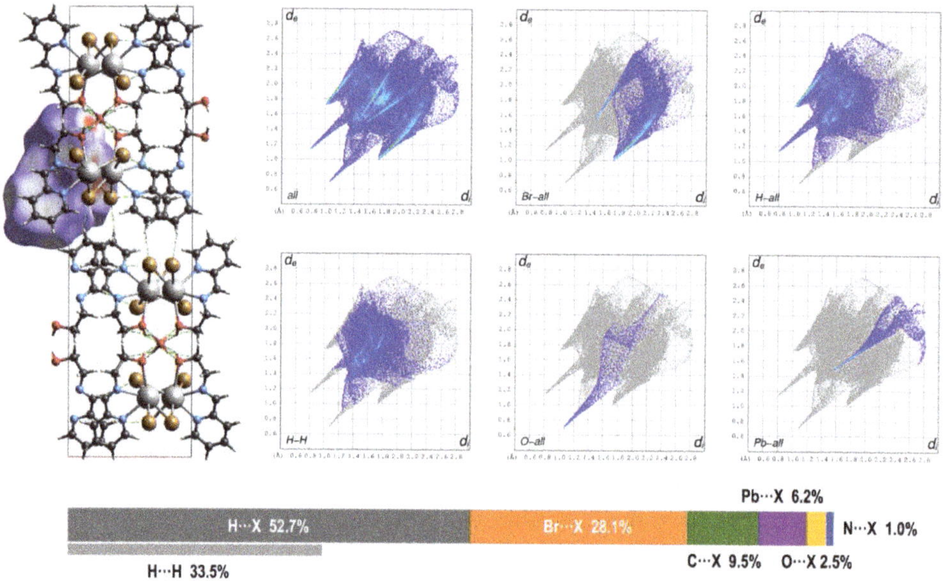

Figure 8. Top left: fragment of the crystal structure of **2** viewed down the *c* axis showing the HS of the fragment $[PbBr_2(L2)]$. Top right: selected fingerprint plots. Bottom: contributions of the specific $X_i \cdots X_e$ contacts.

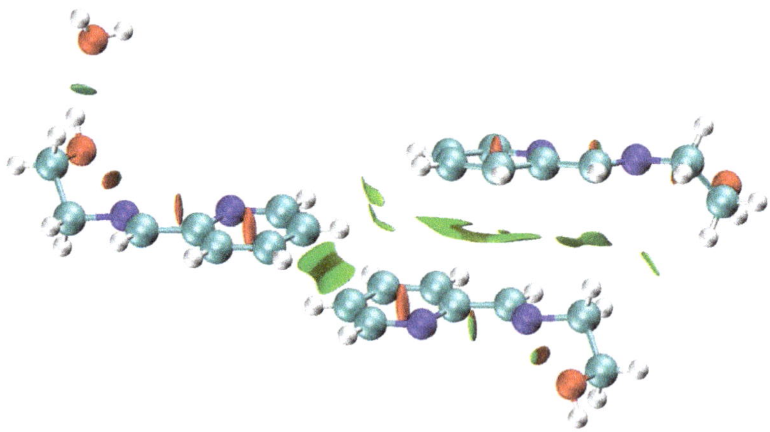

Figure 9. Reduced density gradient surfaces (isovalue of 0.5) showing vdW interactions (green area) between aromatic fragments and strong hydrogen bonds involving water molecule for compound **2**.

The dimeric molecules of **3** create a network of non-covalent interactions, of which the highest contribution (40.9%) to the HS is from numerous H···H contacts (Figures 10 and S6). The latter form a broad surface with no distinct directions (Figure 11), except of the C8–H8A···H8A[1]–C8[1] ([1] = 2 − *x*, −*y*, −*z*) interaction between methylene groups [*d*(C···C) = 4.379(15) Å] and a pair of weak interactions between pyridine rings C13–H13A···H14A–C14[1] and C14–H14A···H13A–C13[1] [[1] = 2 − *x*, 1 − *y*, 1 − *z*; *d*(C···C) = 4.072(18) Å]. As can be seen from the particular Hirshfeld surface, the Br···H and H···Br contacts are mostly located in the *ac* plane (Figure 11). Since the molecules of **3** has no electrical charge, the strengths of these contacts can be estimated [42] from the $\rho_{BCP}(\mathbf{r})$ electron density with sufficient precision (Table 4).

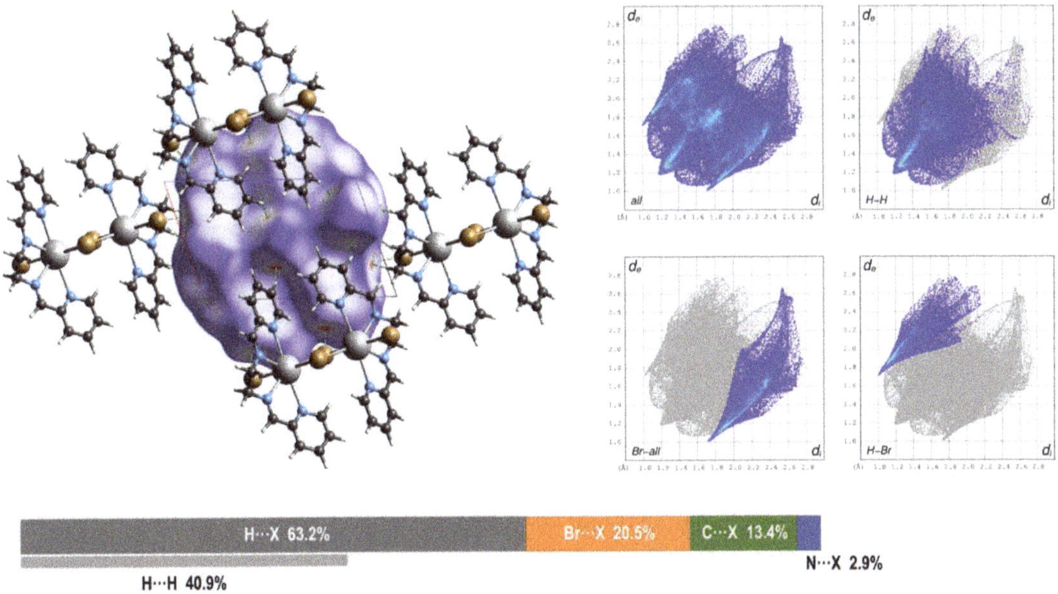

Figure 10. Left: Hirshfeld surface of the dinuclear molecule of **3** surrounded by the closest molecules (view along the *b* axis). Right: selected fingerprint plots of the respective surface. Bottom: contributions (in %) of specific $X_i \cdots X_e$ interactions.

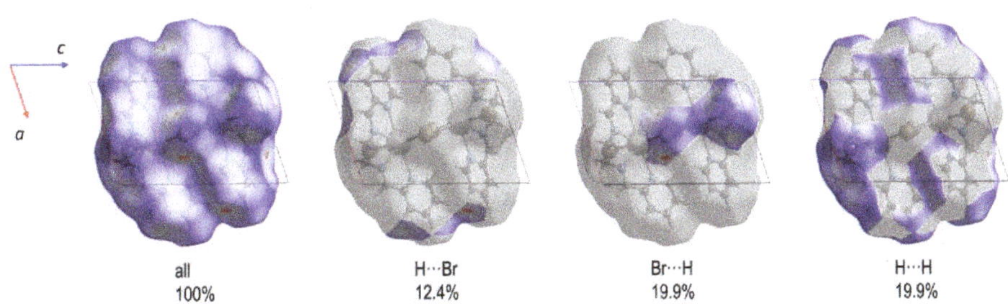

Figure 11. Hirshfeld surface of the dinuclear molecule of **3** highlighting regions of specific interactions.

Table 4. Electron density at the H···Br BCP and respective binding energies for selected intermolecular contacts in the structure **3** [a].

Contact	ρ_{BCP}(r), a.u.	Binding energy, kJ mol^{-1}
C2–H2A···Br1 [14] [b]	8.79×10^{-3}	–5.10
C4–H4B···Br2 [3]	6.01×10^{-3}	–2.50
C6–H6A···Br2 [−5]	5.32×10^{-3}	–1.86
C7–H7A···Br2 [−5]	5.38×10^{-3}	–1.92
C11–H11A···Br2 [15]	6.19×10^{-3}	–2.67
C13–H13A···Br1 [13]	7.71×10^{-3}	–4.09

[a] binding energies were estimated according to the equation for non-charged assemblies [42]; [b] symmetry operations have the same numbering as those in Table 3.

Inspection of the geometry and Hirshfeld surface suggests the existence of the contact C8–H8B···C6[1] [[1] = 1 – x, 1 – y, –z; *d*(C···C) = 3.722(10) Å, ∠(C–H···C) = 161.1°] between methylene CH_2 group and imine carbon atom of the neighboring molecules. However, despite the hydrogen atom being clearly oriented towards C6 one, the respective bond critical point is located rather between H8B and C5[1] atoms (Figure 12). Further analysis of the RDG isosurfaces disclosed the existence of a broad non-covalent interaction region between H8B donor and C4[1], C5[1] and C6[1] ([1] = 1 – x, 1 – y, –z) atoms as acceptors containing 0.0054 electrons (Figure 12). The negative q_{bind} index [43] of this domain (–0.00046383) accounts for a weak attractive interaction. The aromatic moieties of the ligand L3 in the molecule of **3** contact each other through vdW interactions (Figure S7).

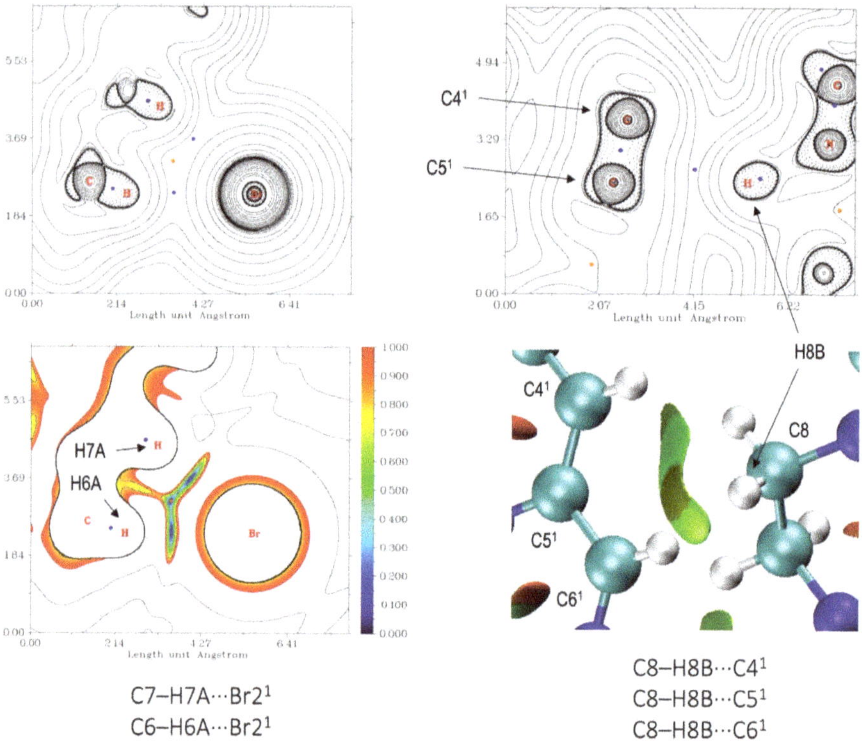

Figure 12. Top: plots of the Laplacian of the electron density in the real space, $\nabla^2\rho$(r), for selected weak contacts. Bottom left: 2D cut plane of the RDG function for the same projection as in the top left plot. Bottom right: 3D isosurface of the RDG (isovalue of 0.5) illustrating non-covalent interactions in the H8B—C4[1]/C5[1]/C6[1] groups of atoms ([1] = 1 – x, 1 – y, –z).

3.4. Optical Properties of 1–3

The optical diffuse reflectance spectra of all the compounds were measured using powder samples at r.t. (Figure 13a). According to the analysis of the spectra by the Tauc plot based on the assumption of direct band gap, $[F(R)h\nu]^2$ vs. $h\nu$ (where $F(R)$ denotes the Kubelka–Munk function, $h\nu$ is photon energy in eV), the optical band gap values of **1**, **2** and **3** were estimated to be 3.36, 3.13 and 2.96 eV, respectively (Figure 13b). The values of **1** and **2** are comparable to those of reported 1D hybrid lead perovskites, such as $[H_2bpp]Pb_2Cl_6$ (3.55 eV), $[H_2bpp]Pb_2Br_6$ (3.30 eV; bpp = 1,3-bis(4-pyridyl)-propane) [38], $[2,6\text{-dmpz}]_3Pb_2Br_{10}$ (3.12 eV; 2,6-dmpz = 2,6-dimethylpiperazine), $[hex]PbBr_3$ (3.41 eV; hex = hexamethyleneimine), $[hep]PbBr_3$ (3.50 eV; hep = heptamethyleneimine) [39]. Using density functional theory calculations, $[2,6\text{-dmpz}]_3Pb_2Br_{10}$, $[hex]PbBr_3$ and $[hep]PbBr_3$ were found to be direct band gap semiconductors [39]. Their valence band maximum (VBM) is composed of mixed Br-p and Pb-s orbitals, while empty Pb-p orbitals constitute conduction band minimum (CBM). In the case of two H_2bpp hybrid lead halides, the VBM was derived from hybridization of Pb-6p and Cl-3p or Br-4p orbitals, and CBM mainly originated from the C-2p/N-2p states of the bicyclic organic cations [38].

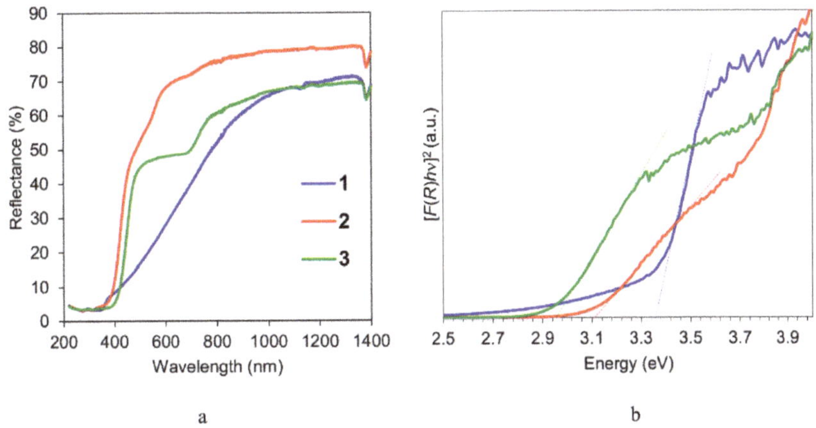

Figure 13. Diffuse reflectance spectra measured using powder samples of **1–3** at r.t. (**a**) and the Tauc plots in the range 2.5–4 eV based on the assumption of direct band gap calculated from the reflectance spectra of **1–3** (**b**).

Within the series of 1D and 2D lead bromide hybrids [39], the electronic structure was suggested to depend on the connectivity mode of $PbBr_6$ octahedra, where the corner-connected compound $[4\text{-amp}]PbBr_4$ [4-amp-4-(aminomethyl)piperidine] exhibits the smallest band gap value (2.93 eV). The face-sharing compounds $[hex]PbBr_3$ and $[hep]PbBr_3$ possess larger band gaps than $[2,6\text{-dmpz}]_3Pb_2Br_{10}$ with the edge- and corner-connecting structure. The band gaps successively going up from "corner-sharing" to "edge-sharing" and further "face-sharing" have also been observed in the lead iodide perovskite-derived organic–inorganic hybrids [44]. The band gaps of edge-sharing **1** and corner-sharing **2** perovskites follow the general trend.

4. Conclusions

In this study, we aimed to explore the templating effect of imidazo[1,5-*a*]pyridinium-based cation with hydroxyl functionality on the dimensionality of the lead halide anionic framework and connectivity modes of the $[PbHal_6]^{4-}$ octahedra, as well as H-bonding interactions in the resulting hybrid lead halide perovskite. To this aim, Ea·HCl was used in the developed protocol for the synthesis of substituted imidazo[1,5-*a*]pyridinium cations by the acid catalyzed oxidative cyclocondensation of FA, amine and 2-PCA. The obtained in situ $[L1]^+$ cation directed the formation of the 1D chloroplumbate $[L1]_{2n}[Pb_2Cl_6]_{n\infty}{\cdot}nH_2O$ (**1**), revealing a new type of the $[Pb_2Cl_6]_\infty$ twin chain constructed from three-edge- and five-edge-sharing $PbCl_6$ octahedra. In the case of $[PbBr_2(L2)]_{n\infty}{\cdot}0.5nH_2O$ (**2**) and $[PbBr_2(L3)]_2$ (**3**), the introduction of HBr—the necessary acid component and source of bromide anions—directly to the reaction media subverted the expected oxidation cyclization and instead facilitated the amine-aldehyde condensation between 2-PCA and Ea or En, respectively. The isolated hybrid lead bromide **2** features a chain of corner-sharing $PbBr_3N_2O$ octahedra, decorated with the tridenate Schiff base ligands on both sides. In the 0D dimer **3**, two irregular $PbBr_3N_4$ pentagonal bipyramids are connected by edge sharing. The Hirshfeld surface and QTAIM analyses disclosed the presence of numerous weak interactions between molecular and polymeric fragments in the structures of **1**–**3**. The band gaps of **1** (3.36 eV) and **2** (3.13 eV) with different connectivity follow the general trend of edge-sharing perovskites exhibiting larger band gaps than corner-sharing structures.

Supplementary Materials: The following supporting information can be downloaded at: https://www.mdpi.com/article/10.3390/cryst13020307/s1, Figure S1: IR spectrum of $[L1]_{2n}[Pb_2Cl_6]_{n\infty}{\cdot}nH_2O$ (**1**); Figure S2: IR spectrum of $[PbBr_2(L2)]_{n\infty}{\cdot}0.5nH_2O$ (**2**); Figure S3: IR spectrum of $[PbBr_2(L3)]_2$ (**3**); Figure S4: Selected fingerprint plots of the (N12, N13A) and (N22, N23A, major component) $[L1]^+$ cations in the crystal structure of **1** showing $X_i{\cdots}X_e$ interactions (where X = Cl, O, C or H); Figure S5: Selected fingerprint plots of the fragment $[PbBr_2(L2)]$ in the crystal structure of **2** showing $X_i{\cdots}X_e$ interactions (where X = Br, O, C or H); Figure S6: Selected fingerprint plots of the dinuclear molecule of **3** showing $X_i{\cdots}X_e$ interactions (where X = Br, N, C or H); Figure S7: Reduced density gradient surfaces (isovalue of 0.5) showing weak vdW interactions (green area) between aromatic fragments of the ligands L3 in the molecular structure of **3** (lead and bromine atoms were excluded from calculations).

Author Contributions: Conceptualization, O.Y.V.; methodology, O.Y.V. and D.S.N.; investigation, O.Y.V., E.A.B., O.V.N. A.N.S. and D.S.N.; writing—original draft preparation, O.Y.V. and D.S.N.; writing—review and editing, O.Y.V. All authors have read and agreed to the published version of the manuscript.

Funding: The work was performed with the support of the Ministry of Education and Science of Ukraine (project 22BP037-13, grant for the perspective development of the scientific direction 'Mathematical sciences and natural sciences' at the Taras Shevchenko National University of Kyiv) and the Fundação para a Ciência e Tecnologia (FCT), Portugal (projects UIDB/00100/2020, UIDP/00100/2020, and LA/P/0056/2020 of Centro de Química Estrutural, contracts IST-ID/086/2018 and IST-ID/117/2018).

Data Availability Statement: Crystallographic data for the structures reported can be obtained free of charge from the Cambridge Crystallographic Data Centre via www.ccdc.cam.ac.uk/data-request/cif (accessed on 27 January 2023) quoting the deposition numbers CCDC 2217143 (**1**), 2223148 (**2**) and 2223147 (**3**).

Conflicts of Interest: The authors declare no conflict of interest.

References

1. Green, M.A.; Hishikawa, Y.; Warta, W.; Dunlop, E.D.; Levi, D.H.; Hohl-Ebinger, J.; Ho-Baillie, A.W. Solar cell efficiency tables (version 50). *Prog. Photovolt. Res. Appl.* **2017**, *25*, 668–676. [CrossRef]
2. Sutherland, B.R.; Sargent, E.H. Perovskite photonic sources. *Nat. Photon.* **2016**, *10*, 295–302. [CrossRef]
3. Saparov, B.; Mitzi, D.B. Organic-Inorganic Perovskites: Structural Versatility for Functional Materials Design. *Chem. Rev.* **2016**, *116*, 4558–4596. [CrossRef] [PubMed]
4. Smith, M.D.; Jaffe, A.; Dohner, E.R.; Lindenberg, A.M.; Karunadasa, H.I. Structural origins of broadband emission from layered Pb–Br hybrid perovskites. *Chem. Sci.* **2017**, *8*, 4497–4504. [CrossRef]
5. Yuan, Z.; Zhou, C.; Tian, Y.; Shu, Y.; Messier, J.; Wang, J.C.; Van De Burgt, L.J.; Kountouriotis, K.; Xin, Y.; Holt, E.; et al. One-dimensional organic lead halide perovskites with efficient bluish white-light emission. *Nat. Commun.* **2017**, *8*, 14051. [CrossRef]
6. Peng, Y.; Yao, Y.; Li, L.; Wu, Z.; Wang, S.; Luo, J. White-light emission in a chiral one-dimensional organic–inorganic hybrid perovskite. *J. Mater. Chem. C* **2018**, *6*, 6033–6037. [CrossRef]
7. Buvaylo, E.A.; Kokozay, V.N.; Linnik, R.P.; Vassilyeva, O.Y.; Skelton, B.W. Hybrid organic–inorganic chlorozincate and a molecular zinc complex involving the in situ formed imidazo [1,5-*a*]pyridinium cation: Serendipitous oxidative cyclization, structures and photophysical properties. *Dalton Trans.* **2015**, *44*, 13735–13744. [CrossRef]

8. Vassilyeva, O.Y.; Buvaylo, E.A.; Linnik, R.P.; Nesterov, D.S.; Trachevsky, V.V.; Skelton, B.W. Synthetic strategy towards halometallates with imidazo [1,5-*a*]pyridinium-based counterions. *CrystEngComm* **2020**, *22*, 5096–5105. [CrossRef]
9. Vassilyeva, O.Y.; Buvaylo, E.A.; Kokozay, V.N.; Skelton, B.W.; Sobolev, A.N. Crystal structures of an imidazo[1,5-*a*] pyridinium-based ligand and its $(C_{13}H_{12}N_3)_2[CdI_4]$ hybrid salt. *Acta Cryst.* **2019**, *75*, 1209–1214. [CrossRef]
10. Vassilyeva, O.; Buvaylo, E.; Kokozay, V.; Skelton, B.W.; Rajnák, C.; Titis, J.; Boca, R. Long magnetic relaxation time of tetracoordinate Co^{2+} in imidazo [1,5-*a*]pyridinium-based $(C_{13}H_{12}N_3)_2[CoCl_4]$ hybrid salt and $[Co(C_{13}H_{12}N_3)Cl_3]$ molecular complex. *Dalton Trans.* **2019**, *48*, 11278–11284. [CrossRef] [PubMed]
11. Vassilyeva, O.Y.; Buvaylo, E.A.; Lobko, Y.V.; Linnik, R.P.; Kokozay, V.N.; Skelton, B.W. Organic–inorganic hybrid tetrachlorocadmates as promising fluorescent agents for cross-linked polyurethanes: Synthesis, crystal structures and extended performance analysis. *RSC Adv.* **2021**, *11*, 7713–7722. [CrossRef] [PubMed]
12. Vassilyeva, O.Y.; Buvaylo, E.A.; Kokozay, V.N.; Skelton, B.W. Organic-inorganic hybrid mixed-halide Zn^{II} and Cd^{II} tetrahalometallates with the 2-methylimidazo[1,5-*a*]pyridinium cation. *Acta Cryst.* **2022**, *78*, 359–364. [CrossRef] [PubMed]
13. Lyons, A.S.; Ingersoll, M.A.; Muniyan, S.; D'Cunha, N.; Robinson, T.; Hoelting, K.; Dwyer, J.G.; Biu, X.R.; Batra, S.K.; Lin, M.F. Novel imidazopyridine derivatives possess anti-tumor effect on human castration-resistant prostate cancer cells. *PLoS ONE* **2015**, *10*, e0131811.
14. Song, G.J.; Bai, S.Y.; Dai, X.; Cao, X.Q.; Zhao, B.X. A ratiometric lysosomal pH probe based on the imidazo[1,5-*a*]pyridine–rhodamine FRET and ICT system. *RSC Adv.* **2016**, *6*, 41317–41322. [CrossRef]
15. Volpi, G.; Garino, C.; Fresta, E.; Casamassa, E.; Giordano, M.; Barolo, C.; Viscardi, G. Strategies to increase the quantum yield: Luminescent methoxylated imidazo[1,5-*a*]pyridines. *Dyes Pigment.* **2021**, *192*, 109455. [CrossRef]
16. Yagishita, F.; Nii, C.; Tezuka, Y.; Tabata, A.; Nagamune, H.; Uemura, N.; Yoshida, Y.; Mino, T.; Sakamoto, M.; Kawamura, Y. Fluorescent N-heteroarenes having large stokes shift and water solubility suitable for bioimaging. *Asian J. Org. Chem.* **2018**, *7*, 1614–1619. [CrossRef]
17. Kortüm, G.; Braun, W.; Herzog, G. Principles and techniques of diffuse-reflectance spectroscopy. *Angew. Chem. Int. Ed.* **1963**, *2*, 333–341. [CrossRef]
18. Landi, S., Jr.; Segundo, I.R.; Freitas, E.; Vasilevskiy, M.; Carneiro, J.; Tavares, C.J. Use and misuse of the Kubelka-Munk function to obtain the band gap energy from diffuse reflectance measurements. *Solid State Commun.* **2022**, *341*, 114573. [CrossRef]
19. Spackman, P.R.; Turner, M.J.; McKinnon, J.J.; Wolff, S.K.; Grimwood, D.J.; Jayatilaka, D.; Spackman, M.A. CrystalExplorer: A program for Hirshfeld surface analysis, visualization and quantitative analysis of molecular crystals. *J. Appl. Cryst.* **2021**, *54*, 1006–1011. [CrossRef]
20. Sheldrick, G.M. Crystal structure refinement with SHELXL. *Acta Cryst.* **2015**, *71*, 3–8.
21. Neese, F. Software update: The ORCA program system—Version 5.0. *Wil. Interdiscip. Rev. Comput. Mol. Sci.* **2022**, *12*, e1606. [CrossRef]
22. Neese, F. The ORCA program system. *Wil. Interdiscip. Rev. Comput. Mol. Sci.* **2012**, *2*, 73–78. [CrossRef]
23. Mardirossian, N.; Head-Gordon, M. ωB97M-V: A combinatorially optimized, range-separated hybrid, meta-GGA density functional with VV10 nonlocal correlation. *J. Chem. Phys.* **2016**, *144*, 214110. [CrossRef] [PubMed]
24. Weigend, F.; Ahlrichs, R. Balanced basis sets of split valence, triple zeta valence and quadruple zeta valence quality for H to Rn: Design and assessment of accuracy. *Phys. Chem. Chem. Phys.* **2005**, *7*, 3297–3305. [CrossRef]
25. Pantazis, D.A.; Neese, F. All-electron scalar relativistic basis sets for the 6p elements. *Theor. Chem. Acc.* **2012**, *131*, 1292. [CrossRef]
26. Weigend, F. Accurate Coulomb-fitting basis sets for H to Rn. *Phys. Chem. Chem. Phys.* **2006**, *8*, 1057–1065. [CrossRef] [PubMed]
27. Stoychev, G.L.; Auer, A.A.; Neese, F. Automatic Generation of Auxiliary Basis Sets. *J. Chem. Theor. Comput.* **2017**, *13*, 554–562. [CrossRef]
28. van Wüllen, C. Molecular density functional calculations in the regular relativistic approximation: Method, application to coinage metal diatomics, hydrides, fluorides and chlorides, and comparison with first-order relativistic calculations. *J. Chem. Phys.* **1998**, *109*, 392. [CrossRef]
29. Barone, V.; Cossi, M. Quantum calculation of molecular energies and energy gradients in solution by a conductor solvent model. *J. Phys. Chem. A* **1998**, *102*, 1995–2001. [CrossRef]
30. Lu, T.; Chen, F.W. Multiwfn: A multifunctional wavefunction analyzer. *J. Comput. Chem.* **2012**, *33*, 580–592. [CrossRef]
31. Johnson, E.R.; Keinan, S.; Mori-Sanchez, P.; Contreras-Garcia, J.; Cohen, A.J.; Yang, W.T. Revealing Noncovalent Interactions. *J. Am. Chem. Soc.* **2010**, *132*, 6498–6506. [CrossRef] [PubMed]
32. Humphrey, W.; Dalke, A.; Schulten, K. VMD—Visual Molecular Dynamics. *J. Mol. Graph.* **1996**, *14*, 33–38. [CrossRef] [PubMed]
33. Dobrzycki, L.; Wozniak, K. Inorganic–organic hybrid salts of diaminobenzenes and related cations. *CrystEngComm* **2008**, *10*, 577–589. [CrossRef]
34. Barkaoui, H.; Abid, H.; Yangui, A.; Triki, S.; Boukheddaden, K.; Abid, Y. Yellowish white-light emission involving resonant energy transfer in a new one-dimensional hybrid material: $(C_9H_{10}N_2)PbCl_4$. *J. Phys. Chem. C* **2018**, *122*, 24253–24261. [CrossRef]
35. Groom, C.R.; Bruno, I.J.; Lightfoot, M.P.; Ward, S.C. The Cambridge Structural Database. *Acta Cryst.* **2016**, *72*, 171–179. [CrossRef]
36. Wu, G.; Zhou, C.; Ming, W.; Han, D.; Chen, S.; Yang, D.; Besara, T.; Neu, J.; Siegrist, T.; Du, M.H.; et al. A one-dimensional organic lead chloride hybrid with excitation-dependent broadband emissions. *ACS Energy Lett.* **2018**, *3*, 1443–1449. [CrossRef]
37. Martinez, C.R.; Iverson, B.L. Rethinking the term "pi-stacking". *Chem. Sci.* **2012**, *3*, 2191–2201. [CrossRef]

38. Sun, X.Y.; Yue, M.; Jiang, Y.X.; Zhao, C.H.; Liao, Y.Y.; Lei, X.W.; Yue, C.Y. Combining dual-light emissions to achieve efficient broadband yellowish-green luminescence in one-dimensional hybrid lead halides. *Inorg. Chem.* **2021**, *60*, 1491–1498. [CrossRef]
39. Mao, L.; Guo, P.; Kepenekian, M.; Hadar, I.; Katan, C.; Even, J.; Schaller, R.D.; Stoumpos, C.C.; Kanatzidis, M.G. Structural diversity in white-light-emitting hybrid lead bromide perovskites. *JACS* **2018**, *140*, 13078–13088. [CrossRef]
40. Roy, S.; Choubey, S.; Bhar, K.; Khan, S.; Mitra, P.; Ghosh, B.K. Syntheses, structures, and luminescence behavior of terephthalate bridged lead(II) complexes with tetradentate N-donor Schiff bases. *J. Mol. Struct.* **2013**, *1051*, 328–335. [CrossRef]
41. Hashemi, L.; Morsali, A. Sonochemical synthesis of nano-structured lead(II) complex: Precursor for the preparation of PbO nano-structures. *J. Coord. Chem.* **2011**, *64*, 4088–4097. [CrossRef]
42. Emamian, S.; Lu, T.; Kruse, H.; Emamian, H. Exploring Nature and Predicting Strength of Hydrogen Bonds: A Correlation Analysis between Atoms-in-Molecules Descriptors, Binding Energies, and Energy Components of Symmetry-Adapted Perturbation Theory. *J. Comput. Chem.* **2019**, *40*, 2868–2881. [CrossRef] [PubMed]
43. Contreras-García, J.; Yang, W.; Johnson, E.R. Analysis of Hydrogen-Bond Interaction Potentials from the Electron Density: Integration of Noncovalent Interaction Regions. *J. Phys. Chem. A* **2011**, *115*, 12983–12990. [CrossRef] [PubMed]
44. Kamminga, M.E.; de Wijs, G.A.; Havenith, R.W.; Blake, G.R.; Palstra, T.T. The role of connectivity on electronic properties of lead iodide perovskite-derived compounds. *Inorg. Chem.* **2017**, *56*, 8408–8414. [CrossRef] [PubMed]

crystals

MDPI

Article

Supramolecular Structure of Tris(1,10-phenanthroline)zinc(II)-Cation and N,N′,N″-tris(carboxymethyl)-1,3,5-benzenetricarboxamide-Anion: Synthesis, Crystal Structure, Vibrational Spectra, and Theoretical Investigations

Niels-Patrick Pook

Institute of Inorganic and Analytical Chemistry, Clausthal University of Technology, Paul-Ernst-Str. 4, D-38678 Clausthal-Zellerfeld, Germany; niels-patrick.pook@tu-clausthal.de; Tel.: +49-5323-72-2887

Abstract: The present work reports on the synthesis, structural, spectroscopic, and theoretical studies of a new solid state ionic compound mainly composed of tris(1,10-phenanthroline)zinc(II) cations and N,N′,N″-tris(carboxymethyl)-1,3,5-benzenetricarboxamide anions. Colorless and well-shaped crystals were obtained from an alkaline aqueous methanolic solution, and single-crystal X-ray diffraction revealed a distinct supramolecular network. Powder diffraction techniques and Rietveld analysis confirmed the phase purity of the crystalline probes. The compound crystallizes in the orthorhombic space group *Pbca* with a cell volume of 9517.0 Å^3. The complex cations $[Zn(phen)_3]^{2+}$ are interconnected via π–π-interactions and form a cationic layer network with holes. The organic counterion, as a dianion, forms dimeric units through π–π-interactions and hydrogen bonds, which also form an anionic layer network with honeycomb-like holes through cooperative classical hydrogen bonds of the O···H–O and O···H–N type with attractive secondary electrostatic interactions. Using the holes, the resulting supramolecular framework can be described as an interpenetrated network of separate anionic and cationic layers linked by further weaker non-covalent interactions such as C–H···π and lone-pair···π interactions. DFT calculations confirmed the experimentally observed spectroscopic (IR and Raman) findings. For a deeper insight into the structural arrangement in the crystal, the different Hirshfeld surfaces of the cation and anion, the pairwise interaction energies as well as the energy framework were calculated, supporting the dominance of attractive and repulsive electrostatic forces between the ions.

Keywords: crystal structure; phenanthroline ligand; supramolecular interactions; π-stacking; Hirshfeld surface analysis; IR and Raman spectroscopy; self-assembly; DFT calculations; hydrogen bonding; noncovalent interactions

Citation: Pook, N.-P. Supramolecular Structure of Tris(1,10-phenanthroline)zinc(II)-Cation and N,N′,N″-tris(carboxymethyl)-1,3,5-benzenetricarboxamide-Anion: Synthesis, Crystal Structure, Vibrational Spectra, and Theoretical Investigations. *Crystals* **2023**, *13*, 569. https://doi.org/10.3390/cryst13040569

Academic Editor: Waldemar Maniukiewicz

Received: 7 March 2023
Revised: 21 March 2023
Accepted: 24 March 2023
Published: 27 March 2023

1. Introduction

Intermolecular interactions are of crucial importance and play a central key role in a multitude of biologically and medically relevant processes. Impressive examples of these biologically essential processes and structures from nature include the assembly of the DNA [1–4] and the associated reproduction of genetic information, the bio-enzymatic catalysis of proteins, protein folding itself, or the impressive self-assembly of the tobacco mosaic virus [5,6]. Underlying many of these fascinating mechanisms of assembly and reaction sequences is a high degree of efficiency in which the involved molecules recognize each other over a complex interplay of several relatively weak non-covalent forces, leading in a controlled manner to complex structures with often remarkable stability.

The associative self-assembly in the DNA double helical structure is based on the formation of classical hydrogen bonds with attractive secondary interactions [7–11] between the different nucleic base pairs of two DNA single strands as well as π–π stacking interactions between the superimposed base pairs [4], which provide an additional contribution to the overall stability of the system.

In addition to the intermolecular forces mentioned so far, other important non-covalent interactions exist such as the attractive and repulsive electrostatic interactions between ions, the non-classical hydrogen bonds, the lone-pair···π and the C–H···π interactions [3,12–20]. Of further central importance to supramolecular chemistry are the characteristic coordinative bonds between a metal atom and organic ligands known from transition metal complexes. One biologically important representative is adenosylcobalamine, also known as coenzyme B12 [1,21,22]. It belongs to the group of cobalamins and is involved in various metabolic functions of the organism. The basic structure consists of a cobalt complex, in which the cobalt shows an octahedral coordination environment. Four of the six coordinative bonds are formed by a corrin ring system. This complex exhibits very high stability. A related structure is found in the heme unit of hemoglobin. This iron porphyrin complex is responsible for oxygen transport in vertebrates [23]. There are many other impressive examples of structures and steering forces and the chemical and biological recognition of molecules. The insights and principles of molecular recognition and self-assembly form the basis of crystal engineering of artificial systems in the field of new effective drugs and functional supramolecular materials with applications in a wide variety of fields. Thus, supramolecular chemistry establishes an interdisciplinary research field of the departments of chemistry, physics, biology, and materials science.

However, with the intention to prepare new metal–organic coordination polymers (CPs) with scintillation effect [24–30] based on zinc as the metal center, 1,10-phenanthroline and N,N′,N″-tris(carboxymethyl)-1,3,5-benzenetricarboxamide (TCMBT) as ligands, the present compound was synthesized. Although the compound does not exhibit a scintillation effect under the chosen excitation conditions [31,32], it is a structurally highly interesting compound of academic interest. Several aspects of self-assembly were considered in the synthesis strategy. On the one hand, the use of the nitrogen-containing 1,10-phenanthroline as an electron-poor aromatic ligand, which is known for π-π stacking [33–36] and lone-pair···π [3,15,18,19,37] interactions, and on the other hand, the organic ligand N,N′,N″-tris(carboxymethyl)-1,3,5-benzenetricarboxamide (TCMBT) which forms very stable dimers [38–43] via π–π stacking and cooperative hydrogen bonds. An overview of the complexes synthesized in which TCMBT acts as bridging and coordinating ligand can be found in the literature [44]. Due to the postulated scintillation mechanism of metal–organic compounds [24,26,27], the use of metal centers with high Z for the high energy conversion process and aromatic compounds (delocalized π-electrons) for an induced fluorescence by the attenuated radiation is preferred.

In the present compound, TCMBT is a doubly deprotonated counter anion and does not coordinate the metal center as shown in Scheme 1. The anions are connected by a distinct network of cooperative hydrogen bonds and hydrogen bonds with attractive secondary electrostatic interactions. They form layers with honeycomb-like holes. Through a SciFinder research, we assume that this coordination mode of TCMBT is structurally described for the first time. The cationic complexes are connected by π–π interactions and also form layers with holes. An interpenetrating network of the separated anion and cation layers is constructed over these holes. The layers are linked by further weaker non-covalent interactions such as C–H···π, lone-pair···π and C–H···O hydrogen bonds.

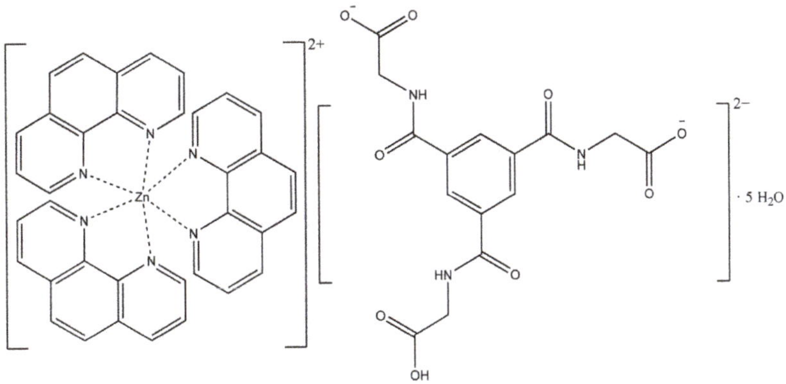

Scheme 1. Schematic representation of the structure of tris(1,10-phenanthroline)zinc(II)-cation and N,N′,N″-tris(carboxymethyl)-1,3,5-benzenetricarboxamide dianion, and water molecules $[Zn(C_{12}H_8N_2)_3]C_{15}H_{13}N_3O_9 \cdot 5H_2O$.

2. Materials and Methods

2.1. Synthesis of the Compound

Cesium carbonate (3 mmol), 1,10-phenanthroline (1 mmol) and N,N′,N″-tris(carboxymethyl)-1,3,5-benzenetricarboxamide (1 mmol) were dissolved in a 1:1 (v/v) mixture of water and methanol (60 mL). The mixture was heated and stirred for 1 h. After cooling to room temperature, 10 mL of an aqueous solution of zinc(II) nitrate (1 mmol) was slowly added under continuous stirring. Good quality colorless block-shaped crystals of the compound suitable for X-ray crystallography could be obtained after five days by slow evaporation under ambient conditions. $C_{51}H_{47}N_9O_{14}Zn$ Yield: 62%; Elemental analysis: Calc.: C 56.96 H 4.41 N 11.44; Found: C 56.41 H 4.31 N 11.28.

2.2. Single-Crystal X-ray Crystallography

The data set for a well-shaped single crystal selected under a polarization microscope was obtained at 223 K on a Stoe IPDS II diffractometer using a graphite monochromated MoK_α radiation (λ = 0.71073 Å). The crystal was separated from the mother liquor and placed in a glass fiber with perfluorinated oil. The structure was solved by direct methods and refined using least-square methods on F^2 with the SHELX-2018 [45] software package. The final structure was checked and validated with PLATON [46–49]. A validation report and a checkCIF file can be found in the Supplementary Information (SI). The atomic displacement parameters for non-hydrogen atoms were refined anisotropically, and all C-bound and N-bound H atoms were set to idealized geometry and refined isotropically with Uiso(H) = 1.2 Ueq(C and N), C–H(aromatic) = 0.94 Å and C–H(methylene) = 0.98 Å N–H(amide) = 0.87 using the riding model. The H atoms of the water molecules were located in a difference-Fourier map and were also refined isotropically with Uiso(H) = 1.5 Ueq(O). Some water molecules were restrained with the DFIX command to 0.87Å which led to a stable hydrogen bonding network. Details of the restrain can be found in the embedded INS-file of the CIF-file. A numerical absorption correction was applied using the STOE software X-Area and X-Red. Figures were prepared with DIAMOND [50] and POV-RAY [51]. Further details of the compound may be obtained free of charge from the Cambridge Crystallographic Data Centre (https://www.ccdc.cam.ac.uk/structures/) on quoting the deposition number CCDC 2234251.

2.3. Powder X-ray Diffraction (PXRD)

The phase purity of the crystalline powder probes of the newly synthesized compound were confirmed by powder diffraction techniques before all other measurements were carried out. PXRD data were collected on a STOE STADI P X-ray powder diffractometer in the

Debye–Scherrer geometry with a position-sensitive detector and a germanium monochromated Cu-K_{α} radiation (λ = 1.54056 Å) at room temperature. Standard measurements were carried out with a generator voltage of 40 kV and a current of 30 mA and a two theta step size of 0.01° with a step width of 0.7° over an angular range of 3–91°. A measuring time of 50 s for each step leads to a total time of two hours for a single range. Eight ranges were measured and added to lower the background noise. Data were collected, analyzied, and visualized with the Stoe WINX^{POW} software, and the JANA2006 program package [52] was used for the leBail and Rietveld refinement procedure.

2.4. *Infrared and Raman Spectroscopy*

FT–IR and FT–Raman spectra were recorded with a Bruker VERTEX 70V. A U-shaped silicon carbide piece was used for infrared excitation, and a DLaTGS detector served as the detector unit. For the measurement of the Raman spectrum a Nd:YAG laser (1064 nm) with a maximum power of 1000 mW and a germanium detector was used. The analysis of the recorded IR and Raman spectra were performed using the Bruker OPUS software package.

2.5. *Computional Chemistry*

Theoretical calculations of the molecular energies, the structure optimization and the IR and Raman spectra were performed with the *ORCA* 4.0.1 program package [53]. The Hirshfeld surfaces [54–56] with the associated fingerprint plots [57,58], the pairwise interaction energies and the energy framework [59,60] were generated with the CrystalExplorer program package [61]. The calculations of the pairwise non-covalent interaction energies between the phenanthroline ligands of the cationic complexes, the phenanthroline ligands and the TCMBT dianion, and the TCMBT dianions were performed using the symmetry-adapted perturbation theory (SAPT) of the PSI4 software [62] with the basis sets of jun-cc-pVDZ and aug-cc-pVDZ, respectively. Further details on the used functionals and basis sets can be found in the associated sections. For all theoretical calculations the initial atomic coordinates were taken from the CIF-file.

2.6. *Thermal and Elemental Analysis*

The thermogravimetric analysis was performed using a TGA951 from TA Instruments Inc. The temperature range for the TGA measurements was from room temperature to 920 °C, with a heating rate of 10 K/min. The samples weighed 4.4090 mg and were measured in a platinum crucible under an argon atmosphere with a flow rate of 50 mL/min. For the elemental analysis, approximately 2 mg of the solid sample were weighed into a thin-walled tin boat, enclosed, and introduced into the Vario EL elemental analyzer from Elementar Analysesysteme GmbH.

3. Results

3.1. *Description of the Crystal Structure*

The single-crystal X-ray crystallography of a colorless block-shaped crystal grown in an alkaline aqueous-methanol solution revealed a complex cation $[Zn(phen)_3]^{2+}$ with a zinc(II) coordination center of a distorted octahedral coordination environment consisting of three bidentate 1,10-phenanthroline ligands, a non-coordinating doubly deprotonated N,N′,N″-tris(carboxymethyl)-1,3,5-benzenetricarboxamide (TCMBT) counter anion, and five water molecules as depicted in Figure 1. The compound crystallizes in the primitive centrosymmetric orthorhombic space group *Pbca* (No. 61) with eight formulas units per unit cell and a cell volume of 9517.0 Å^3. Further details of the single-crystal X-ray analysis and the structure refinement are listed in Table 1. Selected bond length and angles of the known cationic complex and the dianion can be found in the Tables S1 and S2 of the SI, respectively. The values are comparable to those found in the literature [39,63].

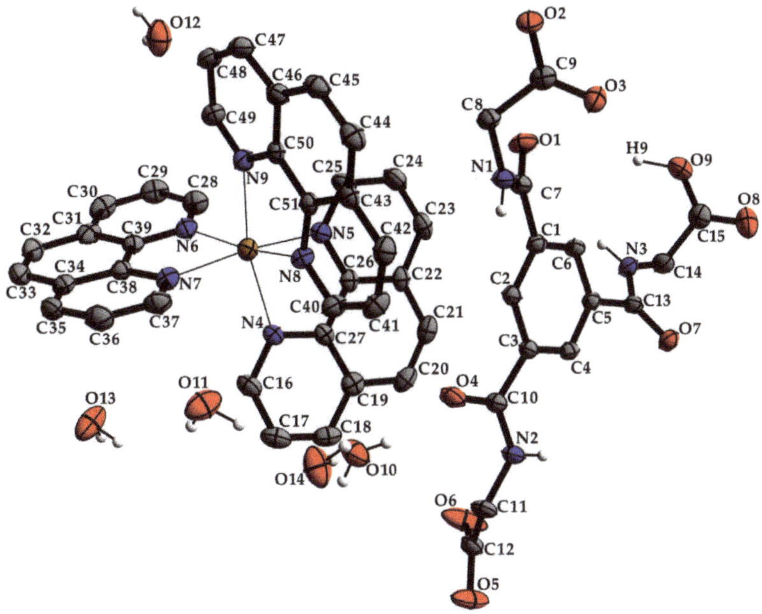

Figure 1. The molecular entities of the zinc(II) complex and the dianion as well as water molecules with atom labels and displacement ellipsoids of non-H atoms drawn at the 40% probability level. The hydrogen atoms of the carbons have been omitted for clarity (see Tables S1 and S2 for details).

The two deprotonated carboxylate groups of the dianion show bond lengths of C9–O2 = 1.267 (5) and C9–O3 = 1.259 (5) Å as well as C12–O5 = 1.248 (6) and C12–O6 = 1.234 (6) Å and bond angles of O2–C9–O3 with 125.7 (4)° as well as O5–C12–O6 with 124.5 (4)°, respectively. This arrangement indicate a delocalized anionic charge. It becomes clearer by the comparison to the protonated carboxylate group which shows a signifcant different bond distance of C15–O8 = 1.218 (6) and C15–O9 = 1.311 (6) Å which are in the expected ranges [64] and identify a double and a single bond. The bond angle of O8–C15–O9 with 121.0 (4)° is decreased compared to the deprotonated carboxylate groups. The illustration of Figure 2, Figure 3 and Figure S1 and Video S1 display the involvement of the carboxylates, amides and several C–H groups of the TCMBT in classical and non-classical hydrogen bonding interactions. A detailed overview of the hydrogen bond geometry is summarized in Table S3 of the SI. As seen, the dianions form dimeric units by π–π stacking and hydrogen bonding. These dimeric units are linked through further hydrogen bonds of a cooperative nature and with attractive secondary electrostatic interactions (Figure 3), leading to a supramolecular anionic layer with honeycomb-like holes (Figure 4). The hydrogen bonds of the anion linkage of the protonated and deprotonated carboxylates and N–H group of O5···H3–N3 and O6···H9–O9, have donor and acceptor distances of 1.926 (4) and 1.35 (6) Å, and angles of O5–H3–N3 and O6–H9–O9 with 158.4 (3)° and 164.8 (5)°, respectively. The short O6···O9 distance of 2.544 (6) Å and the oxygen-hydrogen distance of O9–H9 of 1.21 (6) Å, which is greatly elongated, indicate a strong hydrogen bond (Figure 3). This is known from the literature [65–72]. The diagonal relationships of the secondary stabilizing interactions show distances of H3···O6 and H9···O5 with 2.887 (4) and 2.77 (6) Å, respectively. These attractive diagonal interaction between the parallel hydrogen bonded bridges of O5···H3–N3 and O6···H9–O9 with a DD/AA motif supports and strengthens the hydrogen bonding of the anionic connection. These type of secondary interactions are observed in the hydrogen bonding of the nucleic base pairs of the double-stranded DNA [7–11,73]. In addition to the classical and non-classical hydrogen bonding, a further important role in the architecture of the present supramolecular structure is played by the different π-stacking interactions. As mentioned earlier, two TCMBT molecules interact with a π–π stacking in nearly perfect alignment of a face-to-face arrangement, showing a relatively short *Cg1*···*Cg1* centroid-to-centroid distance of 3.5382 (2) Å [4,13,14,33–35]. A detailed overview of the definition of the centroids is shown in Figure S2 of the SI. Furthermore, the cationic complexes also exhibit π-stacking interactions, resulting in three-dimensional linkage (Figure S3) and the formation of cationic layers with holes (Figure 4). A supramolecular framework is formed over the holes, which can be described as an interpenetrating supramolecular network of separate anionic and cationic layers. These layers are linked via further non-covalent interactions such as lone-pair···π (Figure 3) and C–H···π (Figure S4). Compared to the π–π stacking of the dianions, the cationic stacking show an offset or parallel displaced motif. The centroid-to-centroid distance of *Cg3*···*Cg3* of two phenanthrolines 1 is 3.5544 (1) Å, which is slightly shorter than the distance of *Cg6*···*Cg9* of phenanthroline 2 and phenanthroline 3 with a value of 3.5686 (2) Å. A closer look at the π-stacking interactions of the phenanthrolines reveals that the π–π repulsion, which is more pronounced in the dianion stacking, was minimized by the slipped arrangement of the phenanthrolines, and the π–σ attraction is favored due to the positively charged ring carbons or the C-bounded hydrogen atoms [4,33]. Basically, a transition from π–π stacking to C–H···π interaction can be observed. The nitrogen-containg ligands as electron poor aromatic compounds are predestinated for π-stacking. When a metal atom is coordinated by the phenanthroline, the electron density of the aromatic system decreases even more due to the electron-withdrawing property of the metal, and partial charge separation occurs. This favors the van der Waals attraction of dipole–dipole interactions. For this reason, the phenanthroline ligands are usually stacked with a 180° rotation.

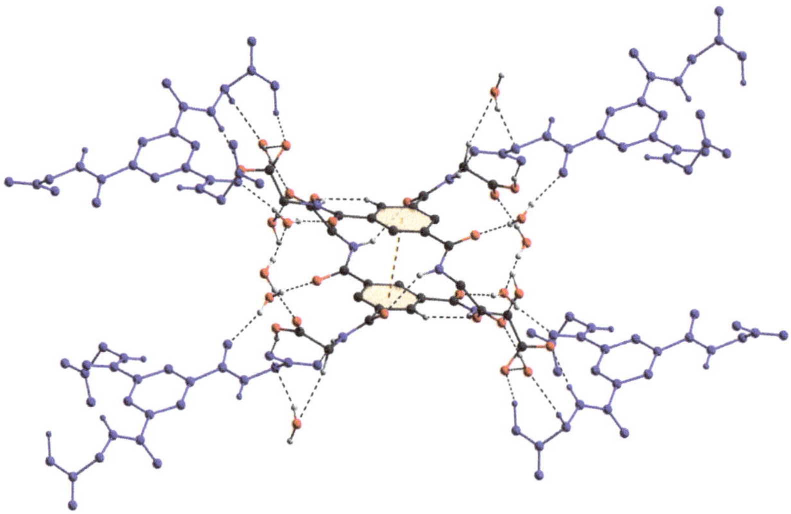

Figure 2. The molecular connections of the dianions in the crystal via hydrogen bonds, cooperative hydrogen bonds with attractive secondary electrostatic interactions, and π–π stacking of the anions. The hydrogen atoms which are not involved in hydrogen bonding have been omitted for clarity.

Table 1. Crystal data and structural refinement parameters of the $[Zn(C_{12}H_8N_2)_3]C_{15}H_{13}N_3O_9 \cdot 5H_2O$.

Compound	$[Zn(C_{12}H_8N_2)_3]C_{15}H_{13}N_3O_9 \cdot 5H_2O$
Empirical formula	$C_{51}H_{47}N_9O_{14}Zn$
Formula weight	1075.34
Temperature (K)	223
Diffractometer	Stoe IPDS II
Wavelength (Å)	0.71073
Crystal system	orthorhombic
Color	colorless
Space group	*Pbca* (No. 61)
a, b, c (Å)	14.643(1), 20.095(1), 32.345(2)
α, β, γ (°)	90
V (Å^3)	9517.0(10)
Z	8
D_{calc} (g·cm^3)	1.501
μ (mm^{-1})	0.597
F(000)	4464.0
Crystal size (mm)	0.250, 0.290, 0.170
θ Range (°)	1.832–25.100
Index ranges	$-17 \leq h \leq 16$ $-23 \leq k \leq 18$ $-38 \leq l \leq 35$
Reflection collected/unique	24712/8439
Completeness to θ (%)	0.995
Absorption correction	Numerical; X-AREA, X-RED (2008)
Max. and min. transmission	0.8040/0.9159
Refinement method	Full-matrix least-squares on F^2
Data/parameters/restrains	8439/709/4
Goodness-of-Fit on F^2	0.908
R_1 [$I \geq 2\sigma(I)$]/R_1 (all data)	0.0536/0.1162
wR_2 [$I \geq 2\sigma(I)$]/wR_2 (all data)	0.0948/0.1277
Largest diff. peak and hole (e·Å^3)	0.280/−0.579
Deposition number	2,234,251

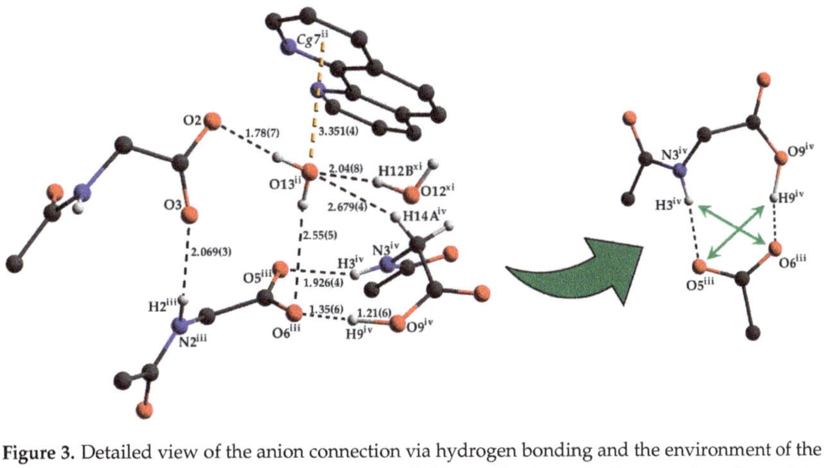

Figure 3. Detailed view of the anion connection via hydrogen bonding and the environment of the O13 water molecule with lone-pair···π interaction (left) as well as the attractive secondary electrostatic interactions (right). The hydrogen atoms which are not involved in hydrogen bonding have been omitted for clarity. Bond lengths are given in Å. Symmetry codes: (ii) −x + 1/2, y + 1/2, z; (iii) −x, −y + 1, −z + 1; (iv) x − 1/2, −y + 3/2, −z + 1; (xi) −x + 1, y, z (see Tables S1–S3 for details).

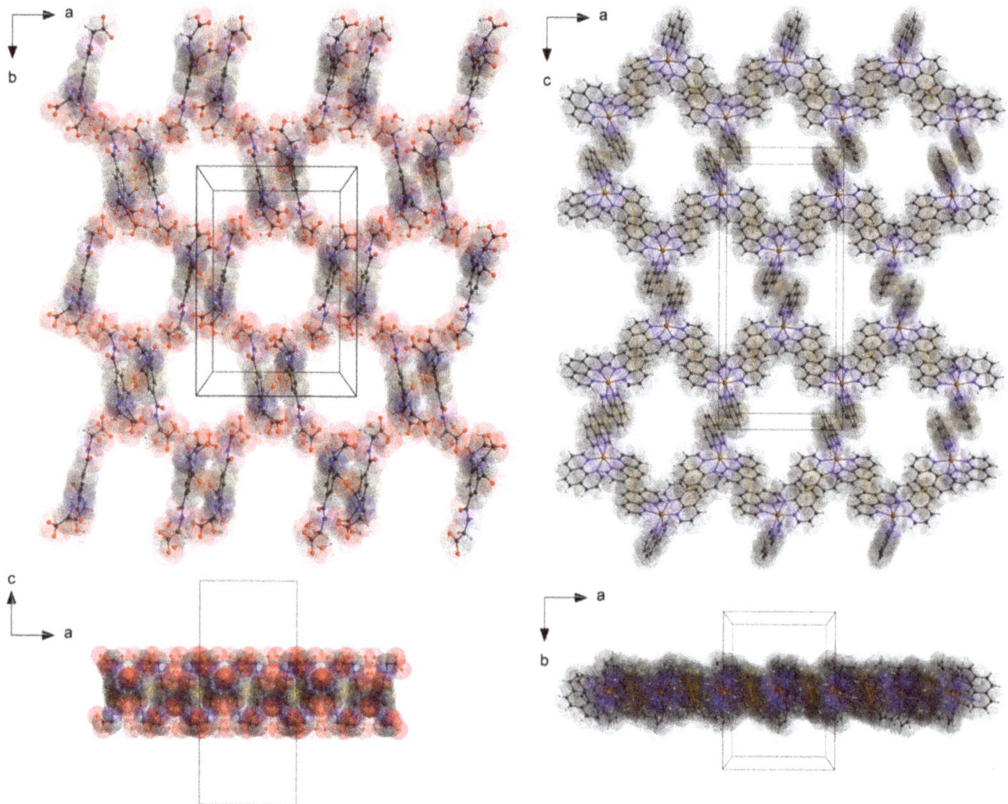

Figure 4. The perspective supramolecular network of the monolayers of the anions and cations connected via hydogen bonding and π–π-interactions. Atoms drawn with transparent van der waals radii. Atom color codes: bronze (Zn), red (O), blue (N), dark grey (C), and white (H).

However, the distance of the hydrogen atom H21 of the *Cg3* centroid of the phenanthroline 1 and the centroid *Cg1* of the dianion (Figure S4) of 3.3033 (1) Å indicate rather an attractive interaction of the positively charged C-bounded hydrogen atom and the aromatic system and thus a C–H···π interaction. This is supported by a comparison of the distance of the centroids *Cg1* and *Cg3* with a distance of 4.4404 (2) Å, which is no longer in the usual range of distances of π–π interactions. The observed geometry of the C–H···π interactions is in good agreement with the literature values [36,74]. The pairwise energy calculations based on SAPT of the PSI4 program package [62] show the largest energy gain in the interaction of phenanthroline 1 and the dianion. The pairwise interaction energy calculation results for the phenanthrolines and the phenanthrolines with the dianion can be found in Table S4 of the SI.

The lone-pair of the O13 water molecule interact with the aromatic system of the centroid *Cg7* of the phenantroline 2. The measured distance between O13 and the centroid *Cg7* is 3.351 (4) Å and comparable with the literature [15,18]. Due to the cooperative nature of this water molecule's involvement in hydrogen bonding and further weak interactions, the hydrogen bond distance of H13B to the hydrogen bond acceptor is greatly extended (Figure 3).

3.2. Hirshfeld Surface Analysis, Pairwise Interaction Energies and Energy Framework

In order to describe and understand the composition of the different weak interactions in supramolecular architectures of crystals, Hartree–Fock (HF) and Density Functional Theory (DFT) can be used. These quantum mechanical modelling methods are employed in chemistry and physics as well as in related scientific research fields of material science for calculating electronic structures, spectra, and energies of molecules. The visualization of the results in form of orbitals of the different theories (e.g., MOs, NBOs, etc.) and the molecule surfaces [55,75] gain a deeper insight into the weaker interactions in the crystals, and the findings can be utilized in crystal engineering. Based on the mentioned methods, the CrystalExplorer [61] software offers the possibility of generating such surfaces and newly calculate of the pairwise interaction energies which form the basis for the visualization of the energy framework of the crystal [59,60]. To investigate the close contacts between the ionic molecules in the present structure, the Hirshfeld surfaces (HS) mapped over the d_{norm}, the electrostatic potential energy and the shape index with the associated fingerprint plots (FP) have been generated with CrystalExplorer using TONTO with standard settings and for the electrostatic potential energy with the dispersion corrected hybird functional B3LYP and the 6-311G(d,p) basis set. To allow the viewing of all atoms and molecules around which the surface is calculated, all the surfaces are drawn transparently.

The three-dimensional HS mapped over d_{norm} of the cationic complex in the range −0.1974 to 1.5761 and for the dianion in the range −0.8003 to 1.6868 are shown in Figures 5 and 6, respectively. The blue areas of the HS surfaces represent the longer distances than the van der Waals radii, and the white areas the close contacts equal to the sum of van der Waals radii. The red spots display closer distances than the van der Waals radii and thus indicate donors and acceptors of classical and non-classical hydrogen bonding. The big bright-red spots reveal the types of O–H···O as the stronger classical hydrogen bonds of the dianion, whereas the smaller and lighter red spots reveal the non-classical hydrogen bonds of the type C–H···O of the aromatic C-bounded hydrogens of the cationic complex and the dianion.

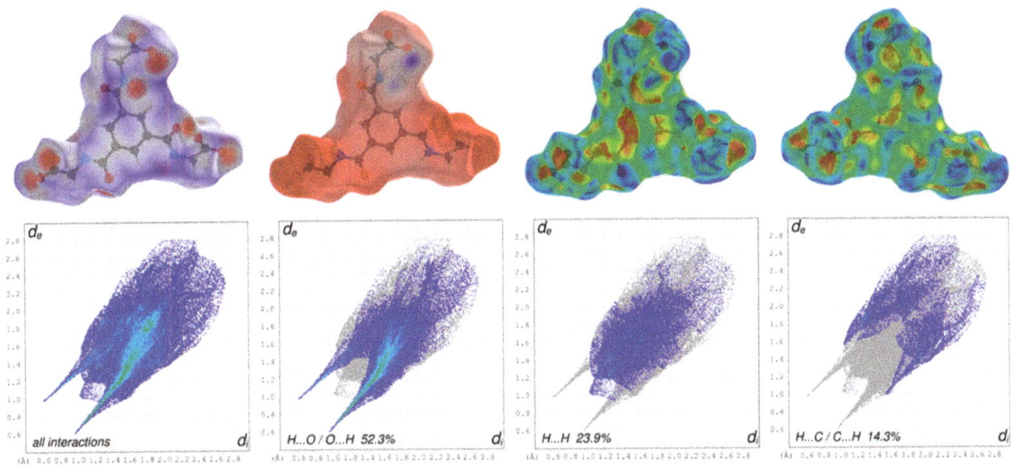

Figure 5. Graphical representations of the Hirshfeld surfaces of the dianion drawn with transparency and mapped over the d_{norm}, the electrostatic potential energy and the shape-index of both sides of the anion (upper part; from left to right) and the assoiated two-dimensional fingerprint plots of all and main interactions (lower part).

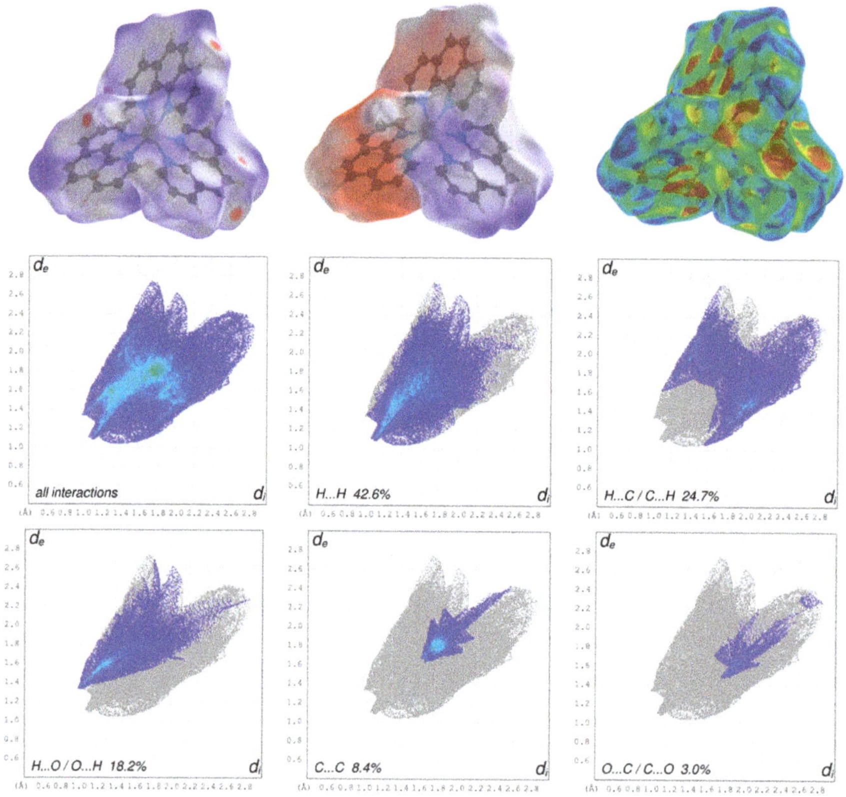

Figure 6. Graphical representations of the Hirshfeld surfaces of the cationic complex drawn with transparency and mapped over the d_{norm}, the electrostatic potential energy and the shape-index (upper part, from left to right) and the associated two-dimensional fingerprint plots of all and main interactions (lower part).

The appearance of the small single spike in the corresponding FP of the cationic complex in the region de ~1.35 Å/di~1.00 Å as well as the larger and long spikes of dianion in the region de ~0.98 Å/di~0.61 Å and de ~0.61 Å/di~0.98 Å confirm the presence of the hydrogen bonding, respectively. The d_i and the d_e are defined as the internal and external distances between the HS and are given in angstrom. The absence of the second spike and the blue region in the FP of the close contacts of hydrogens and oxygens of the cationic complex reveals that the cationic complex only acts as a hydrogen bond donor. The HS mapped over the electrostatic potential energy verifies the role of donor and acceptor functions in the hydrogen bonding network. The transparent HS of the complex cation in the range –0.0320 to 0.0471 and the dianion in the range –0.3274 to 0.1925 are depicted in Figures 5 and 6, respectively. The blue spots are assigned to the hydrogen bond donors of the aromatic C-bounded hydrogens of the cationic complex. The region around the protonated oxygens of the carboxylate group of the dianion and the nitrogen of the amides show also big blue spots and, therefore, indicates the hydrogen bond donor function. The big red spots of the HS in the region around the deprotonated oxygens of the dianion can be assigned to the hydrogen bond acceptors. In addtion, the three-dimensional deformation density maps of the dianion and the phenathrolines were also plotted using CrystalExplorer and are illustrated in Figure S6 of the SI.

The deformation energy may reveal the presence of charge depletion directed to the region of charge concentration. In a comparison of the main contributions to the HS of the complex cation and dianion, it is striking that the close contacts of the cation, which are attributed to the non-classical hydrogen bonding of H···O/O···H (18.2%), are decreased to the values of the dianion of classical and non-classical hydrogen bonding of H···O/O···H (52.3%) with a simultaneously increasing of the close contacts of H···H (42.6%) of cationic complex and H···H (23.9%) of the dianion. This difference is due to the fact, that the hydrogen bonding of the cationic complex has only a donor nature and is more involved in π–π stacking interactions as well as the high number of the H atoms of the phenantroline ligands of the cationic complex. Therefore, the close contacts of C···C (8.4%) of the cationic complex is increased to the C···C (5.6%) close contacts of the dianion. In general, the blue–red triangles in the three-dimensional HS mapped over the shape index indicate π–π stacking interactions of aromatic compounds and moities. Figure 5 shows a blue–red triangle of the HS mapped ovre the shape index on one side of the dianion and an orange deformation of the surface on the other side. This corresponds to the π–π stacking interactions of two dianions leading to the dimeric anion units. The deformation of the HS is caused by the hydrogen atom of the phenanthroline 1 which is involved in C–H···π interaction with the dianion (Figure S4). The mentioned triangles are also observed by the phenantrholines of the cationic complex. For a better overview, the HS mapped over the shape index for all phenanthrolines and with both sides are given in Figure S5 of the SI. A detailed look shows that all phenanthroline ligands have blue–red triangles and, therefore, are involved in π–π stacking interactions. Phenanthroline 2 and phenanthroline 3 have a orange colored deformation on the surface. This indicates the C–H···π interaction of the phenanthrolines and is in agreement with the single-crystal structure analysis (Figure S4).

CrystalExplorer has also been used for the intermolecular energy calculation based on the CE-B3LYP/6-31G(d,p) and CE-B3LYP/DGDZVP level of theory for the crystal monomers, respectively, and to visualize the pairwise interaction energies and the three-dimensional energy frameworks of the present compound (Figure 7). The calculated pairwise interaction energy details are summarized in Table S5 of the SI, and the energy framework divided into the electrostic, dispersion and total energy is depicted in Figure S7. The total energies are only reported for two benchmarked energy models which are the sum of the four energy components and scaled appropriately [60]. The pairwise interaction energies and the associated energy framework exhibit the dominance of attractive and repulsive electrostatic forces between the counter ions.

3.3. Vibrational and Theoretical Spektra

For all the theoretical calculations, the density functional programs provided by the ORCA 4.0.1. [53] package was used. Before the spectra were calculated, the structures were optimized to avoid a convergence to a saddle-point leading to negative frequencies. The optimization results are presented in Figures 8 and 10 wherein the original structures of the cationic complex and dianion based on the crystal structure coordinates are colored in yellow, and the optimized structures were colored in red. All calculations were carried out using the pure density, local, and gradient corrected functionals of Becke and Perdew, which is known as BP86 [76] and provide excellent geometries and vibrational spectra in most cases. The def2 basis sets of the Karlsruhe group of TZVPP and TZVP [77] were employed for the infrared and Raman spectra calculations, and the def2/J auxiliary basis [78], including the RI approximation, was used for the coulomb fitting, respectively, featuring a valence triple-zeta basis set with polarization functions on all atoms. For the unrestricted theoretical calculations of the anion spectra, the diffuse and minimally augmented basis sets were used [79]. The resulting theoretical spectra are compared with the measured ones as depicted in Figures 9 and 11, wherein the lines of the theoretical spectra of the cationic complexes are colored in red and the lines of the dianion are colored in green. The room temperature baseline-corrected FT–MIR spectrum was measured in the range of 3600 and 400 cm^{-1}, and the Raman spectrum was measured in the range of 3600 and 0 cm^{-1}, respectively. The lines of the measured spectra are colored in black.

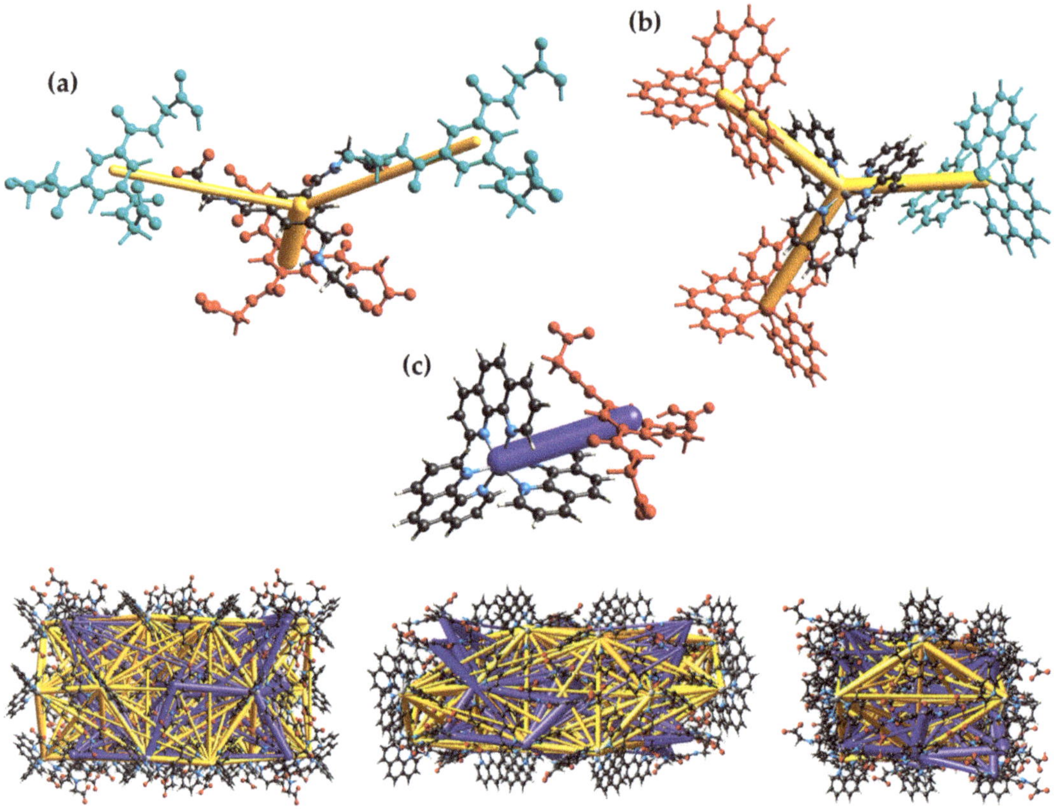

Figure 7. Visualization of the pairwise interaction energies (gold and blue bars) between the anions (**a**), the complex cations (**b**), anion and cation (**c**) with color code as well as the energy framework in the crystal (bottom) along the *a* axis (left), *b* axis (middle) and *c* axis (right).

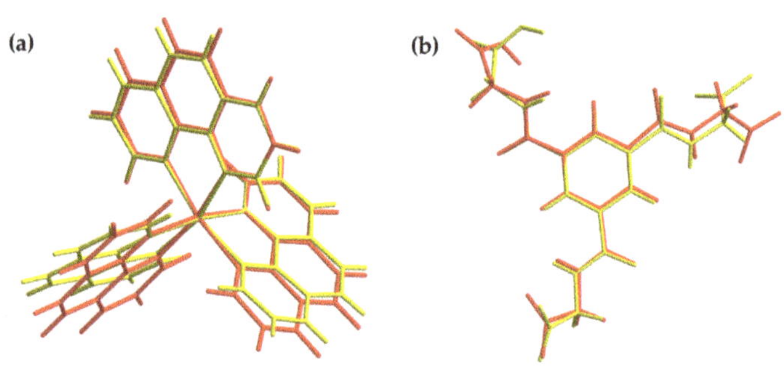

Figure 8. Overlay view of the structures in the crystal (yellow) and optimized structures (red) of the cationic complex (**a**) and the dianion (**b**). Optimization based on BP86/def2-TZVPP (cation) and BP86/ma-def2-TZVPP (anion) level of theory, respectively.

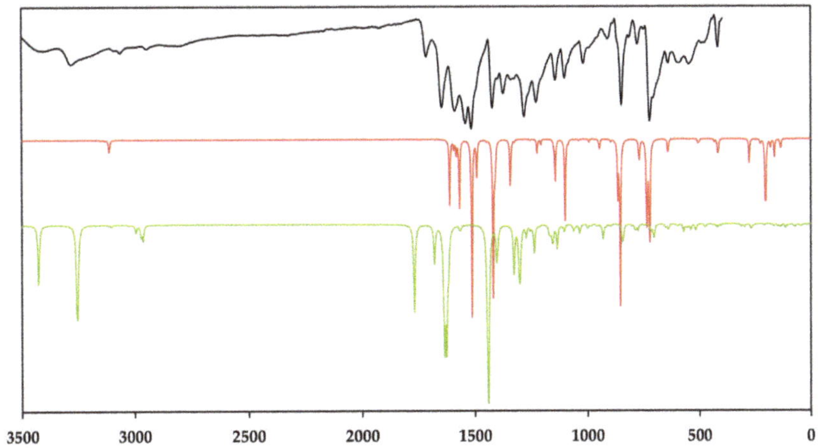

Figure 9. The baseline corrected room temperature FT–MIR spectrum (black) of the crystalline $[Zn(C_{12}H_8N_2)_3]C_{15}H_{13}N_3O_9{\cdot}5H_2O$ and the calculated IR spectra of the cationic complex (red) and the dianion (green). Theoretical spectra were calculated using BP86/def2-TZVPP (cation) and BP86/ma-def2-TZVPP (anion), respectively. The intensity of the normalizied IR spectra is given in transmittance.

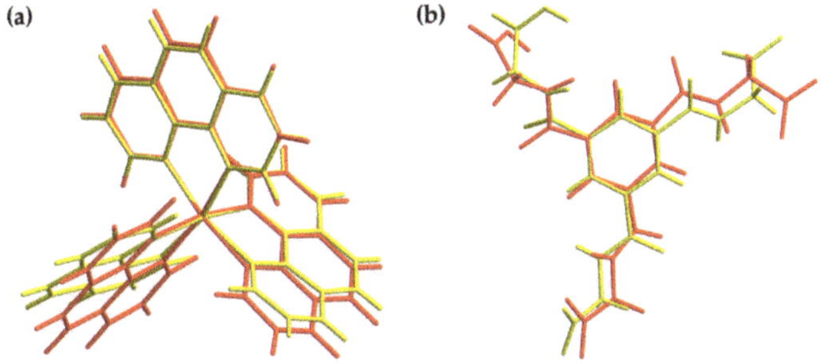

Figure 10. Overlay view of the structures in the crystal (yellow) and optimized structures (red) of the cationic complex (**a**) and the dianion (**b**). Optimization based on BP86/def2-TZVP (cation) and UKS BP86/ma-def2-TZVP (anion) level of theory, respectively.

As depicted in Figure 9, the measured infrared spectrum shows two broad absorption bands in the region of 3404 and 3278 cm^{-1} with a sharper intensity due to the asymmetric and symmetric O–H stretching vibrations of the lattice water and the secondary amide of the amide groups. It is evident that these bands have a sharp character in the theoretical spectrum. This fact can be explained by the absence of the hydrogen bonds between the lattice water and the amines of different strength, which broaden these bands [80,81]. The aromatic C–H stretching vibrations of TCMBT and the phenanthroline occur in the range of 3109 and 3089 cm^{-1} and followed by the asymmetric and symmetric stretching vibrations of the methylene group of the side chain of the TCMBT in the region of 2969 and 2946 cm^{-1} in the infrared and Raman spectra. As a result of the strong hydrogen bonding of the carboxylate groups, the very weak broad absorption at 2806 cm^{-1} in the infrared spectrum may thatindicate the O–H stretching vibrations of the carboxylate group, which is underpinned by non-appearance in the calculated spectrum of the dianion [82]. A further indication is provided by the occurrence of the vibration bands located at 1715 cm^{-1} of a medium intensity in the infrared spectrum and at 1719 cm^{-1} of a very weak intensity in the Raman spectrum, which derive from the out-of-phase and in-phase C=O stretch in dimers [83], or it may indicate symmetric C=O stretching vibrations [82]. The appearance of the absorption at 1646 cm^{-1} in the infrared and 1625 cm^{-1} in the Raman spectrum is associated with the C=O stretching of the carbonyl and/or carboxylate group.

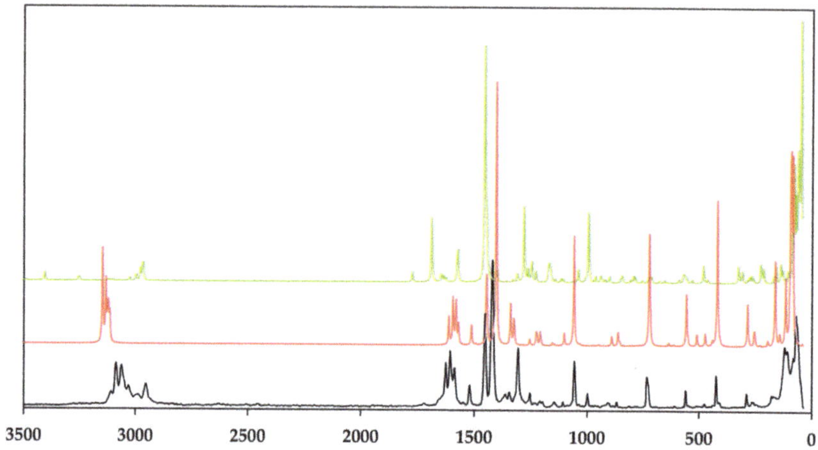

Figure 11. Room temperature raman spectrum of the crystalline $[Zn(C_{12}H_8N_2)_3]C_{15}H_{13}N_3O_9{\cdot}5H_2O$ (black) and the calculated Raman spectra of the cationic complex (red) and the dianion (green). Theoretical spectra were calculated using BP86/def2-TZVP (cation) and UKS BP86/ma-def2-TZVP (anion), respectively. Raman intensity of the normalized spectra is given in arbitrary units.

However, the spectra are dominated by the vibrations of the aromatic rings and the associated groups. The metal-nitrogen stretches are found in the region of 400–180 cm^{-1}, and the assignments can be found in the literature [84,85]. The vibrational modes below 180 cm^{-1} are caused by the lattice vibrations. The full vibrational assignments of the phenanthroline have been described earlier [86–89]. Due to the symmetry of the cationic complex, the Raman spectrum is dominated by the vibrations of this complex. It must be noted that the findings from the single-crystal structure analysis correspond very well with those of the measured and calculated spectra.

3.4. *Rietveld Refinement*

For the Rietveld refinement procedure and visualization of the refinement results the JANA2006 program package [52] was used. Figure 12 gives a graphically summarized overview of the Rietveld refinement results. For the refinement the CIF file informations of the single-crystal X-ray crystallography loaded with JANA 2006 and during the LeBail fitting and profile matching the parameters of the scale factor, background, unit cell parameters, as well as peak shape, width, and peak symmetry were refined. The pseudo-Voigt function was used for the description of the line shape of the diffraction peaks, and the low theta peak asymmetry were corrected by the divergence correction method. The asymmetry fitting results are presented in the small window of Figure 12 and show a very good agreement with the peak shape of the measured powder diffraction pattern. All mentioned parameters were refined during the following Rietveld refinement procedure expecting the anisotropic displacement parameters, atomic distances, and angles which were fixed to the ones of the single-crystal structure refinement. As shown in the difference profile of the Rietveld refinement (black line), the calculated and measured intensities exhibit a very good correlation. Additionally, all intensities were included in the refinement. Based on the standard quality criteria of the R-values and the goodness-of-fit, the Rietveld refinement yields excellent agreement with the results of the single-crystal structure determination.

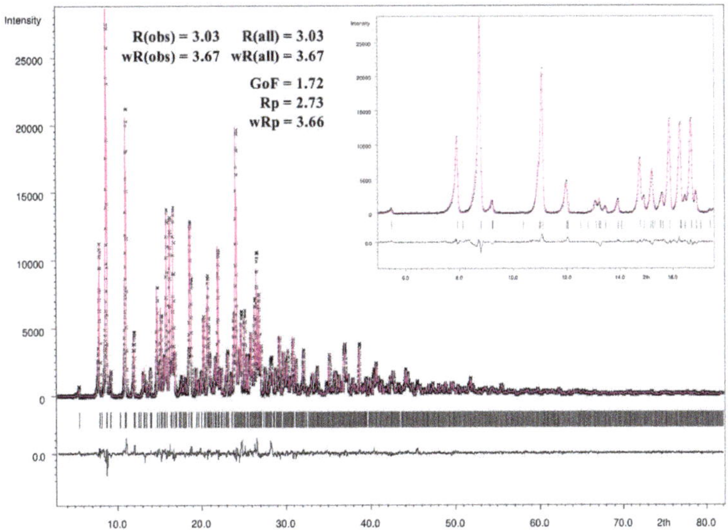

Figure 12. Observed (black crosses) and calculated (magenta line) X-ray powder diffraction pattern with difference profile of the Rietveld refinement (black line) as well as the positions of Bragg reflections (black bars) of a powder sample of the crystalline $[Zn(C_{12}H_8N_2)_3]C_{15}H_{13}N_3O_9 \cdot 5H_2O$.

3.5. Thermal Analysis

To evaluate the thermal stability of the new compound, its thermal behavior was examined using thermogravimetric analysis (TGA) (Figure 13). A weight loss of 8.7% was observed in the temperature range of 78–142 °C, which corresponds to the removal of the water molecules. Subsequently, a weight loss of 38.2% was observed in the temperature range of 245–378 °C, indicating the decomposition of the non-coordinating organic dianion (TCMBT). The additional weight loss of the sample can be attributed to the decomposition of the tris(1,10-phenanthroline)zinc(II) cation. Based on the TGA results, the tris(1,10-phenanthroline)zinc(II) cation exhibits the greatest stability.

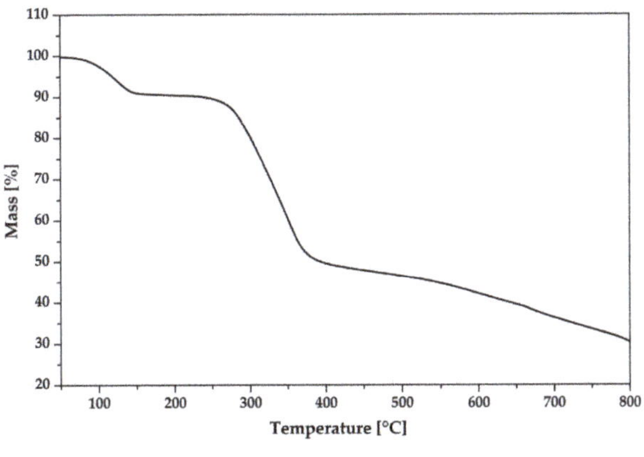

Figure 13. The thermogravimetric analysis of the crystalline $[Zn(C_{12}H_8N_2)_3]C_{15}H_{13}N_3O_9 \cdot 5H_2O$.

4. Conclusions

In summary, the reaction of N,N′,N″-tris(carboxymethyl)-1,3,5-benzenetricarboxamide (TCMBT), 1,10-phenathroline and zinc(II) nitrate in an alkaline aqueous methanolic solution leads to colorless block-shaped crystals. The single-crystal X-ray structure analysis revealed an extended supramolecular architecture mainly composed of the complex cations $[Zn(C_{12}H_8N_2)_3]^{2+}$ and TCMBT dianions. The cationic complexes are linked via π–π stacking and form a three-dimensional cationic layer network with holes. Based on hydrogen bonding and π-stacking, the dianions form dimeric units. These units are connected through cooperative hydrogen bonds with attractive secondary electrostatic interactions, resulting in an anionic layer network with honeycomb-like holes. Based on the holes, the resulting supramolecular framework can be described as an interpenetrating network of separate anionic and cationic layers connected by further weaker non-covalent interactions such as C–H···π and lone-pair···π interactions. The theoretical infrared and Raman spectra were calculated with the DFT methods and confirm the experimentally observed spectroscopic findings. The analysis of the different Hirshfeld surfaces proved the key role of the hydrogen bonding and the π-interactions in the composition of the supramolecular architecture in the crystal. Moreover, the pairwise interaction energies and associated energy framework reveal the dominance of attractive and repulsive electrostatic forces between the counterions.

Supplementary Materials: The following supporting information can be downloaded at: https://www.mdpi.com/article/10.3390/cryst13040569/s1, Table S1: Selected bond lengths (Å) and bond angle (°) of the cationic complex; Table S2: Selected bond lengths (Å) and bond angle (°) of the dianion; Table S3: Hydrogen bond geometry (Å, °) for $[Zn(C_{12}H_8N_2)_3]C_{15}H_{13}N_3O_9 \cdot 5H_2O$; Figure S1: View of the dimeric anions with hydrogen bonds (black dashed lines) and π–π-interaction (yellow dashed lines). The carbon hydrogen atoms have been omitted for clarity. Symmetry codes: (ii) −x + 1/2, y + 1/2, z; (iii) −x, −y + 1, −z + 1; Figure S2: Defintion of centroids of the phenantrholines of the cationic complex and the dianion with numbering, respectively. The hydrogen atoms have been omitted for clarity; Figure S3: View of the three-dimensional cationic π–π-interactions (yellow dashed lines). The hydrogen atoms have been omitted for clarity. Symmetry codes: (i) −x + 1, −y + 1, −z + 1; (ix) x + 1/2, y, −z + 1/2; (x) x 1/2, y, −z + 1/2; Figure S4: Representation of the C–H···π interaction of the phenanthroline 1 and the dianion (left) as well as the π stacking and C–H···π interactions of the phenanthroline 2 and 3 (right). The hydrogen atoms have been omitted for clarity. Bond lengths are given in Å. Symmetry code: (ix) x + 1/2, y, −z + 1/2; Table S4: Pairwise interaction energies (kJ/mol) calculated with PSI4 (SAPT0); Figure S5: Graphical representations of the Hirshfeld surfaces of the phenanthrolin ligands of the cationic complex drawn with transparency and mapped over the shape index; Figure S6: Graphical representations of the Hirshfeld surfaces of the dianion (upper part) and the phenanthrolin ligands of the cationic complex (lower part) mapped over the deformation density; Figure S7: Visualization of the energy framework in the crystal along the a axis (left), b axis (middle) and c axis (right) of the dispersion energy (top), the coulomb energy (middle), and the total energy (bottom); Table S5. Pairwise interaction energies of the cationic complexes and anions with color code. Interaction energy is given in kJ/mol,m and R is the distance between molecular centroids (mean atomic position) in Å; The CIF, the checkCIF and Platon output files as well as the Video S1 are included in the Supplementary Materials.

Funding: The author acknowledges financial support from the Open Access Publishing Fund of Clausthal University of Technology.

Data Availability Statement: On quoting the deposition number 2234251 at the Cambridge Crystallographic Data Centre (https://www.ccdc.cam.ac.uk/structures/) the single-crystal crystallo-graphic data set of the present compound are available free of charge.

Acknowledgments: The author is deeply indebted to A. Adam and M. Gjikaj for their support and helpful suggestions. The author thanks Karin Bode for the measurement of the IR and Raman spectra, thermal analysis and elemental analysis.

Conflicts of Interest: The authors declare no conflict of interest.

References

1. Steed, J.W.; Atwood, J.L. *Supramolecular Chemistry*, 2nd ed.; Wiley: Chichester, UK, 2009; ISBN 978-0-470-51234-0.
2. Cerný, J.; Hobza, P. Non-covalent interactions in biomacromolecules. *Phys. Chem. Chem. Phys.* **2007**, *9*, 5291–5303. [CrossRef] [PubMed]

3. Chawla, M.; Chermak, E.; Zhang, Q.; Bujnicki, J.M.; Oliva, R.; Cavallo, L. Occurrence and stability of lone pair-π stacking interactions between ribose and nucleobases in functional RNAs. *Nucleic Acids Res.* **2017**, *45*, 11019–11032. [CrossRef]
4. Martinez, C.R.; Iverson, B.L. Rethinking the term "pi-stacking". *Chem. Sci.* **2012**, *3*, 2191. [CrossRef]
5. Lehn, J.-M. *Supramolecular Chemistry: Concepts and Perspectives*; VCH: Weinheim, UK, 1995; ISBN 978-3-527-29311-7.
6. Culver, J.N. Tobacco mosaic virus assembly and disassembly: Determinants in pathogenicity and resistance. *Annu. Rev. Phytopathol.* **2002**, *40*, 287–308. [CrossRef]
7. Kawahara, S.; Uchimaru, T. Secondary Interaction Contribution in Hydrogen-Bonded Complex: Theoretical Model Study in Hydrogen Fluoride Trimer. *J. Comput. Chem. Jpn.* **2004**, *3*, 41–48. [CrossRef]
8. Zimmerman, S.C.; Murray, T.J. Hydrogen bonded complexes with the AA·DD, AA·DDD, and AAA·DD motifs: The role of three centered (bifurcated) hydrogen bonding. *Tetrahedron Lett.* **1994**, *35*, 4077–4080. [CrossRef]
9. Jorgensen, W.L.; Pranata, J. Importance of secondary interactions in triply hydrogen bonded complexes: Guanine-cytosine vs uracil-2,6-diaminopyridine. *J. Am. Chem. Soc.* **1990**, *112*, 2008–2010. [CrossRef]
10. Hisamatsu, Y.; Shirai, N.; Ikeda, S.; Odashima, K. A new quadruple hydrogen-bonding module with a DDAA array: Formation of a stable homodimer without competition from undesired hydrogen-bonded dimers. *Org. Lett.* **2009**, *11*, 4342–4345. [CrossRef]
11. van der Lubbe, S.C.C.; Zaccaria, F.; Sun, X.; Guerra, C.F. Secondary Electrostatic Interaction Model Revised: Prediction Comes Mainly from Measuring Charge Accumulation in Hydrogen-Bonded Monomers. *J. Am. Chem. Soc.* **2019**, *141*, 4878–4885. [CrossRef]
12. Hobza, P.; Müller-Dethlefs, K. *Non-Covalent Interactions: Theory and Experiment*; Royal Society of Chemistry: Cambridge, UK, 2010; ISBN 978-1847558534.
13. Schneider, H.-J. Binding mechanisms in supramolecular complexes. *Angew. Chem. Int. Ed. Engl.* **2009**, *48*, 3924–3977. [CrossRef]
14. Meyer, E.A.; Castellano, R.K.; Diederich, F. Interactions with aromatic rings in chemical and biological recognition. *Angew. Chem. Int. Ed Engl.* **2003**, *42*, 1210–1250. [CrossRef]
15. Jain, A.; Ramanathan, V.; Sankararamakrishnan, R. Lone pair ... pi interactions between water oxygens and aromatic residues: Quantum chemical studies based on high-resolution protein structures and model compounds. *Protein Sci.* **2009**, *18*, 595–605. [CrossRef] [PubMed]
16. Novotný, J.; Bazzi, S.; Marek, R.; Kozelka, J. Lone-pair-π interactions: Analysis of the physical origin and biological implications. *Phys. Chem. Chem. Phys.* **2016**, *18*, 19472–19481. [CrossRef]
17. Wan, C.-Q.; Chen, X.-D.; Mak, T.C.W. Supramolecular frameworks assembled via intermolecular lone pair-aromatic interaction between carbonyl and pyridyl groups. *CrystEngComm* **2008**, *10*, 475. [CrossRef]
18. Egli, M.; Sarkhel, S. Lone pair-aromatic interactions: To stabilize or not to stabilize. *Acc. Chem. Res.* **2007**, *40*, 197–205. [CrossRef] [PubMed]
19. Jia, C.; Miao, H.; Hay, B.P. Crystal Structure Evidence for the Directionality of Lone Pair−π interactions: Fact or Fiction? *Cryst. Growth Des.* **2019**, *19*, 6806–6821. [CrossRef]
20. Nishio, M. The CH/π hydrogen bond in chemistry. Conformation, supramolecules, optical resolution and interactions involving carbohydrates. *Phys. Chem. Chem. Phys.* **2011**, *13*, 13873–13900. [CrossRef]
21. Brenig, C.; Prieto, L.; Oetterli, R.; Zelder, F. A Nickel(II)-Containing Vitamin B12 Derivative with a Cofactor-F430-type π-System. *Angew. Chem. Int. Ed. Engl.* **2018**, *57*, 16308–16312. [CrossRef]
22. Eschenmoser, A. Vitamin B12: Experiments Concerning the Origin of Its Molecular Structure. *Angew. Chem. Int. Ed. Engl.* **1988**, *27*, 5–39. [CrossRef]
23. Franck, B.; Nonn, A. Novel Porphyrinoids for Chemistry and Medicine by Biomimetic Syntheses. *Angew. Chem. Int. Ed. Engl.* **1995**, *34*, 1795–1811. [CrossRef]
24. Doty, F.P.; Bauer, C.A.; Skulan, A.J.; Grant, P.G.; Allendorf, M.D. Scintillating Metal-Organic Frameworks: A New Class of Radiation Detection Materials. *Adv. Mater.* **2009**, *21*, 95–101. [CrossRef]
25. Allendorf, M.D.; Bauer, C.A.; Bhakta, R.K.; Houk, R.J.T. Luminescent metal-organic frameworks. *Chem. Soc. Rev.* **2009**, *38*, 1330–1352. [CrossRef] [PubMed]
26. Wang, C.; Volotskova, O.; Lu, K.; Ahmad, M.; Sun, C.; Xing, L.; Lin, W. Synergistic assembly of heavy metal clusters and luminescent organic bridging ligands in metal-organic frameworks for highly efficient X-ray scintillation. *J. Am. Chem. Soc.* **2014**, *136*, 6171–6174. [CrossRef] [PubMed]
27. Wang, X.; Wang, Y.; Wang, Y.; Liu, H.; Zhang, Y.; Liu, W.; Wang, X.; Wang, S. Color-tunable X-ray scintillation based on a series of isotypic lanthanide-organic frameworks. *Chem. Commun.* **2019**, *56*, 233–236. [CrossRef]
28. Lecoq, P. Development of new scintillators for medical applications. *Nucl. Instrum. Methods Phys. Res. Sect. A Accel. Spectrometers Detect. Assoc. Equip.* **2016**, *809*, 130–139. [CrossRef]
29. Perry IV, J.J.; Feng, P.L.; Meek, S.T.; Leong, K.; Doty, F.P.; Allendorf, M.D. Connecting structure with function in metal–organic frameworks to design novel photo- and radioluminescent materials. *J. Mater. Chem.* **2012**, *22*, 10235. [CrossRef]
30. Pook, N.-P. A Scintillating One-Dimensional Coordination Polymer Based on Cadmium(II), N,N′-(1,4-Phenylenedicarbonyl)diglycinate, and 2,2′-Bipyridine: Crystal Structure, Hirshfeld Surface Analysis, and Luminescence Lifetime Properties †. *Solids* **2021**, *2*, 371–384. [CrossRef]
31. Pook, N.-P.; Fruhner, C.-J.; Franzl, T.; Denzer, U.; Adam, A. Instrumentation for X-ray Excited and Laser Induced Fluorescence Lifetime Spectroscopy Using Two-Dimensional Photon Counting. *IEEE Trans. Nucl. Sci.* **2012**, *59*, 2319–2323. [CrossRef]

32. Pook, N.-P.; Fruhner, C.-J.; Franzl, T.; Denzer, U.; Adam, A. Further performance tests of a picosecond X-ray and laser induced streak camera system with fast scintillation materials. *Radiat. Meas.* **2013**, *56*, 281–284. [CrossRef]
33. Janiak, C. A critical account on π–π stacking in metal complexes with aromatic nitrogen-containing ligands †. *J. Chem. Soc. Dalton Trans.* **2000**, *21*, 3885–3896. [CrossRef]
34. Salonen, L.M.; Ellermann, M.; Diederich, F. Aromatic rings in chemical and biological recognition: Energetics and structures. *Angew. Chem. Int. Ed. Engl.* **2011**, *50*, 4808–4842. [CrossRef]
35. Riwar, L.-J.; Trapp, N.; Kuhn, B.; Diederich, F. Substituent Effects in Parallel-Displaced π-π Stacking Interactions: Distance Matters. *Angew. Chem. Int. Ed. Engl.* **2017**, *56*, 11252–11257. [CrossRef] [PubMed]
36. Lu, Z.; Gamez, P.; Mutikainen, I.; Turpeinen, U.; Reedijk, J. Supramolecular Assemblies Generated from Both Lone-Pair···π and C−H···π Binding Interactions. *Cryst. Growth Des.* **2007**, *7*, 1669–1671. [CrossRef]
37. Li, W.; Usabiaga, I.; Calabrese, C.; Evangelisti, L.; Maris, A.; Favero, L.B.; Melandri, S. Characterizing the lone pairπ-hole interaction in complexes of ammonia with perfluorinated arenes. *Phys. Chem. Chem. Phys.* **2021**, *23*, 9121–9129. [CrossRef] [PubMed]
38. Bose, P.P.; Drew, M.G.B.; Das, A.K.; Banerjee, A. Formation of triple helical nanofibers using self-assembling chiral benzene-1,3,5-tricarboxamides and reversal of the nanostructure's handedness using mirror image building blocks. *Chem. Commun.* **2006**, *30*, 3196–3198. [CrossRef] [PubMed]
39. Gong, B.; Zheng, C.; Yan, Y. Structure of N,N′,N′′-tris(carboxymethyl)-1,3,5benzenetricarboxamide trihydrate. *J. Chem. Crystallogr.* **1999**, *29*, 649–652. [CrossRef]
40. Dokorou, V.N.; Powell, A.K.; Kostakis, G.E. Two pseudopolymorphs derived from alkaline earth metals and the pseudopeptidic ligand trimesoyl-tris-glycine. *Polyhedron* **2013**, *52*, 538–544. [CrossRef]
41. Raynal, M.; Li, Y.; Troufflard, C.; Przybylski, C.; Gontard, G.; Maistriaux, T.; Idé, J.; Lazzaroni, R.; Bouteiller, L.; Brocorens, P. Experimental and computational diagnosis of the fluxional nature of a benzene-1,3,5-tricarboxamide-based hydrogen-bonded dimer. *Phys. Chem. Chem. Phys.* **2021**, *23*, 5207–5221. [CrossRef]
42. Sun, R.; Li, Y.-Z.; Bai, J.; Pan, Y. Synthesis, Structure, Water-Induced Reversible Crystal-to-Amorphous Transformation, and Luminescence Properties of Novel Cationic Spacer-Filled 3D Transition Metal Supramolecular Frameworks from N, N′, N′′-Tris(carboxymethyl)-1,3,5-benzenetricarboxamide. *Cryst. Growth Des.* **2007**, *7*, 890–894. [CrossRef]
43. Sun, R.; Wang, S.; Xing, H.; Bai, J.; Li, Y.; Pan, Y.; You, X. Unprecedented 4(2)6(4) topological 2-D rare-earth coordination polymers from a flexible tripodal acid with additional amide groups. *Inorg. Chem.* **2007**, *46*, 8451–8453. [CrossRef]
44. Loukopoulos, E.; Michail, A.; Kostakis, G. A 12-Fold ThSi2 Interpenetrated Network Utilizing a Glycine-Based Pseudopeptidic Ligand. *Crystals* **2018**, *8*, 47. [CrossRef]
45. Sheldrick, G.M. Crystal structure refinement with SHELXL. *Acta Cryst C* **2015**, *71*, 3–8. [CrossRef] [PubMed]
46. Spek, A.L. Structure validation in chemical crystallography. *Acta Crystallogr. D Biol. Crystallogr.* **2009**, *65*, 148–155. [CrossRef] [PubMed]
47. Spek, A.L. Checkcif validation ALERTS: What they mean and how to respond. *Acta Crystallogr. E Crystallogr. Commun.* **2020**, *76*, 1–11. [CrossRef] [PubMed]
48. Spek, A.L. What makes a crystal structure report valid? *Inorg. Chim. Acta* **2018**, *470*, 232–237. [CrossRef]
49. Spek, A.L. Single-crystal structure validation with the program PLATON. *J. Appl. Crystallogr* **2003**, *36*, 7–13. [CrossRef]
50. Crystal Impact—Dr. H. Putz & Dr. K. Brandenburg GbR, Kreuzherrenstr. 102, 53227 Bonn, Germany. Diamond—Crystal and Molecular Structure Visualization. Available online: http://www.crystalimpact.com/diamond (accessed on 6 March 2023).
51. Persistence of Vision Raytracer (Version 3.6)[Computer Software]. Persistence of Vision Pty. Ltd. 2004. Available online: http://www.povray.org/download/ (accessed on 6 March 2023).
52. Petříček, V.; Dušek, M.; Palatinus, L. Crystallographic Computing System JANA2006: General features. *Z. Für Krist. —Cryst. Mater.* **2014**, *229*, 345–352. [CrossRef]
53. Neese, F. Software update: The ORCA program system, version 4.0. *WIREs Comput. Mol. Sci.* **2018**, *8*, e1327. [CrossRef]
54. McKinnon, J.J.; Mitchell, A.S.; Spackman, M.A. Hirshfeld Surfaces: A New Tool for Visualising and Exploring Molecular Crystals. *Chem. Eur. J.* **1998**, *4*, 2136–2141. [CrossRef]
55. Spackman, M.A.; Jayatilaka, D. Hirshfeld surface analysis. *CrystEngComm* **2009**, *11*, 19–32. [CrossRef]
56. Spackman, M.A.; McKinnon, J.J.; Jayatilaka, D. Electrostatic potentials mapped on Hirshfeld surfaces provide direct insight into intermolecular interactions in crystals. *CrystEngComm* **2008**, *10*, 377–388. [CrossRef]
57. McKinnon, J.J.; Jayatilaka, D.; Spackman, M.A. Towards quantitative analysis of intermolecular interactions with Hirshfeld surfaces. *Chem. Commun.* **2007**, *37*, 3814–3816. [CrossRef]
58. Parkin, A.; Barr, G.; Dong, W.; Gilmore, C.J.; Jayatilaka, D.; McKinnon, J.J.; Spackman, M.A.; Wilson, C.C. Comparing entire crystal structures: Structural genetic fingerprinting. *CrystEngComm* **2007**, *9*, 648. [CrossRef]
59. Turner, M.J.; Thomas, S.P.; Shi, M.W.; Jayatilaka, D.; Spackman, M.A. Energy frameworks: Insights into interaction anisotropy and the mechanical properties of molecular crystals. *Chem. Commun.* **2015**, *51*, 3735–3738. [CrossRef]
60. Mackenzie, C.F.; Spackman, P.R.; Jayatilaka, D.; Spackman, M.A. CrystalExplorer model energies and energy frameworks: Extension to metal coordination compounds, organic salts, solvates and open-shell systems. *IUCrJ* **2017**, *4*, 575–587. [CrossRef]
61. Spackman, P.R.; Turner, M.J.; McKinnon, J.J.; Wolff, S.K.; Grimwood, D.J.; Jayatilaka, D.; Spackman, M.A. CrystalExplorer: A program for Hirshfeld surface analysis, visualization and quantitative analysis of molecular crystals. *J. Appl. Crystallogr.* **2021**, *54*, 1006–1011. [CrossRef]

62. Smith, D.G.A.; Burns, L.A.; Simmonett, A.C.; Parrish, R.M.; Schieber, M.C.; Galvelis, R.; Kraus, P.; Kruse, H.; Di Remigio, R.; Alenaizan, A.; et al. Psi4 1.4: Open-source software for high-throughput quantum chemistry. *J. Chem. Phys.* **2020**, *152*, 184108. [CrossRef]
63. Wei, D.-Y.; Zheng, Y.-Q.; Lin, J.-L. Self—Assembly of Zinc Ions with Suberic Acid and Phenanthroline—Crystal Structures of five New Zinc Phenanthroline Suberato Complexes. *Z. Anorg. Allg. Chem.* **2002**, *628*, 2005–2012. [CrossRef]
64. Pook, N.-P.; Adam, A.; Gjikaj, M. Crystal structure and Hirshfeld surface analysis of (μ-2-{4-(carboxyl-atometh-yl)carbamo-ylbenz-amido}-acetato-κ2O:O′)bis-bis-(1,10-phenanthroline-κ2N,N′)copper(II) dinitrate N,N′-(1,4-phenyl-enedicarbon-yl)diglycine monosolvate octa-hydrate. *Acta Crystallogr. E Crystallogr. Commun.* **2019**, *75*, 667–674. [CrossRef]
65. Emsley, J. Very strong hydrogen bonding. *Chem. Soc. Rev.* **1980**, *9*, 91. [CrossRef]
66. Feng, G.; Favero, L.B.; Maris, A.; Vigorito, A.; Caminati, W.; Meyer, R. Proton transfer in homodimers of carboxylic acids: The rotational spectrum of the dimer of acrylic acid. *J. Am. Chem. Soc.* **2012**, *134*, 19281–19286. [CrossRef]
67. Guo, J.; Tolstoy, P.M.; Koeppe, B.; Golubev, N.S.; Denisov, G.S.; Smirnov, S.N.; Limbach, H.-H. Hydrogen bond geometries and proton tautomerism of homoconjugated anions of carboxylic acids studied via H/D isotope effects on 13C NMR chemical shifts. *J. Phys. Chem. A* **2012**, *116*, 11180–11188. [CrossRef]
68. Lin, J.; Pozharski, E.; Wilson, M.A. Short Carboxylic Acid-Carboxylate Hydrogen Bonds Can Have Fully Localized Protons. *Biochemistry* **2017**, *56*, 391–402. [CrossRef]
69. Silvi, B.; Ratajczak, H. Hydrogen bonding and delocalization in the ELF analysis approach. *Phys. Chem. Chem. Phys.* **2016**, *18*, 27442–27449. [CrossRef]
70. Steiner, T. The Hydrogen Bond in the Solid State. *Angew. Chem. Int. Ed. Engl.* **2002**, *41*, 48–76. [CrossRef]
71. van der Lubbe, S.C.C.; Fonseca Guerra, C. The Nature of Hydrogen Bonds: A Delineation of the Role of Different Energy Components on Hydrogen Bond Strengths and Lengths. *Chem. Asian J.* **2019**, *14*, 2760–2769. [CrossRef]
72. Sarkhel, S.; Desiraju, G.R. N-H ... O, O-H ... O, and C-H ... O hydrogen bonds in protein-ligand complexes: Strong and weak interactions in molecular recognition. *Proteins* **2004**, *54*, 247–259. [CrossRef]
73. Lüning, U.; Kühl, C.; Uphoff, A. Four Hydrogen Bonds−DDAA, DADA, DAAD and ADDA Hydrogen Bond Motifs. *Eur. J. Org. Chem.* **2002**, *2002*, 4063–4070. [CrossRef]
74. Kumar Seth, S.; Dey, B.; Kar, T.; Mukhopadhyay, S. Experimental observation of supramolecular carbonyl–π/π–π/π–carbonyl assemblies of CuII complex of iminodiacetate and dipyridylamine. *J. Mol. Struct.* **2010**, *973*, 81–88. [CrossRef]
75. Hirshfeld, F.L. Bonded-atom fragments for describing molecular charge densities. *Theoret. Chim. Acta* **1977**, *44*, 129–138. [CrossRef]
76. Becke, A.D. Density-functional exchange-energy approximation with correct asymptotic behavior. *Phys. Rev. A Gen. Phys.* **1988**, *38*, 3098–3100. [CrossRef] [PubMed]
77. Weigend, F.; Ahlrichs, R. Balanced basis sets of split valence, triple zeta valence and quadruple zeta valence quality for H to Rn: Design and assessment of accuracy. *Phys. Chem. Chem. Phys.* **2005**, *7*, 3297–3305. [CrossRef] [PubMed]
78. Weigend, F. Accurate Coulomb-fitting basis sets for H to Rn. *Phys. Chem. Chem. Phys.* **2006**, *8*, 1057–1065. [CrossRef] [PubMed]
79. Zheng, J.; Xu, X.; Truhlar, D.G. Minimally augmented Karlsruhe basis sets. *Theoret. Chim. Acta* **2011**, *128*, 295–305. [CrossRef]
80. Hesse, M.; Maier, H.; Zeeh, B. *Spektroskopische Methoden in der Organischen Chemie: 96 Tabellen*, 4th ed.; Thieme: Stuttgart, Germany, 1991; ISBN 3-13-576104-5.
81. Nakamoto, K. *Infrared and Raman Spectra of Inorganic and Coordination Compounds*, 6th ed.; Wiley-Blackwell: Oxford, UK, 2008; ISBN 978-0-471-74493-1.
82. Socrates, G. *Infrared and Raman Characteristic Group Frequencies: Tables and Charts*, 3rd ed.; Wiley: Chichester, UK, 2001; ISBN 0-471-85298-8.
83. Larkin, P.J. *Infrared and Raman Spectroscopy: Principles and Spectral Interpretation*; Elsevier: Amsterdam, The Netherlands, 2011; ISBN 978-0-12-386984-5.
84. Krishnan, K.; Plane, R.A. Raman and infrared spectra of o-phenanthroline and its complexes with Zn(II) and Hg(II). *Spectrochim. Acta Part A Mol. Spectrosc.* **1969**, *25*, 831–837. [CrossRef]
85. Thornton, D.A.; Watkins, G.M. The Infrared Spectra (4000–50 cm^{-1}) of Complexes of 2,2′-Bipyridine, 1,10-Phenanthroline and their Perdeuterated Analogues with Metal(II) Perchlorates of the First Transition Series. *J. Coord. Chem.* **1992**, *25*, 299–315. [CrossRef]
86. Altmann, V.W.; Perkampus, H.-H. Normalkoordinatenanalyse der phenanthroline teil. I. Die planaren grundschwingungen. *Spectrochim. Acta Part A Mol. Spectrosc.* **1979**, *35*, 253–257. [CrossRef]
87. Perkampus, H.-H.; Rother, W. Die Infrarot- und Ramanspektren der Phenanthroline. *Spectrochim. Acta Part A Mol. Spectrosc.* **1974**, *30*, 597–610. [CrossRef]
88. Reiher, M.; Brehm, G.; Schneider, S. Assignment of Vibrational Spectra of 1,10-Phenanthroline by Comparison with Frequencies and Raman Intensities from Density Functional Calculations. *J. Phys. Chem. A* **2004**, *108*, 734–742. [CrossRef]
89. Thornton, D.A.; Watkins, G.M. A full vibrational assignment (4000-50 cm^{-1}) of 1, 10-phenanthroline and its perdeuterated analogue. *Spectrochim. Acta Part A Mol. Spectrosc.* **1991**, *47*, 1085–1096. [CrossRef]

crystals

MDPI

Article

Novel Mononuclear Tetrabromonitrosylrhenate(II) Complexes Containing Azole-Type Ligands: Magnetostructural Characterization through Hirshfeld Surfaces Analysis

Mario Pacheco [1,*], Javier González-Platas [2], Carlos Kremer [1], Miguel Julve [3], Francesc Lloret [3,*] and Alicia Cuevas [1]

1 Área de Química Inorgánica, Departamento Estrella Campos, Facultad de Química, Universidad de la República, Av. General Flores 2124, Montevideo CP 11800, Uruguay; acuevas@fq.edu.uy (A.C.)
2 Departamento de Física, Instituto Universitario de Estudios Avanzados en Física Atómica, Molecular y Fotónica (IUDEA), MALTA Consolider Team, Universidad de La Laguna, Avenida Astrofísico Fco. Sánchez s/n, E-38204 La Laguna, Tenerife, Spain
3 Instituto de Ciencia Molecular (ICMol)-Departamento de Química Inorgánica, Universidad de Valencia, C/Catedrático José Beltrán 2, 46980 Paterna, Valencia, Spain; miguel.julve@uv.es
* Correspondence: mpacheco@fq.edu.uy (M.P.); francisco.lloret@uv.es (F.L.)

Abstract: Our research group has made incursions into the scarcely known coordination chemistry of rhenium(II). The literature shows that Re(II) mononuclear complexes are attractive in molecular magnetism due to high magnetic anisotropy because of a significant spin-orbit coupling, making them a potential source for new molecule-based magnets. In this work, we present the preparation of four novel Re(II) compounds of general formula $NBu_4[Re(NO)Br_4(L)]$ [NBu_4^+ = tetra-*n*-butylammonium: L = imidazole (**1**), pyrazole (**2**), 1,2,4-triazole (**3**) and 1H-tetrazole (**4**)]. The four compounds were fully characterized by single-crystal X-ray diffraction, infrared spectroscopy, and cryomagnetic measurements in the temperature range of 1.8–300 K. Their crystal structures consist of mononuclear $[Re(NO)Br_4(L)]^-$ complex anions and NBu_4^+ cations. Each Re(II) ion is six-coordinate with a linear nitrosyl group and one monodentate nitrogen-donor (L), which are trans-positioned, plus four bromide groups, building a tetragonally distorted octahedral surrounding. The inter-anionic contacts were thoroughly analyzed using Hirshfeld surface analyses (plots over the d_{norm}, shape index, and 2D fingerprints). Cryomagnetic measurements show that these complexes behave as quasi-magnetically isolated spin doublets with weak antiferromagnetic interactions at low temperatures. The magnetic behavior of Re(II) was modeled by the influence of the ligand field, tetragonal distortion, spin-orbit coupling, and covalence effects. In addition, the antiferromagnetic exchange coupling was correlated to the nature of the intermolecular interactions.

Keywords: X-ray structure; rhenium(II); triazole; pyrazole; imidazole; tetrazole; magnetic properties

Citation: Pacheco, M.; González-Platas, J.; Kremer, C.; Julve, M.; Lloret, F.; Cuevas, A. Novel Mononuclear Tetrabromonitrosylrhenate(II) Complexes Containing Azole-Type Ligands: Magnetostructural Characterization through Hirshfeld Surfaces Analysis. *Crystals* **2023**, *13*, 658. https://doi.org/10.3390/cryst13040658

Academic Editor: Waldemar Maniukiewicz

Received: 24 March 2023
Revised: 5 April 2023
Accepted: 6 April 2023
Published: 11 April 2023

1. Introduction

In the last twelve years, mononuclear rhenium(II) complexes have garnered interest due to their various properties, such as catalysis [1,2], substance delivery [3], and photochemistry [4]. In the field of molecular magnetism, the Re(II) ion's high anisotropy, resulting from significant spin-orbit coupling, makes its complexes a potential source for discovering new molecule-based magnets [5,6]. In this respect, an illustrative example is the single-molecule magnet (SMM) behavior reported for $[\{MnCl\}_4\{Re^{II}(triphos)(CN)_3\}_4]$ [triphos = 1,1,1-tris(diphenylphosphinomethyl)ethane] by Dunbar's group [7,8]. Since then, our research team has started a systematic magnetostructural study of new mononuclear rhenium(II) complexes containing potentially bridging diazine-type ligands that can generate polynuclear complexes [9–12]. Tetrabromonitrosylrhenate(II) complexes have proven themselves to be suitable metalloligands towards first-row transition metal ions whose coordination sphere is partially blocked. The mononuclear Re(II) complexes behave like quasi-magnetically isolated spin doublets with weak antiferromagnetic interactions across short Re-X···X-Re intermolecular contacts. Some of these complexes have been studied as salts of small diamagnetic and paramagnetic ions, affecting the through-space magnetic coupling between the $[Re(NO)Br_4(L)]^-$ complex anions [13–15].

A detailed magnetic description of this type of compounds must be accompanied by a deep chemical characterization of the compound, particularly with regard to the electronic structure of the spin-carrying centers, their tridimensional arrangement, and connectivity, in order to formulate coherent Hamiltonian formalisms consistent with the number and type of metal centers in the crystal [16]. Herein, the observed magnetic exchange interactions occur in interionic or intermolecular forms. This behavior can be investigated through Hirshfeld surfaces analysis, which is a simple yet powerful tool that allows us to determine and describe semi-quantitatively the intermolecular surrounding in the considered species [17,18]. In this work, we focus on the preparation and characterization of a new set of Re(II) mononuclear complexes containing azole-type ligands that could potentially act as precursors of polynuclear compounds by using them as metalloligands [19–21]. Specifically, we report the synthesis, chemical, and cryomagnetic characterization of four novel compounds of $NBu_4[Re(NO)Br_4(L)]$ (NBu_4^+ = tetra-*n*-butylammonium: L = imidazole (**1**), pyrazole (**2**), 1,2,4-triazole (**3**) and 1*H*-tetrazole (**4**). Additionally, we performed a detailed study of the crystal packing using Hirshfeld surfaces analysis and paid particular attention to the Br···Br contacts, $\pi-\pi$ stacking, C−H···O, and conventional hydrogen bonds as potential pathways for through-space magnetic interactions [22].

2. Materials and Methods

2.1. Materials

All manipulations were performed under aerobic conditions, using materials as received (reagent grade). The precursor $NBu_4[Re(NO)Br_4(EtOH)]$ was prepared from $KReO_4$ and NO, as previously reported [9].

2.2. Physical Methods

Elemental analyses (C, H, N) were performed with a CE Instruments CHNS 1100 Elemental Analyzer [samples of 25 (**1**), 20 (**2**), 25 (**3**), and 23 mg (**4**)] by the Central Service for the Support to Experimental Research (SCSIE) at the University of Valencia. The IR spectra were recorded on an FTIR Shimadzu Prestige-21 spectrophotometer in the range 4000–400 cm^{-1}. In addition, magnetic susceptibility measurements were carried out on polycrystalline samples with a Superconducting Quantum Interference Design (SQUID) magnetometer in the 1.9–300 K temperature range. In order to avoid saturation phenomena, external dc magnetic fields of 250 G ($T < 50$ K) and 5000 G ($T \geq 50$ K) were used. Diamagnetic corrections from the constituent atoms were estimated from Pascal constants [23]. Corrections for the magnetization of the sample holder (a plastic bag) measured in the same conditions were also applied.

2.3. Synthesis of the Complexes

Complexes **1–4** were prepared by using a common synthetic procedure: 0.91 mmol of L was added to a solution of $NBu_4[Re(NO)Br_4(EtOH)]$ (0.060 mmol, 50 mg) dissolved in 20 cm^3 of EtOH. The reaction mixture was stirred at room temperature for three hours, leading to a yellow-greenish solution. Then, it was filtered to remove any small solid particles and allowed to evaporate slowly at room temperature. X-ray-quality crystals, as green polyhedrons, were grown after ten days. These were filtered off and dried in the air. Yield (based on Re): ca. 48 (**1**), 41 (**2**), 35 (**3**), and 41% (**4**). Calc. for $C_{19}H_{40}Br_4N_4ORe$ (**1**/**2**): C, 27.0; H, 4.76; N, 6.62. Found: C, 27.3; H, 4.85; N, 6.44% (**1**). C, 27.3; H, 4.76; N, 6.48% (**2**). Anal. Calc. for $C_{18}H_{39}Br_4N_5ORe$ (**3**): C, 25.5; H, 4.64; N, 8.26. Found: C, 25.2; H, 4.64; N, 8.26% (**3**). Anal. Calc. for $C_{17}H_{38}Br_4N_6ORe$ (**4**): C, 24.1; H, 4.51; N, 9.91. Found: C, 24.0; H, 4.51; N, 9.91% (**4**). IR (KBr disk/cm^{-1}): ν(NO) is located at 1765(s), 1769(s), 1769(s) and 1765(s) for **1–4**, respectively. For other IR absorption bands, refer to Table S1 [24–26].

2.4. X-ray Data Collection and Structure Refinement

X-ray diffraction data were collected for single crystals of **1–4** using Agilent SuperNOVA (for **1**, **2**, and **4**) and Bruker D8 venture (for **3**) diffractometers at 150 K (for **1** and **3**) and 293 K (for **2** and **4**), with microfocus X-ray Mo Kα radiation (λ = 0.71073 Å) used in both diffractometers. CrysAlisPro v2021 [27] software (**1**, **2**, and **4**) and Apex2 [28] (**3**) were employed to collect, index, scale, and apply a numerical absorption correction based on Gaussian integration over a multifaceted crystal model. The structures were solved by applying the dual-space algorithm implemented in the SHELXT-2014 program [29]. Fourier recycling and least-squares refinement were used for the model completion with SHELXL-2018 [30]. All non-hydrogen atoms were refined with anisotropic thermal parameters through full-matrix least-squares procedures on F^2. All hydrogen atoms were allowed to ride on their parent atoms with $U_{iso}(H) = 1.2U_{eq}(C)$. Geometrical analysis of the interactions in the structures was performed with PLATON [31] and Olex2 [32] programs. Only in structure **4** did we apply restraints to the C-C distances for one of the NBu_4^+ cations since it presented a slight disorder. The CIF files contain the corresponding refinement details. Crystal data, collection procedures, and refinement results are summarized in Table S2, while selected bond lengths and angles are listed in Table 1. Mercury software (2022.1.0 version) was used for crystal structure visualizations [33]. The crystallographic data for the structures reported in this paper have been deposited with the Cambridge Crystallographic Data Centre as supplementary publication 2247513–2247516. Copies of the data can be obtained free of charge by applying to the CCDC, Cambridge, UK (http://www.ccdc.cam.ac.uk (accessed on 6 April 2023).

Table 1. Selected bond distances (Å) and angles (°) for **1–4**.

		Bond Lengths (Å)		
	1	**2**	**3**	**4A/4B**
$Re1-N2_L$	2.173(4)	2.191(3)	2.194(4)	2.171(12)/2.183(12)
$Re1-N1_{NO}$	1.736(4)	1.732(4)	1.743(4)	1.735(13)/1.759(14)
Re1−Br1	2.5628(5)	2.5335(5)	2.5558(5)	2.5013(16)/2.5020(18)
Re1−Br2	2.5347(5)	2.5346(5)	2.5290(4)	2.5197(17)/2.5286(18)
Re1−Br3	2.5103(5)	2.5250(5)	2.5041(4)	2.5172(15)/2.5017(13)
Re1−Br4	2.5274(5)	2.5130(5)	2.5228(5)	2.5145(16)/2.5070(16)
$N1_{NO}-O1$	1.185(5)	1.176(5)	1.179(5)	1.139(13)/1.100(14)
		Bond angles (°)		
$N1_{NO}-Re1-N2_L$	178.69(14)	177.6(1)	177.9(1)	178.8(5)/179.2(6)
$Re1-N_{NO}-O1$	177.8(4)	177.5(3)	176.8(3)	177.3(13)/177.6(16)
Br1−Re1−Br3	172.141(19)	173.98(2)	171.41(2)	171.11(6)/171.99(6)
Br2−Re1−Br4	175.197(18)	171.72(2)	173.71(2)	173.20(6)/172.77(7)
$N1_{NO}-Re1-Br2$	90.93(12)	96.5(1)	94.1(1)	91.7(4)/93.0(4)
$N2_L-Re1-Br2$	88.13(9)	85.87(9)	87.03(9)	87.3(3)/87.0(3)

2.5. Hirshfeld Surface Analysis

Hirshfeld surfaces [34] were constructed using the Crystal Explorer 17 software [35]. In particular, the intermolecular interactions in the solid state were studied by mapping the surface over the d_{norm} along with the two-dimensional fingerprint plots and over the shape index surface [36,37]. The Hirshfeld surfaces mapped over d_{norm} were calculated with the default setting of arbitrary units range; the rotation of the generated plot enables the identification of regions of interest. The colors employed to refer to the shortest and longest d_{norm} are red and blue, respectively. Red spots on the surface correspond to the shortest contacts within the surface, indicating the formation of intermolecular bonds. Contacts close in length to the van der Waals limit are white-colored [18]. Moreover, 2D fingerprint plots were computed for each interatomic contact and overall interaction. In the computation of individual interatomic contacts, the reciprocal interaction of each interatomic contact was also included.

3. Results

3.1. Details on the Preparation and Characterization of the Complexes

All of the complexes were prepared via ligand substitution from the anionic precursor $[Re(NO)Br_4(EtOH)]^-$ as shown in the following chemical equation:

$$[Re(NO)Br_4(EtOH)]^- + L \rightarrow [Re(NO)Br_4L]^- + EtOH$$

This reactant complex has been previously used as a precursor for other tetrabromonitrosylrhenate(II) complexes [9,38]. The $\{Re(NO)Br_4\}^-$ fragment appears to be highly stable. The ethanol ligand is an easy-leaving group upon its substitution by other nitrogen donors from aromatic ligands under mild conditions because of the trans effect produced by the NO group. Higher yields are achieved in the presence of an excess of the substituent ligand. The IR spectra of these metal complexes are dominated by the characteristic stretching mode of the nitrosyl group at approximately 1750–1770 cm^{-1}. This absorption peak experiences a mild shift when the ligand substitution occurs, making it helpful in monitoring the ligand exchange in the aforementioned reaction.

3.2. Crystal Structure and Hirshfeld Analysis

Single crystals of **1–4** were grown directly from the reaction solution and were used for the X-ray data collection. We observed how **1–3** crystallize in the monoclinic crystal system, space group $P2_1/n$, whereas **4** crystallizes in the monoclinic space group $P2_1/c$. They consist of discrete $[Re(NO)Br_4(L)]^-$ complex anions and bulky n-$(C_4H_9)_4N^+$ organic cations in a 1:1 molar ratio, which are kept by electrostatic and van der Waals interactions.

Perspective views of the $[Re(NO)Br_4(L)]^-$ complex anion of **1–4** are presented in Figure 1, and the main bond lengths and angles are summarized in Table 1. Each $[Re(NO)Br_4(L)]^-$ unit comprises a rhenium(II) central ion surrounded by four bromide ligands in an equatorial plane, one NO group, and an azole-type ligand in the axial position. The geometry of the complex can be qualitatively described as a distorted octahedron and it is quantitatively confirmed by continuous symmetry measures (Table S3, [39]). The L ligands are coordinated to the rhenium(II) center through a nitrogen atom. The rhenium atom is slightly displaced from the mean equatorial plane towards the NO group, with the values of the *d*(Re–plane (centroid)) being equal to 0.138 (**1**), 0.156 (**2**), 0.161 (**3**), 0.171 (**4A**) and 0.166 Å (**4B**). The Re–NO group is practically linear, and the $Re–N_{NO}$ and N–O distances agree with those previously observed in similar complexes [11,14] (see Table 1). The Re-Br bond lengths are comparable, covering the narrow range 2.5017–2.5628 Å, similar to previously reported values. The fact that the values of the Re1–Br1 bond distance observed in **1–3** are slightly longer than the other Re–Br bonds is due to the participation of Br1 in Br···Br or Br···H intermolecular contacts in their respective compounds, as we will analyze later in this work. The planarity of the L ligands in **1–4** is preserved upon coordination, and their C–C and C–N bond lengths are almost identical to those observed in the corresponding free molecules in the solid state.

The analysis of the crystal packing of the complex anions in **1–3** reveals a similar pattern between them. Infinite supramolecular chains of $[Re(NO)Br_4(L)]^-$ units are observed along the crystallographic *b*-axis, connected by bromo and hydrogen intermolecular contacts as well as π-stacking interactions [40,41], as shown in Figures 2, 3 and S2. These shortest intermolecular Br···Br distances are 3.7387(6) ($Br2\cdots Br3^i$ in **1**), 3.8811(7) ($Br1\cdots Br4^i$ in **2**) and 3.7817(5) Å ($Br4\cdots Br3^{ii}$ in **3** [symmetry code: $(i) = \frac{1}{2} - x, -\frac{1}{2} + y, \frac{1}{2} - z$; $(ii) = 1.5 - x, -1/2 + y, 1.5 - z$]. They are slightly greater than the sum of the van der Waals radii (ca. 3.7 Å) [42–44] but considered in any matter as weak intermolecular contacts [45]. Additionally, π-stacking interactions [40] are observed between the pairs of HIm ligands in **1** [$d(C\cdots C) = 3.516$ Å; <P_CC = 27.171°], the Hpz ligands in **2** [$d(C\cdots C) = 3.910$ Å; <P_CC = 31.97°] and the Htz ligands in **3** [$d(C\cdots C) = 3.677$ Å; <P_CC = 30.72°], along the crystallographic *b* axis [<P_CC is the slip angle between the normal of the pyridine plane (P) and the centroid vector (C)]. Finally, it is worth mentioning the observation

of two weak hydrogen bonds in **1** between the Br1 and Br2 atoms from an anionic unit and the H3 and H2 atoms, respectively, from the imidazole ligand from another complex anion unit [*d*(Br1···H3) = 2.7029 Å; *d*(N3···Br1) = 3.531(4) Å; <(N3–H3···Br1) = 161.97°; *d*(Br2···H2) = 3.0119 Å; *d*(N2···Br2) = 3.678(5) Å; <(C2–H2···Br2) = 129.88°]. Similar intermolecular hydrogen bonds are observed in **3** between Br1 and the H1 atom from the Htz ligand of an adjacent complex anion [*d*(Br1···H1) = 2.6246 Å; *d*(C1···Br1) = 3.534(5) Å; <(C1–H1···Br1) =165.6°].

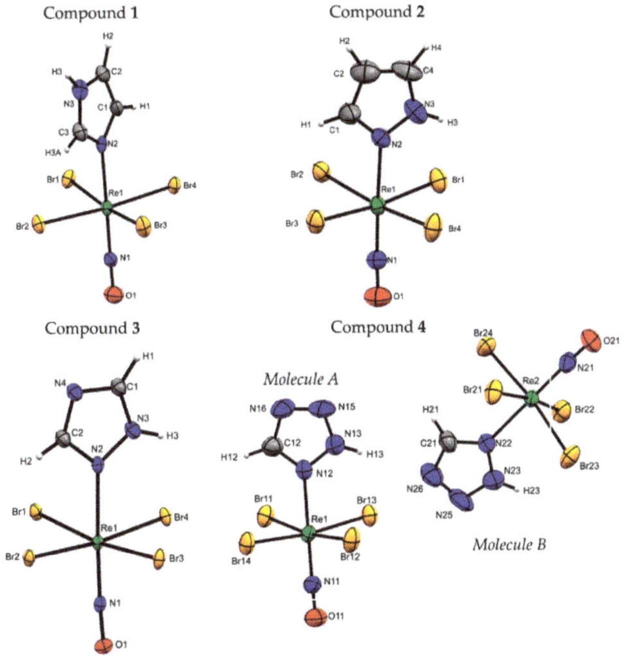

Figure 1. Perspective drawings of **1–4** showing the atom numbering. Thermal ellipsoids are plotted at a 30% probability level. The Bu_4N^+ cation was omitted for clarity. A and B stand for the two crystallographically independent complex anions in **4**.

The crystal packing in **4** shows a different pattern, as shown in Figure 3a. A supramolecular corrugated sheet parallel to the *bc* plane is formed through intermolecular N-H···H hydrogen bonds and weak Br···Br and Br···N-type contacts. The hydrogen bonds correspond to the N13–H13···N26 set of atoms [*d*(N26···H13) = 2.55 Å; *d*(N13···N26) = 3.41(2) Å; <(N13–H13···N26) = 176°]. The Br···Br contacts observed involve the Br11 and $Br22^{iii}$ atoms [symmetry code: (*iii*) = x, $1.5 - y$, $\frac{1}{2} + z$], the distance between them being 3.574(3) Å, a value which is 3.4% smaller than the sum of the van der Waals radii. The Br···N contacts concern the Br13 and $N26^{iv}$ atoms [symmetry code: (*iv*) = x, y, z], the value of the interatomic distance being 3.32(2) Å. The bulky $NBu_4{}^+$ cations separate these supramolecular 2D motifs from each other well, and no intermolecular interactions are observed between the anionic sheets. Furthermore, no π-π stacking is observed in the crystal structure. It is worth pointing out that the $[Re(NO)Br_5L]^-$ anions are arranged similarly in **1–3**, that is, with all of the anionic units orientated antiparallel in the same direction, whereas the $[Re(NO)Br_4(Htrz)]^-$ complex anions display different orientations in the crystal of **4**, as shown in Figure 3b.

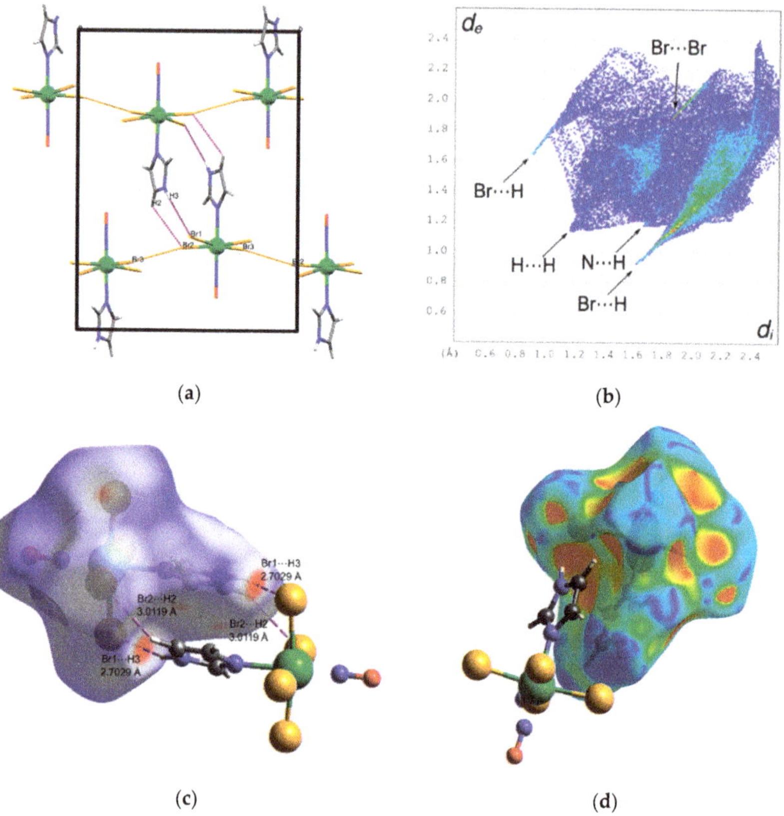

Figure 2. (**a**) Packing diagram down the crystallographic *a* axis, (**b**) fingerprint plot, (**c**) d_{norm} surface, and (**d**) shape index surface of the $[Re(NO)Br_4(HIm)]^-$ anionic units in **1**. Short Br···Br (yellow), Br···H (purple) contacts and π-π stacking interactions (grey ovals) are shown.

The crystal packing in **4** shows a different pattern, as shown in Figure 3a. A supramolecular corrugated sheet parallel to the *bc* plane is formed through intermolecular N–H···H hydrogen bonds and weak Br···Br and Br···N-type contacts. The hydrogen bonds correspond to the N13–H13···N26 set of atoms [*d*(N26···H13) = 2.55 Å; *d*(N13···N26) = 3.41(2) Å; <(N13–H13···N26) = 176°]. The Br···Br contacts observed involve the Br11 and $Br22^{iii}$ atoms [symmetry code: (*iii*) = *x*, 1.5 − *y*, $\frac{1}{2}$ + *z*], the distance between them being 3.574(3) Å, a value which is 3.4% smaller than the sum of the van der Waals radii. The Br···N contacts concern the Br13 and $N26^{iv}$ atoms [symmetry code: (*iv*) = *x*, *y*, *z*], the value of the interatomic distance being 3.32(2) Å. The bulky NBu_4^+ cations separate these supramolecular 2D motifs from each other well, and no intermolecular interactions are observed between the anionic sheets. Furthermore, no π-π stacking is observed in the crystal structure. It is worth pointing out that the $[Re(NO)Br_5L]^-$ anions are arranged similarly in **1–3**, that is, with all of the anionic units orientated antiparallel in the same direction, whereas the $[Re(NO)Br_4(Htrz)]^-$ complex anions display different orientations in the crystal of **4**, as shown in Figure 3b.

In order to gain a deeper understanding of the intermolecular contacts between the complex anions in **1–4**, we conducted a Hirshfeld surface analysis on this family of compounds. While a direct inspection of the crystal structure usually reveals the intermolecular contacts, Hirshfeld surface analysis allows us to qualitatively correlate the strength of intermolecular contacts with the intensity of the red spots in the d_{norm} plots [46]. In **1** and

3, the strongest inter-anion interactions are due to Br-H contacts. However, no solid red spots due to anion–anion interactions are observed in the Hirshfeld surface over the d_{norm} in **2**. White-colored surface areas are present around the shortest Br···Br intermolecular distances in **1–3** due to the values relative to the sum of the van der Waals radii.

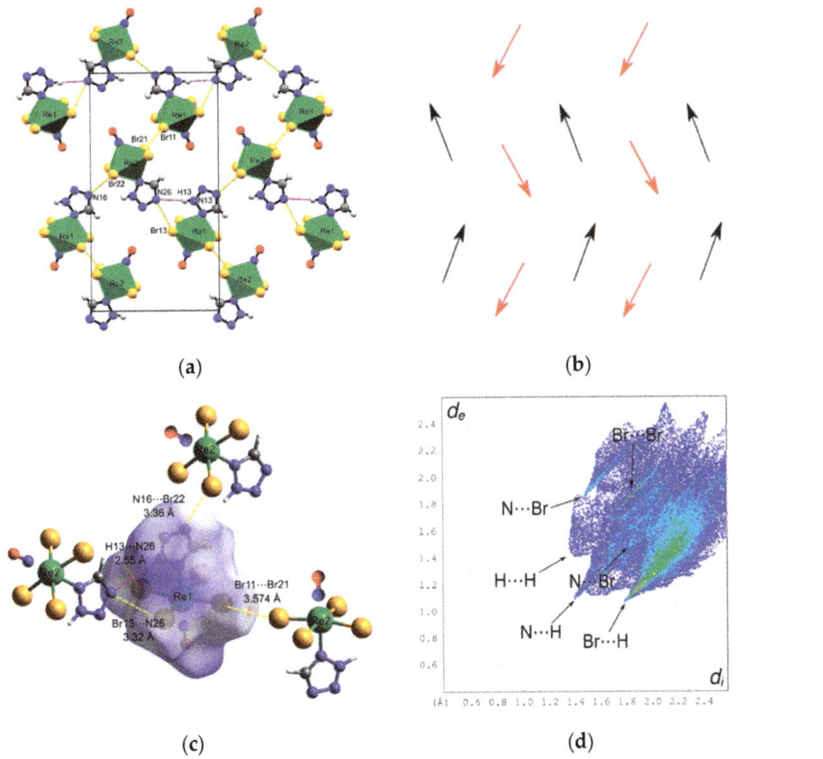

Figure 3. (**a**) Packing diagram down the crystallographic *a* axis; (**b**) complex anion orientation scheme along the crystallographic *a* axis (black and red arrows represent the A and B molecules, respectively); (**c**) d_{norm} surface and (**d**) shape index surface of the $[Re(NO)Br_4(Htrz)]^-$ units in **4**. Short through space Br···Br (yellow) and Br···N (yellow) contacts and N–H···N hydrogen bonds (purple) are also shown.

Conversely, the Hirshfeld plot over the d_{norm} map for **4A** reveals the presence of equally significative Br···Br and Br···N through space contacts and N–H···N hydrogen bonds. Notably, in contrast to what usually constitutes the focus of discussion, conventional hydrogen bonding often makes relatively small percentage contributions to the overall surface. The HS analysis of anion B in **4** reveals that the interaction for this molecule only varies in proportion (less than 1%) and not in nature, compared to the case of the A anion. In all compounds, the other red spots observed that we do not point out are due to less-significant cation-anion interactions (magnetically speaking). The supramolecular 2D fingerprint plots analysis shown in Figures 2, 3b, S2 and S3b, and the quantitative analysis of the contributions of the different intermolecular contacts to the HS of the anions shown in Figure 4, reveals that the main contribution to the HS is H-involving contacts, mainly Br···H (50% to 53%) and H···H contacts (11% to 22%). This is expected due to the bulky bromide ligands of the complex anion and the presence of the large NBu_4^+ cation. The exclusively anion–anion interactions (of magnetic interest) that contribute to the HS are the Br···N and Br···Br-type contacts. However, they contribute less than 5% and 3% to the total HS, respectively.

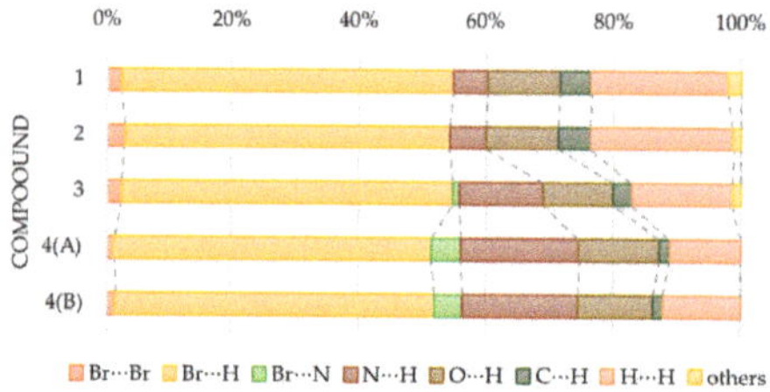

Figure 4. Relative contributions to the Hirshfeld surface area of the overall close molecular contacts for the complex anions in **1–4**.

The HS surfaces plotted over the shape index [Figures 2, 3d, S2 and S3d] show triangular-shaped regions around the aromatic rings, indicating that $\pi\cdots\pi$ stacking interactions occur in the crystal packing. However, it is found that these stacking interactions are very weak because the values of the centroid-to-centroid separation range from 4.755(3) to 5.963(1) Å. This can also be seen as brighter regions near the center of the fingerprint plot, just in the vicinity of (d_i, d_e) 1.8–2.0 Å, a range of values typical of the interplanar spacing of polycyclic aromatic hydrocarbons.

3.3. Magnetic Properties

The temperature dependence of the $\chi_M T$ product per Re(II) ion for **1–4** is shown in Figure 5. A quasi-Curie law is observed in a wide range of temperatures with a $\chi_M T$ value around 0.43 cm^3 K mol^{-1} at 295 K and a slight constant decrease in the $\chi_M T$ value, which is as expected for a local spin doublet (S_{Re} = 1/2) with a temperature-independent paramagnetism (TIP). However, upon cooling below c.a. 15 K, $\chi_M T$ decreases significantly due to the occurrence of antiferromagnetic interactions between the paramagnetic anionic units, the smallest value at 2.0 K being 0.207 cm^3 mol^{-1} K for **2**.

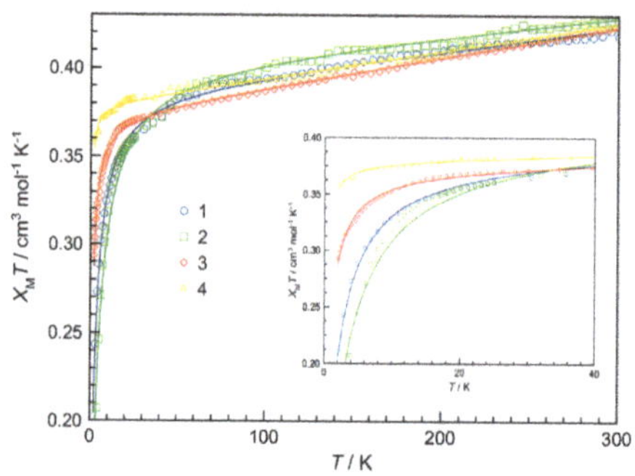

Figure 5. Thermal dependence of the $\chi_M T$ product for **1–4**. The inset shows a detail of the low-temperature domain.

As previously reported, the theoretical magnetic behavior of magnetically non-interacting six-coordinate Re(II) complexes in an octahedral environment can be reproduced by considering a sequence of perturbations made up by the spin-orbit coupling (SOC), the tetragonal distortion, and the magnetic field [9,10]. Consequently, having in mind these considerations and the crystal structures of **1–4**, their magnetic susceptibility data were analyzed through the following Hamiltonian [Equation (1)]:

$$\hat{H}_{Re} = k\lambda\hat{L}\hat{S} + \Delta\left(\hat{L}_z^2 - 2/3\right) + \beta H\left(k\hat{L} + g_e\hat{S}\right). \quad (1)$$

The first term in Equation (1) considers the spin-orbit coupling between the $S = 1/2$ spin and the $L = 1$ effective angular moment for the $^2T_2$ ground term of Re(II) due to the isomorphism | |T$_2$| | = -| |P| | [9,47]. λ is the spin-orbit coupling parameter, and κ is the orbital reduction factor. In the second term, Δ represents the energy gap between the ^2E and $^2B_2$ levels arising from splitting the $^2T_2$ ground state in octahedral geometry (axial distortion) [48]. Finally, the last term accounts for the Zeeman effects, g_e being the Landé factor for the free electron. In the absence of SOC, the tetragonal distortion (considering the C_{4v} point group) splits the $^2T_2$ term into an orbital singlet ($^2B_2$) and an orbital doublet (^2E) separated by an energy gap (Δ), which is defined as positive if the singlet is the lowest level. Both perturbations must be considered simultaneously due to the strong spin-orbit coupling ($\lambda \approx -2000$ cm^{-1}) and ligand field operating in the 5*d* metal ions, which results in six double-degenerated Kramer doublets, as shown in the splitting diagram in Figure 6 [9].

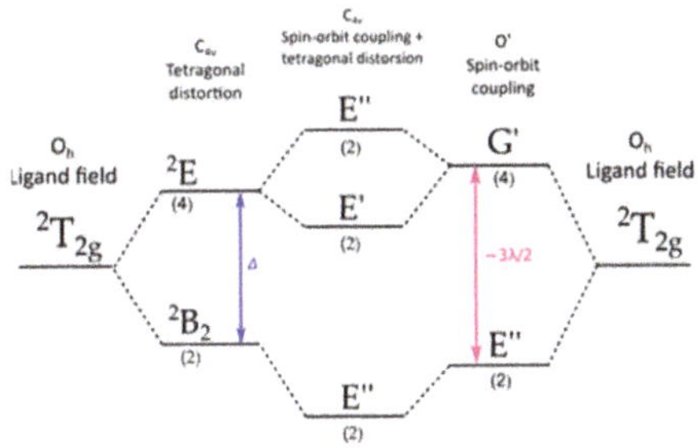

Figure 6. Splitting of the $^2T_2$ term by spin-orbit coupling and a tetragonal distortion. Numbers in parentheses correspond to the degree of orbital degeneracy.

The energy values associated with the wave functions of the $^2T_2$ term are calculated using the Figgis [49] and Mabbs [50] methodology. These values are the following: $E_1 = E_2 = \kappa\lambda(\nu - 0{,}5)$, $E_3 = E_4 = \frac{\kappa\lambda}{\sqrt{2}}a$ and $E_5 = E_6 = \frac{\kappa\lambda}{\sqrt{2}}b$, ν being the tetragonal distortion parameter defined as $\nu = \Delta/\kappa\lambda$ with $a = \frac{1}{\sqrt{2}}[(\nu + 0{,}5) - Z]$, $b = \frac{1}{\sqrt{2}}[(\nu + 0{,}5) + Z]$ and $Z = \sqrt{\nu^2 + \nu + 2{,}25}$. As the system is anisotropic, the effect of the magnetic field on the parallel direction (*z*) is different from that on the perpendicular ones (*x* and *y*). Under an axial symmetry, there is a parallel and a perpendicular component of the magnetic susceptibility [Equation (2)]:

$$\chi_{\parallel} = \frac{N_A\beta^2}{4kT}\left[\frac{2(\kappa + 1 - a^2)}{(1 + a^2)}\right]^2 - \frac{2N_A\beta^2(\kappa + 1 - ab)}{\kappa\lambda(1 + a^2)(1 + b^2)Z} \tag{2}$$

$$\chi_{\perp} = \frac{N_A\beta^2}{4kT}\left[\frac{2\left(\kappa a\sqrt{2} - a\right)}{(1 + a^2)}\right]^2 - \frac{2N_A\beta^2}{\kappa\lambda}\left\{\frac{\left[k(a + b) - ab\sqrt{2}\right]^2}{2(1 + a^2)(1 + b^2)Z} - \frac{\left(\sqrt{2} - \kappa a\right)^2}{(1 + a^2)(\nu - 1{,}5 - Z)}\right\}$$

Given that the magnetic susceptibility was measured on a powder sample, it must obey the expression $\chi_M = \frac{\chi_{\parallel} + 2\chi_{\perp}}{3}$. In the case where the intermolecular interactions are not considered, it is expected that a polycrystalline sample of any of the mononuclear Re(II) compounds would exhibit a Curie-law behavior with a total spin $S_{eff} = \frac{1}{2}$, ($\chi_M T = C = N\beta^2 g_{av}^2/4k$). If the temperature-independent paramagnetism is considered, the lineal equation $\chi_M T = A + BT$ is derived [Equations (3)–(5)]:

$$\chi_M T = N\beta^2 g_{av}^2/4k + \chi_{TIP}^{av}\, T \tag{3}$$

$$g_{av}^2 = \left(g_{\parallel}^2 + 2g_{\perp}^2\right)/3 \tag{4}$$

$$\chi_{TIP}^{av} = \left(\chi_{TIP}^{\parallel} + 2\chi_{TIP}^{\perp}\right)/3 \tag{5}$$

The values of χ_{TIP}^{u} y g_u are deduced from Equation (2) and summarized in Table 2.

Table 2. Values of g and χ in the parallel and perpendicular directions to the magnetic field.

Direction (*u*)	g_u	χ_{TIP}^{u}
Parallel (*z*)	$\left[\frac{2\left(\kappa+1-a^2\right)}{(1+a^2)}\right]$	$-\frac{2N_A\beta^2(\kappa+1-ab)}{\kappa\lambda(1+a^2)(1+b^2)Z}$
Perpendicular (*x* and *y*)	$\left[\frac{2\left(\kappa a\sqrt{2}-a\right)}{(1+a^2)}\right]$	$-\frac{2N_A\beta^2}{\kappa\lambda}\left\{\frac{\left[k(a+b)-ab\sqrt{2}\right]^2}{2(1+a^2)(1+b^2)Z}-\frac{\left(\sqrt{2}-\kappa a\right)^2}{(1+a^2)(v-1{,}5-Z)}\right\}$

It is worth mentioning that a Weiss constant θ (in the form of $T-\theta$) was introduced in the fitting procedure to account for the intermolecular magnetic interactions. The least-squares fit of the magnetic data using matrix-diagonalization techniques through the VPMAG [47] program led to the parameters listed in Table 3.

Table 3. Best-fit parameters for **1–4** [a].

Compound	**Δ**	**λ**	**κ**	**ν**	**TIP** [b]	**θ/K**	$g_{\parallel}$	$g_{\perp}$	g_{av}	$R^{c} \times 10^6$
1	2225	–1847	0.62	–1.94	125	–1.78	1.25	2.33	2.03	0.5
2	2589	–1800	0.63	–2.28	98	–3.12	1.43	2.34	2.08	1.8
3	1850	–1899	0.63	–1.55	169	–0.59	1.01	2.32	1.98	0.2
4 [d]	2016	–1849	0.64	–1.70	155	–0.14	1.07	2.34	2.01	0.2

[a] All energies are given in cm^{-1}. [b] χ_{TIP} = TIP × 10^{-6} cm^3 mol^{-1} and calculated through Equations (2) and (3). [c] *R* is the agreement factor defined as $\sum[(\chi_M T)_{exp} - (\chi_M T)_{calcd}]^2/\sum[(\chi_M T)_{exp}]^2)$. [d] Although there are two crystallographically non-equivalent $[Re(NO)Br_4(Htrz)]^-$ complex anions in **4**, they are structurally quite similar—in particular, the coordination geometry around the rhenium atom (Table 1); therefore, their magnetic behavior was assumed to be identical.

These values are similar to those previously obtained for other related Re(II)-nitrosyl complexes [9,10]. The relatively low values of κ (0.62 to 0.64) and high values of $|\lambda|$ (1800 to 1899 cm^{-1}) are indicative of a significant covalence degree of the rhenium to ligand interaction in **1–4**. The value of Δ (axial distortion) is positive in sign, indicating that the orbital singlet ($^2B_2$) is lower in energy than the doublet (^2E) (Figure 6), in agreement with the electronic configuration expected for a tetragonal Re(II)-nitrosyl complex [51]. A simulation of the dependence of the g_u value with the distortion parameter ν by using the equations in Table 3, shown in Figure 7a, agrees with the experimental $g_{\parallel} < g_{\perp}$ pattern determined experimentally. Additionally, there is a good agreement between the experimental values calculated for the $g_{\parallel} < g_{\perp}$ pattern and those inferred from Figure 7b. ($g_{\perp} = 2.2 - 2.7$ and $g_{\parallel} = 0.9 - 1.4$), considering $\lambda = -2000$ cm^{-1} and values of κ in the range 0.5–0.75.

The calculated values of *g* and TIP are very close to those observed for similar complexes previously reported [9,10,14,52]. Finally, the exchange pathways for the very weak intermolecular magnetic interactions have to be attributed to the through-space Br···Br, Br-H contacts, as well as π-π stacking interactions. Ligands that provide a supramolecular pathway through hydrogen bonding or halide–halide interactions may be essential for promoting significant magnetic interactions among mononuclear complexes in the crystal packing [52]. This is particularly relevant in the magnetochemistry of related mononuclear Re(IV) complexes (t_{2g}^3 electronic configuration), where through-space magnetic interactions can provide the magnetic pathways that lead to magnetic ordering at relatively high temperatures [53]. It is worth noting that although the shortest Br···Br distances in **1** are close to the van der Waals one (ca. 3.7 Å) [42–44], they may still be viewed as likely magnetic exchange pathways.

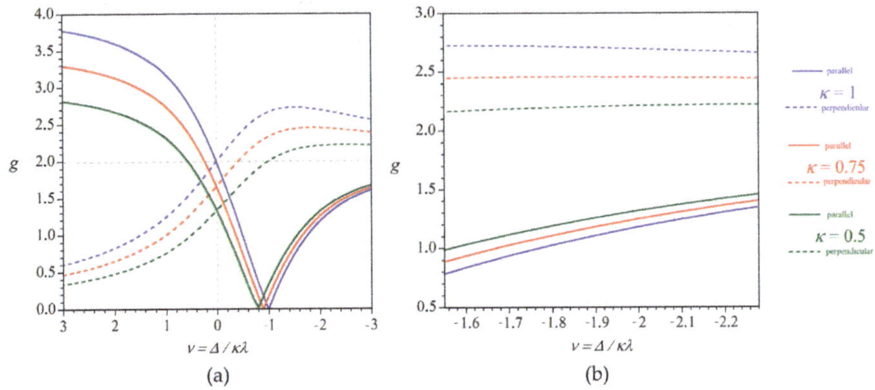

Figure 7. (**a**) Calculated values of the *g* factor for an electronic configuration t_2^5 ($\lambda < 0$) as a function of possible negative and positive values of the distortion factor (ν) under different values of the covalence factor (κ). (**b**) Details of the calculated values of the *g* factor for λ = –2000 cm^{-1} and κ = 0.5 – 0.75.

4. Conclusions

Four mononuclear Re(II)-nitrosyl complexes with a series of azole-type molecules as co-ligands (compounds **1–4**) were isolated and magnetostructurally investigated. A thorough analysis of the influence of the magnetic field, SOC, axial distortion, and covalence effects as variable parameters successfully simulated their cryomagnetic data. Very weak intermolecular antiferromagnetic interactions occur in **1–4**, the through-space Br$\cdots$Br contacts, among other weak intermolecular interactions, providing exchange pathways for these complexes. The use of Hirshfeld surface analysis in the solid-state structure studies of **1–4** enabled the visualization of the proportion and nature of the interactions present in the structures, some of which are not immediately evident through knowledge of the crystal structure alone. Determining these intrinsic parameters and their relation to molecular or crystal structure is critical in providing essential tools for the rational design of molecular materials with interesting magnetic properties.

Supplementary Materials: The following supporting information can be downloaded at: https://www.mdpi.com/article/10.3390/cryst13040658/s1, Table S1. List of the infrared absorption bands observed in compounds **1–4**, and classified upon their origin. (ligand, NBu_4^+ or Re-NO). Table S2. Summary of the crystal data and structure refinement for **1–4**. Table S3. Geometrical distortion from OC-6 calculated by CSM (OC-6) and the calculated tetragonal distortion for several tetrabromonitrosylrhenium(II) complexes. Figure S1. (a) Packing diagram down the crystallographic *a* axis, (b) fingerprint plot, (c) d_{norm} surface, and (d) shape index surface of the $[Re(NO)Br4(Hpz)]^-$ anionic units in **2**. Short Br$\cdots$Br (yellow dashed lines) contacts and π-π stacking interactions (grey ovals) are shown. Figure S2. (a) Packing diagram down the crystallographic *a* axis, (b) fingerprint plot, (c) d_{norm} surface, and (d) shape index surface of the $[Re(NO)Br4(Htz)]^-$ anionic units in **3**. Short Br$\cdots$Br (yellow), Br$\cdots$H (red) contacts and π-π stacking interactions (grey ovals) are shown. File S1. checkCIF/PLATON report.

Author Contributions: M.P.: Methodology, Investigation, Formal analysis, Writing—original Draft, review and Editing, Visualization. J.G.-P.: Investigation, Formal analysis, Resources, Writing—Review. C.K.: Conceptualization. M.J.: Formal analysis, Visualization, Writing—Original Draft, review and Editing, Resources. F.L.: Formal analysis, Visualization, Writing—Original Draft, Review and Editing, Resources. A.C.: Conceptualization, Supervision, Writing—Review, Project administration. All authors have read and agreed to the published version of the manuscript.

Funding: Financial support from PEDECIBA (Uruguay), CSIC (Uruguay, Programa de Apoyo a Grupos de Investigación, Project N°245), MICIU (Spain, Project PID2019-109735GB-I00), the Generalitat Valenciana (Spain, AICO/2020/183 and AICO/2021/295), Agencia Estatal de Investigación under the National Program of Sciences and Technological Materials (Spain, PID2019-106383GB-C44/AEI/10.13039/501100011033) and by EU-FEDER is gratefully acknowledged.

Institutional Review Board Statement: Not applicable.

Informed Consent Statement: Not applicable.

Data Availability Statement: No data provided.

Acknowledgments: M.P. thanks Fundación Carolina (ES), CAP-UdelaR (UY) and ANII (UY) for studentship grants. J.G.-P. thanks to Servicios Generales de Apoyo a la Investigación (SEGAI) at La Laguna University.

Conflicts of Interest: The authors declare no conflict of interest. The funders had no role in the design of the study; in the collection, analyses, or interpretation of data; in the writing of the manuscript; or in the decision to publish the results.

References

1. Dudle, B.; Rajesh, K.; Blacque, O.; Berke, H. Rhenium in Homogeneous Catalysis: [ReBrH(NO)(Labile Ligand)(Large-Bite-Angle Diphosphine)] Complexes as Highly Active Catalysts in Olefin Hydrogenations. *J. Am. Chem. Soc.* **2011**, *133*, 8168–8178. [CrossRef] [PubMed]
2. Jiang, Y.; Blacque, O.; Berke, H. Probing the Catalytic Potential of Chloro Nitrosyl Rhenium(I) Complexes. *Dalton Trans.* **2011**, *40*, 2578–2587. [CrossRef]
3. Zobi, F.; Blacque, O. Reactivity of 17 E − Complex $[ReIIBr_4(CO)_2]^{2-}$ with Bridging Aromatic Ligands. Characterization and CO-Releasing Properties. *Dalton Trans.* **2011**, *40*, 4994–5001. [CrossRef]
4. Adams, J.J.; Arulsamy, N.; Sullivan, B.P.; Roddick, D.M.; Neuberger, A.; Schmehl, R.H. Homoleptic Tris-Diphosphine Re(I) and Re(II) Complexes and Re(II) Photophysics and Photochemistry. *Inorg. Chem.* **2015**, *54*, 11136–11149. [CrossRef]
5. Bar, A.K.; Pichon, C.; Sutter, J.-P. Magnetic Anisotropy in Two- to Eight-Coordinated Transition–Metal Complexes: Recent Developments in Molecular Magnetism. *Coord. Chem. Rev.* **2016**, *308*, 346–380. [CrossRef]
6. Wang, X.-Y.; Avendaño, C.; Dunbar, K.R. Molecular Magnetic Materials Based on 4d and 5d Transition Metals. *Chem. Soc. Rev.* **2011**, *40*, 3213–3238. [CrossRef] [PubMed]
7. Schelter, E.J.; Prosvirin, A.V.; Dunbar, K.R. Molecular Cube of ReII and MnII That Exhibits Single-Molecule Magnetism. *J. Am. Chem. Soc.* **2004**, *126*, 15004–15005. [CrossRef] [PubMed]
8. Schelter, E.J.; Prosvirin, A.V.; Reiff, W.M.; Dunbar, K.R. Unusual Magnetic Metal–Cyanide Cubes of ReII with Alternating Octahedral and Tetrahedral Corners. *Angew. Chem. Int. Ed.* **2004**, *43*, 4912–4915. [CrossRef]
9. Pacheco, M.; Cuevas, A.; González-Platas, J.; Faccio, R.; Lloret, F.; Julve, M.; Kremer, C. Synthesis, Crystal Structure and Magnetic Properties of the Re(Ii) Complexes $NBu_4[Re(NO)Br_4(L)]$ (L = Pyridine and Diazine Type Ligands). *Dalton Trans.* **2013**, *42*, 15361–15371. [CrossRef]
10. Pacheco, M.; Cuevas, A.; González-Platas, J.; Lloret, F.; Julve, M.; Kremer, C. The Crystal Structure and Magnetic Properties of 3-Pyridinecarboxylate-Bridged Re(ii)M(ii) Complexes (M = Cu, Ni, Co and Mn). *Dalton Trans.* **2015**, *44*, 11636–11648. [CrossRef] [PubMed]
11. Pacheco, M.; González-Platas, J.; Julve, M.; Lloret, F.; Kremer, C.; Cuevas, A. Crystal Structure and Magnetic Properties of 3,5-Pyridinedicarboxylate-Bridged Re(II)M(II) Heterodinuclear Complexes (M = Cu, Ni and Co). *Polyhedron* **2021**, *208*, 115414. [CrossRef]
12. Rosas, K.S.; Torres, J.; Pacheco, M.; Ramos, J.; Gancheff, J.S. Proton-Transfer Reactions of Re(II)-Nitrosyl Complexes: Potentiometric Studies, DFT and TD-DFT Calculations. *Results Chem.* **2022**, *4*, 100455. [CrossRef]
13. Pacheco, M.; Cuevas, A.; González-Platas, J.; Gancheff, J.S.; Kremer, C. Complex Salts of $[Re^{II}(NO)Br_4(Pyz)]^-$: Synthesis, Crystal Structures, and DFT Studies. *J. Coord. Chem.* **2014**, *67*, 4028–4038. [CrossRef]
14. Pacheco, M.; Alvarez, N.; Cuevas, A.; Romerosa, A.; Lloret, F.; Kremer, C. Crystal Structure and Magnetic Study of the Complex Salt $[RuCp(PTA)_2$–μ-CN-1κC:2κN–$RuCp(PTA)_2][Re(NO)Br_4(EtOH)_{0.5}(MeOH)_{0.5}]$. *Acta Crystallogr. Sect. E Crystallogr. Commun.* **2021**, *77*, 749–754. [CrossRef]
15. Pacheco, M.; Cuevas, A.; González-Platas, J.; Kremer, C. Synthesis, Spectroscopic Characterization and Crystal Structure of $[ReV(O)_2(Pyz)_4][ReII(NO)Br_4(Pyz)]$ (Pyz = Pyrazine). *Commun. Inorg. Synth.* **2015**, *2*, 20–24. [CrossRef]
16. Kahn, O. *Molecular Magnetism*; Wiley-VCH: New York, NY, USA, 2001; ISBN 978-0-471-18838-4.
17. Barbour, L.J. Single-Crystal X-ray Diffraction. In *Comprehensive Supramolecular Chemistry II*; Elsevier: Amsterdam, The Netherlands, 2017; pp. 23–43. ISBN 978-0-12-803199-5.
18. McKinnon, J.J.; Jayatilaka, D.; Spackman, M.A. Towards Quantitative Analysis of Intermolecular Interactions with Hirshfeld Surfaces. *Chem. Commun.* **2007**, *7*, 3814–3816. [CrossRef]

19. Aromí, G.; Barrios, L.A.; Roubeau, O.; Gamez, P. Triazoles and Tetrazoles: Prime Ligands to Generate Remarkable Coordination Materials. *Coord. Chem. Rev.* **2011**, *255*, 485–546. [CrossRef]
20. Chen, S.-S. The Roles of Imidazole Ligands in Coordination Supramolecular Systems. *Crystengcomm* **2016**, *18*, 6543–6565. [CrossRef]
21. Haasnoot, J.G. Mononuclear, Oligonuclear and Polynuclear Metal Coordination Compounds with 1,2,4-Triazole Derivatives as Ligands. *Coord. Chem. Rev.* **2000**, *200–202*, 131–185. [CrossRef]
22. Li, H.; Zhang, S.-G.; Xie, L.-M.; Yu, L.; Shi, J.-M. π–π Stacking, Hydrogen Bonding and Anti-Ferromagnetic Coupling Mechanism on a Mononuclear Cu(II) Complex. *J. Coord. Chem.* **2011**, *64*, 1456–1468. [CrossRef]
23. Earnshaw, A. *Introduction to Magnetochemistry*; Elsevier Science: Saint Louis, MI, USA, 2013; ISBN 978-1-4832-7069-2.
24. AIST: Spectral Database for Organic Compounds, SDBS. Available online: http://sdbs.db.aist.go.jp/sdbs/cgi-bin/cre_index.cgi (accessed on 15 October 2016).
25. Nakamoto, K. Complexes of Alcoxides, Alcohols, Ethers, Ketones, Aldehydes, Esteres, and Carboxylic Acids. En: Infrared and Raman Spectra of Inorganic and Coordination Compounds. In *Part II: Applications in Coordination, Organometallic, and Bioinorganic*; John Wiley & Sons: Hoboken, NJ, USA, 1997.
26. Billes, F.; Endrédi, H.; Keresztury, G. Vibrational Spectroscopy of Triazoles and Tetrazole. *J. Mol. Struct. THEOCHEM* **2000**, *530*, 183–200. [CrossRef]
27. Agilent Technologies. CrysAlisPro (v42). Agilent Technologies, Oxfordshire, UK. 2021. Available online: https://www.rigaku.com/products/crystallography/crysalis (accessed on 6 April 2023).
28. BRUKER APEX2. Available online: https://www.cif.iastate.edu/x-ray/apexii (accessed on 3 April 2023).
29. Palatinus, L.; Chapuis, G. Superflip—A Computer Program for the Solution of Crystal Structures by Charge Flipping in Arbitrary Dimensions. *J. Appl. Crystallogr.* **2007**, *40*, 786–790. [CrossRef]
30. Sheldrick, G.M. Crystal Structure Refinement with SHELXL. *Acta Cryst. C* **2015**, *71*, 3–8. [CrossRef]
31. Speck, A. *Platon: A Multipurpose Crystalographic Tool*; Utrecht University: Utrecht, The Netherlands, 2002.
32. Dolomanov, O.V.; Bourhis, L.J.; Gildea, R.J.; Howard, J.A.K.; Puschmann, H. OLEX2: A Complete Structure Solution, Refinement and Analysis Program. *J. Appl. Crystallogr.* **2009**, *42*, 339–341. [CrossRef]
33. Macrae, C.F.; Sovago, I.; Cottrell, S.J.; Galek, P.T.A.; McCabe, P.; Pidcock, E.; Platings, M.; Shields, G.P.; Stevens, J.S.; Towler, M.; et al. Mercury 4.0: From Visualization to Analysis, Design and Prediction. *J. Appl. Crystallogr.* **2020**, *53*, 226–235. [CrossRef] [PubMed]
34. Spackman, M.A.; Jayatilaka, D. Hirshfeld Surface Analysis. *CrystEngComm* **2009**, *11*, 19–32. [CrossRef]
35. Turner, M.J.; McKinnon, J.J.; Wolff, S.K.; Grimwood, D.J.; Spackman, P.R.; Jayatilaka, D.; Spackman, M.A. *CrystalExplorer17*; University of Western Australia: Perth, Australia, 2017.
36. Spackman, M.A.; McKinnon, J.J. Fingerprinting Intermolecular Interactions in Molecular Crystals. *CrystEngComm* **2002**, *4*, 378–392. [CrossRef]
37. McKinnon, J.J.; Spackman, M.A.; Mitchell, A.S. Novel Tools for Visualizing and Exploring Intermolecular Interactions in Molecular Crystals. *Acta Crystallogr. Sect. B Struct. Sci.* **2004**, *60*, 627–668. [CrossRef]
38. Ciani, G.; Giusto, D.; Manassero, M.; Sansoni, M. Reactions of Pentachloro- and Pentabromonitrosylrhenate(II) Anions with N-Donor Ligands. *Gazz. Chim. Ital.* **1977**, *107*, 429–430.
39. Steed, J.W.; Atwood, J.L. *Supramolecular Chemistry*, 2nd ed.; Wiley: Chichester, UK, 2009; ISBN 978-0-470-51233-3.
40. Alkorta, I.; Elguero, J.; Frontera, A. Not Only Hydrogen Bonds: Other Noncovalent Interactions. *Crystals* **2020**, *10*, 180. [CrossRef]
41. Cuevas, A.; Chiozzone, R.; Kremer, C.; Suescun, L.; Mombrú, A.; Armentano, D.; De Munno, G.; Lloret, F.; Cano, J.; Faus, J. Rhenium (IV)-Copper (II) Heterobimetallic Complexes with a Bridge Malonato Ligand. Synthesis, Crystal Structure, and Magnetic Properties. *Inorg. Chem.* **2004**, *43*, 7823–7831. [CrossRef] [PubMed]
42. Cuevas, A.; Kremer, C.; Hummert, M.; Schumann, H.; Lloret, F.; Julve, M.; Faus, J. Magnetic Properties and Molecular Structures of Binuclear (2-Pyrazinecarboxylate)-Bridged Complexes Containing Re (IV) and M (II)(M = Co, Ni). *Dalton Trans.* **2007**, *3*, 342–350. [CrossRef] [PubMed]
43. Martínez-Lillo, J.; Armentano, D.; Munno, G.D.; Julve, M.; Lloret, F.; Faus, J. Ferromagnetic Coupling and Spin Canting Behaviour in Heterobimetallic ReIVMII/III (M = CoII/III, NiII) Species. *Dalton Trans.* **2013**, *42*, 1687–1695. [CrossRef]
44. Schiemenz, G.P. The Sum of van der Waals Radii—A Pitfall in the Search for Bonding. *Z. Für Nat. B* **2007**, *62*, 235–243. [CrossRef]
45. Tan, S.L.; Jotani, M.M.; Tiekink, E.R.T. Utilizing Hirshfeld Surface Calculations, Non-Covalent Interaction (NCI) Plots and the Calculation of Interaction Energies in the Analysis of Molecular Packing. *Acta Crystallogr. Sect. E Cryst. Commun.* **2019**, *75*, 308–318. [CrossRef] [PubMed]
46. Lloret, F.; Julve, M.; Cano, J.; Ruiz-García, R.; Pardo, E. Magnetic Properties of Six-Coordinated High-Spin Cobalt(II) Complexes: Theoretical Background and Its Application. *Inorg. Chim. Acta* **2008**, *361*, 3432–3445. [CrossRef]
47. Machura, B. Structural and Spectroscopic Properties of Rhenium Nitrosyl Complexes. *Coord. Chem. Rev.* **2005**, *249*, 2277–2307. [CrossRef]
48. Figgis, B.N. The Magnetic Properties of Transition Metal Ions in Asymmetric Ligand Fields. Part 3.—The Behaviour of Some T52g Complexes. *Trans. Faraday Soc.* **1961**, *57*, 204–209. [CrossRef]
49. Mabbs, F.E.; Machin, D.J. *Magnetism and Transition Metal Complexes*; Springer: Berlin/Heidelberg, Germany, 1973; ISBN 978-1-5041-2035-7.

50. Cano, J. *VPMAG*; University of Valencia: Valencia, Spain, 2003.
51. Schelter, E.J.; Bera, J.K.; Bacsa, J.; Galán-Mascarós, J.R.; Dunbar, K.R. New Paramagnetic Re(II) Compounds with Nitrile and Cyanide Ligands Prepared by Homolytic Scission of Dirhenium Complexes. *Inorg. Chem.* **2003**, *42*, 4256–4258. [CrossRef]
52. Silva, C.P.; Junior, H.C.S.; Santos, I.F.; Bernardino, A.M.R.; Cassaro, R.A.A.; Novak, M.A.; Vaz, M.G.F.; Guedes, G.P. Synthesis, Crystal Structure, Magnetic Properties and DFT Calculations of a Mononuclear Copper(II) Complex: Relevance of Halogen Bonding for Magnetic Interaction. *Inorg. Chim. Acta* **2018**, *482*, 395–401. [CrossRef]
53. Martinez.Lillo, J.; Faus, J.; Lloret., F.; Julve, M. Towards multifunctional magnetic systems through molecular-programmed self assembly of Re(IV) metalloligands. *Coord. Chem. Rev.* **2015**, *289–290*, 215–237. [CrossRef]

crystals MDPI

Article

Structural and Hirshfeld Surface Analysis of Thallium(I) and Indium(III) Complexes of a Soft Scorpionate Ligand

Kiyoshi Fujisawa [1,*], Ayaka Kuboniwa [1], Sang Loon Tan [2] and Edward R. T. Tiekink [2,*]

1 Department of Chemistry, Ibaraki University, Mito 310-8512, Ibaraki, Japan
2 Research Centre for Crystalline Materials, School of Medical and Life Sciences, Sunway University, Bandar Sunway 47500, Selangor Darul Ehsan, Malaysia; alant@sunway.edu.my
* Correspondence: kiyoshi.fujisawa.sci@vc.ibaraki.ac.jp (K.F.); edwardt@sunway.edu.my (E.R.T.T.)

Abstract: Two complexes containing a soft sulfur-substituted tris(pyrazolyl)hydroborate ligand, namely $[Tl^{I}(Tm^{tBu})]_2 \cdot 2H_2O$ and $[In^{III}(Tm^{tBu})_2](InCl_4)$, where Tm^{tBu} is the tris(3-*tert*-butyl-2-sulfanylidene-1*H*-imidazol-1-yl)hydroborate anion, have been characterized. The $\{TlS\}_2$ core of the former has the shape of a diamond. Each S atom of the Tm^{tBu} anion coordinates differently: one S is connected to one Tl atom, one bridges both Tl atoms, while the third S atom connects solely to the second Tl atom. The S_4 donor set defines a seesaw geometry. The independent H_2O molecule forms O–H···S and localized O–H···π(pyrazolyl) contacts. Flattened octahedral geometries defined by S_6 donor sets are noted for the two independent cations in $[In^{III}(Tm^{tBu})_2](InCl_4)$. In the crystal of $[Tl^{I}(Tm^{tBu})]_2 \cdot 2H_2O$, pyrazolyl-C–H···O(water) interactions connect the dimeric units into a linear supramolecular chain, chains pack without directional interactions between them. In the crystal of $[In^{III}(Tm^{tBu})_2](InCl_4)$, alternating rows of independent cations are interspersed by anions. The primary points of contact within a three-dimensional architecture are of the type In–Cl···π(pyrazolyl) and C–H···Cl. The assessment of the molecular packing was complemented by considering the calculated Hirshfeld surfaces and two-dimensional fingerprint plots (overall and delineated into individual contacts).

Keywords: thallium; indium; main group chemistry; tripodal ligand; thiol ligand; crystal structure; hydrogen bonding; Hirshfeld surface analysis

Citation: Fujisawa, K.; Kuboniwa, A.; Tan, S.L.; Tiekink, E.R.T. Structural and Hirshfeld Surface Analysis of Thallium(I) and Indium(III) Complexes of a Soft Scorpionate Ligand. *Crystals* **2023**, *13*, 745. https://doi.org/10.3390/cryst13050745

Academic Editor: Waldemar Maniukiewicz

Received: 10 April 2023
Revised: 26 April 2023
Accepted: 26 April 2023
Published: 29 April 2023

1. Introduction

Tripodal nitrogen-containing ligands, such as tris(pyrazolyl)hydroborate, have been utilized in the fields of inorganic and coordination chemistry [1,2]. One reason why the chemistry of this type of ligand has been studied so extensively relates to the fact that it is relatively facile to introduce substituents in the pyrazolyl rings with varying steric and electronic profiles. Recently, we developed transition metal complexes ligated by tris(pyrazolyl)hydroborate anion and/or their neutral analogues, i.e., tris(pyrazolyl)methanes, to determine how to control small molecule activation and their magnetism [3–6]. The history of the development and use of the tris(pyrazolyl)hydroborate ligand, often referred to as 'scorpionate', has been outlined by the founder of this chemistry, the late Prof. Swiatoslaw Trofimenko [7]. In Trofimenko's historical account, it was noted that new ligand architectures could also be obtained by the introduction of other heteroatoms, such as oxygen, sulfur, and phosphorus [7].

Thallium and indium are toxic metal *p*-block elements [8,9]. Recently, indium(III) oxide has been used as a transparent conductive coating on glass substrates in electroluminescent panels, i.e., ITO [9]. Thallium(I) and indium(III) are stable formal oxidation states and have electron configurations of $[Xe]4f^{14}5d^{10}6s^{2}6p^{0}$ and $[Kr]4d^{10}5s^{0}5p^{0}$, respectively. With respect to tripodal ligands, the introduction of sulfur gives rise to S_3-tripod type ligands, e.g., tris(3-*tert*-butyl-2-sulfanylidene-1*H*-imidazol-1-yl)hydroborate (denoted Tm^{tBu}) (Figure 1, left; R = tBu), being a soft tris(pyrazolyl)hydroborate derivative that readily complexes heavy metal *p*-block elements, such as bismuth(III) [10]; for relevant reviews of the coordination chemistry of *p*-block elements with soft S_3-type ligands, see [11–13].

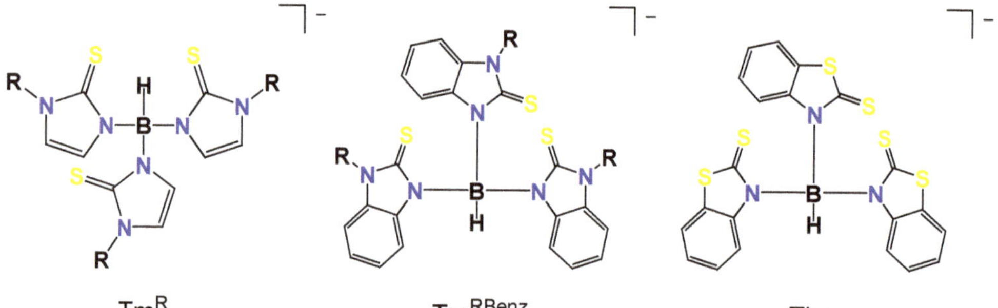

Figure 1. Schematic drawing of the ligands: tris(3-R-2-sulfanylidene-1*H*-imidazol-1-yl)hydroborate (Tm^R), tris(3-R-2-sulfanylidene-1*H*-benzimidazol-1-yl)hydroborate (Tm^{RBenz}), and tris(2-sulfanylidene-1*H*-benzothiazol-1-yl)hydroborate (Tbz).

In continuation of previous work, the crystal and molecular structures, as well as a detailed analysis of the calculated Hirshfeld surfaces, are described for thallium(I), $[Tl^I(Tm^{tbu})]_2$, characterized as a dihydrate, and indium(III), $[In^{III}(Tm^{tBu})_2]^+$, complexes ligated by the same soft tripod sulfur-containing type ligand employed in an earlier study [10], namely Tm^{tBu}. This work compliments the literature precedents of thallium(I), thallium(III), indium(I), and indium(III) complexes ligated by tris(3-R-2-sulfanylidene-1*H*-imidazol-1-yl)hydroborate (Tm^R), tris(3-R-2-sulfanylidene-1*H*-benzimidazol-1-yl)hydroborate (Tm^{RBenz}), and tris(2-sulfanylidene-1*H*-benzothiazol-1-yl)hydroborate (Tbz) ligands (Figure 1)—thallium(I): $[Tl(Tm^{MeBenz})]$ [14], $[Tl(Tm^{tBuiBenz})]\cdot C_6H_6$ [14], $[Tl(Tm^{tBu})]_2$ [15], $[Tl(Tm^{Ph})]_2$ [16], and $[Tl(Tbz)]_\infty\cdot CH_2Cl_2$ [17]; thallium(III): $[Tl(Tm)_2](TlI_4)$ [18], $[Tl(Tm)_2](I)$ [14], and $[Tl(Tm^{Ph})_2](ClO_4)$ [16]; indium(I): $[In(Tm^{tBu})]$ [15], $[In(Tm^{tBu})\{B(C_6F_5)_3\}]$ [15], $[In(Tm^{tBu})(\kappa^2\text{-}S_4)]$ [15]; and $[In(Tm^{tBu})_2](I)$ [15]; indium(III): $[In(Tm^{Ad})_2](InI_4)$ [19], $[In(Tm^{Ad})(\kappa^2\text{-mim}^{Ad})](Cl)$ [19], $[In(Tm^{Ad})\{B(C_6F_5)_3\}](Cl)$ [19], $[In(Tm^{tBu})]$ [15], and $[In(Tm^{Me})_2](I)$ [20].

2. Materials and Methods

2.1. Chemicals and Instrumentation

The preparation and handling of the two complexes were performed under an argon atmosphere using standard Schlenk tube techniques. Dichloromethane was carefully purified by refluxing and distilling under an argon atmosphere over phosphorous pentoxide. Heptane, toluene, and tetrahydrofuran were carefully purified by refluxing and distilling under an argon atmosphere over sodium benzophenone ketyl [21]. Dry ethanol was purchased from Wako Pure Chemical Ind. Ltd. and deoxygenated by purging with argon gas. Deuteriochloroform was obtained from Cambridge Isotope Laboratories, Inc. (Tewksbury, MA, USA). Other reagents were commercially available and used without further purification. The potassium salt of tris(3-*tert*-butyl-2-sulfanylidene-1*H*-imidazol-1-yl)hydroborate (KTm^{tBu}) was prepared by published methods [22–25].

2.2. Instrumentation

IR spectra (4000–400 cm^{-1}) were recorded as KBr pellets using a JASCO FT/IR-6300 spectrophotometer (JASCO, Tokyo, Japan). Raman spectra (4000–200 cm^{-1}) were measured as powders on a JASCO RFT600 spectrophotometer with a YAG laser 650 mW (JASCO, Tokyo, Japan). Abbreviations used in the description of vibrational data are as follows: s, strong; m, medium; w, weak. ^1H-NMR (500 MHz) and ^{13}C-NMR (125 MHz) spectra were obtained on a Bruker AVANCE III-500 NMR spectrometer at room temperature (298 K) in $CDCl_3$ (Bruker Japan, Yokohama, Japan). ^1H and ^{13}C chemical shifts were reported as δ

values relative to residual solvent peaks (7.26 and 77.16 ppm, respectively). UV–Vis spectra (solution CH_2Cl_2, 1050–250 nm) were recorded on an Agilent 8453 UV–visible spectroscopy system (Agilent, Tokyo, Japan). The elemental analyses (C, H, and N) were performed by the Chemical Analysis Center of Ibaraki University.

2.3. Preparation of Complexes

2.3.1. $[Tl(Tm^{tBu})]_2 \cdot 2H_2O$

A solution of $K(Tm^{tBu})$ (0.5114 g, 0.991 mmol) in dichloromethane (15 mL) was added to a solution of thallium(I) acetate (0.2649 g, 1.006 mmol) in degassed ethanol (10 mL). After allowing the reaction to proceed overnight, the solvent was removed under reduced pressure, and the resulting solid was extracted by dichloromethane (15 mL). Colorless crystals were obtained by slow evaporation from a saturated dichloromethane/heptane (1:1 v/v) solution and were characterized crystallographically as $[Tl(Tm^{tBu})]_2 \cdot 2H_2O$ (0.2820 g, 0.410 mmol, yield: 41%). Elemental analysis (bulk material): Anal. Calcd. for $[Tl(Tm^{tBu})] \cdot 1/3H_2O$: C 36.66, H 5.08, N 12.22%; Found: C 36.43, H 4.90 N 11.93%. IR (KBr, cm^{-1}): 3147 w ν(C–H), 2975 m ν(C–H), 2922 w ν(C–H), 2455 w ν(B–H), 1626 m, 1560 w, 1407 m, 1396 m, 1357 s, 1273 m, 1199 s, 1165 m, 1099 m, 981 w, 737 w, 715 m, 682 m. Raman (cm^{-1}): 3188 w (C–H), 3145 w ν(C–H), 2979m ν(C–H), 2959 m ν(C–H), 2922m ν(C–H), 2445 w ν(B–H), 1564 s, 1458 m, 1358 s, 1258 w, 1226 w, 1153 w, 1099 w, 1058 w, 1026 w, 931 w, 821 w, 716 w, 611 w, 564 w, 407 w, 314 w. ^1H NMR ($CDCl_3$, 298 K): 1.79 (s, 27H, CH_3), 6.12 (3H, imidazole H), 6.86 (3H, imidazole H). ^{13}C NMR ($CDCl_3$, 298 K): 28.7 (CCH_3), 58.5 (CCH_3), 115.9 (imidazole C_4 or C_5), 123.1 (imidazole C_4 or C_5), 161.2 (imidazole C=S). UV–vis (CH_2Cl_2, 298 K; λ_{max}, nm (ε, $M^{-1}cm^{-1}$)): 270 (20,700).

2.3.2. $[In(Tm^{tBu})_2](InCl_4)$

A solution of $[Tl(Tm^{tBu})]_2 \cdot 2H_2O$ (0.0706 g, 0.050 mmol) in dichloromethane (5 mL) was added to a solution of $InCl_3 \cdot 4H_2O$ (0.0294 g, 0.100 mmol) in tetrahydrofuran (5 mL). After allowing the reaction to proceed overnight, the solvent was removed under reduced pressure, and the resulting solid was extracted with dichloromethane (10 mL). Colorless crystals were obtained by slow evaporation from a saturated dichloromethane/toluene (1:1 v/v) solution as $[In(Tm^{tBu})_2](InCl_4)$ (0.0586 g, 0.044 mmol, yield: 88%). Elemental analysis: Anal. Calcd. for $[In(Tm^{tBu})_2](InCl_4)$: C 38.03, H 5.17, N 12.67%; Found: C 38.04, H 5.09 N 12.33%. IR (KBr, cm^{-1}): 3181w ν(C–H), 3147 w ν(C–H), 2979 m ν(C–H), 2928 m ν(C–H), 2423 w ν(B–H), 1567 w, 1480 m,1420 s, 1398 m, 1357 s, 1308 m, 1260 w, 1228 m, 1098 s, 1177 s, 1072 w, 822 w, 768 w, 732 m, 687 m, 590 w, 552 w, 496 w, 457 w. Raman (cm^{-1}): 3184 w ν(C–H), 3152 w ν(C–H), 3111 w ν(C–H), 2986 m ν(C–H), 2928 m ν(C–H), 2425 w ν(B–H), 1568 m, 1450 w, 1431 w, 1359 s, 1307 w, 1255 w, 1248 w, 1073 w, 1041 w, 986 w, 932 w, 825 w, 738 w, 637 w, 590 w, 401 w, 320 w. ^1H NMR ($CDCl_3$, 298 K): 1.74 (*s*, 9H, CH_3), 1.79 (*s*, 18H, CH_3), 6.82 (*d*, 1H, J = 2.0 Hz, imidazole *H*), 6.89 (*d*, 2H, J = 2.0 Hz, imidazole *H*), 7.08 (*d*, 1H, J = 2.0 Hz, imidazole *H*), 7.12 (*d*, 2H, J = 2.0 Hz, imidazole *H*). ^{13}C NMR ($CDCl_3$, 298 K): 29.9 (CCH_3), 30.1 (CCH_3), 60.1 (CCH_3), 60.8 (CCH_3), 118.3 (imidazole C_4 or C_5), 123.9 (imidazole C_4 or C_5), 153.2 (imidazole C=S). UV–vis (CH_2Cl_2, 298 K; λ_{max}, nm (ε, $M^{-1}cm^{-1}$)): 268 (17,900), 307 (8000).

2.4. X-ray Crystallography

Colorless crystals of $[Tl(Tm^{tBu})]_2 \cdot 2H_2O$ and $[In(Tm^{tBu})_2](InCl_4)$ were coated with Paratone-N oil (Hampton Research, Aliso Viejo, CA, USA) and mounted on a Dual-Thickness MicroLoop LD (200 μM) (MiTeGen, New York, NY, USA). X-ray intensity data were measured at T = 178 K on a Rigaku/Oxford Diffraction Rigaku XtaLAB P200 diffractometer (Rigaku Oxford Diffraction, Oxfordshire, UK) fitted with MoKα radiation (λ = 0.71073 Å) so that 100% data completeness was achieved at θ_{max} = 25.2°. Data reduction, including empirical absorption correction, was accomplished with CrysAlisPro (Rigaku Oxford Diffraction, Oxfordshire, UK) [26]. The structures were solved by direct methods [27] and refined (anisotropic displacement parameters and C-bound H atoms in the riding model approximation) on F^2 [28]. For **2**, the positions of the water-bound H atoms were idealized based on chemically reasonable positions with the O–H and H···H distances initially refined with restraints 0.840 ± 0.001 and 1.30 ± 0.001 Å, respectively, and fixed at these values in the final cycles of refinement. A weighting scheme of the form $w = 1/[\sigma^2(F_o^2) + (aP)^2 + bP]$, where $P = (F_o^2 + 2F_c^2)/3)$, was applied toward the latter stages of each refinement. At the conclusion of each refinement, relatively large residual electron density peaks were noted; details are given in the respective CIFs. The molecular structure diagrams were generated with ORTEP for Windows [29] with 50% displacement ellipsoids, and the packing diagrams were drawn with DIAMOND [30]. Additional data analysis was made with PLATON [31]. Crystal data and refinement details are given in Table 1.

Table 1. Crystallographic data and refinement details for $[Tl(Tm^{tBu})]_2{\cdot}2H_2O$ and $[In(Tm^{tBu})_2](InCl_4)$.

Complex	$[Tl(Tm^{tBu})]_2{\cdot}2H_2O$	$[In(Tm^{tBu})_2](InCl_4)$
Formula	$C_{42}H_{68}B_2N_{12}S_6Tl_2$, $2(H_2O)$	$C_{42}H_{68}B_2InN_{12}S_6$, $InCl_4$
Molecular weight	1399.86	1326.50
Crystal size/mm^3	0.03 × 0.10 × 0.13	0.13 × 0.22 × 0.26
Colour	colorless	colorless
Crystal system	triclinic	triclinic
Space group	$P\bar{1}$	$P\bar{1}$
a/Å	9.5421(2)	11.1900(1)
b/Å	11.8656(2)	11.4449(2)
c/Å	14.5839(3)	23.2153(3)
a/°	67.963(2)	94.421(1)
β/°	71.203(2)	92.606(1)
γ/°	73.428(2)	90.262(1)
V/Å^3	1423.33(6)	2961.14(7)
Z	1	2
D_c/g cm^{-3}	1.633	1.488
μ/mm^{-1}	5.918	1.212
Measured data	48,875	101474
θ range/°	2.5–29.9	2.6–29.8
Unique data	7633	15775
Observed data ($I \geq 2.0\sigma(I)$)	6786	14551
No. of parameters	298	616
R, obs. data; all data	0.032; 0.038	0.028; 0.030
a; b in weighting scheme	0.055; 1.294	0.041; 2.674
Rw, obs. data; all data	0.085; 0.087	0.075; 0.076
Range of residual electron density peaks/eÅ^{-3}	−1.38–2.58	−0.78–1.76

3. Results and Discussion

3.1. Synthesis and Characterization

The reactions of the ligand $K(Tm^{tBu})$ [22–25] with one equivalent of thallium(I) acetate (TlOAc) were carried out at room temperature, and single crystals of the thallium(I) complex, formulated as $[Tl(Tm^{tBu})]_2{\cdot}2H_2O$, were obtained by slow evaporation of a dichloromethane/ethanol solution at room temperature (Figure 2). The indium(III) complex, $[In(Tm^{tBu})_2](InCl_4)$, was obtained by the reaction of $[Tl(Tm^{tBu})_2]$ with indium(III) chloride $InCl_3{\cdot}4H_2O$. The colorless crystals were obtained from the mixed solution of the saturated dichloromethane/toluene solution (Figure 2).

The expected signals in IR and Raman spectra were obtained for each $[Tl(Tm^{tBu})]_2$ and $[In(Tm^{tBu})_2](InCl_4)$. Noteworthy were the B–H stretching bands at 2455 cm^{-1} for $[Tl(Tm^{tBu})]_2$ and 2423 cm^{-1} for $[In(Tm^{tBu})_2](InCl_4)$, which were clearly evident and red-shifted compared to 2480 cm^{-1} for $K(Tm^{tBu})$ [25] (Figures S1 and S2 of the Supplementary Materials). The ^{1}H-and ^{13}C NMR spectra of $[Tl(Tm^{tBu})]_2$ in $CDCl_3$ occurred at chemical shifts identical to those of KTm^{tBu} (Figures S3–S5 of the Supplementary Materials). For $[In(Tm^{tBu})_2](InCl_4)$, all chemical shifts in the ^{1}H-NMR, and all those except for C=S carbon shifts in the ^{13}C-NMR, were clearly split due to the different structural arrangements, i.e., occupying equatorial and axial coordination sites. The UV–Vis absorption spectra of $[Tl(Tm^{tBu})]_2$ and $[In(Tm^{tBu})_2](InCl_4)$ were also measured (Figure S6 of the Supplementary Materials). Two characteristic absorption bands at 268 and 307 nm were observed for $[In(Tm^{tBu})_2](InCl_4)$, but for $[Tl(Tm^{tBu})]_2$, only a band at 261 nm was noted. From this observation, the high-energy bands at 260 and 268 nm band were assigned to a ligand Tm^{tBu}-based absorption, and the low-energy band for $[In(Tm^{tBu})_2](InCl_4)$ is metal centered. For more detailed assignments, computational chemistry calculations are required, which are beyond the scope of this study. The ^{1}H-NMR and UV–Vis spectroscopy results indicate that the structures remained intact in the solution state.

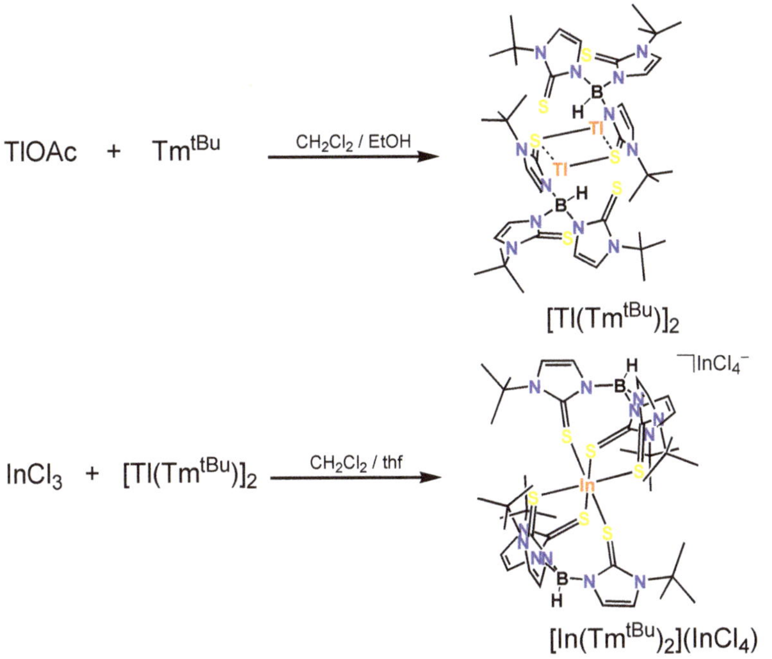

Figure 2. Syntheses of thallium(I) complex, $[Tl(Tm^{tBu})]_2$, and indium(III) complex, $[In(Tm^{tBu})_2](InCl_4)$.

3.2. *Crystal and Molecular Structures*

3.2.1. Molecular Structures

The crystallographic asymmetric unit of $[Tl^{I}(Tm^{tBu})]_2 \cdot 2H_2O$ comprises one-half of a dimeric complex molecule, being located about a center of inversion and a water molecule of crystallization. The complex molecule is shown in Figure 3a and comprises a central, diamond-shaped core with almost equivalent Tl1–S1, S1^{i} bond lengths (Table 2). The four-coordinate geometry for thallium is completed by the thione-S2 and symmetry-related thione-S3^{i} atoms. An indication of the coordination geometry defined by the S_4 donor set is τ_4, which is computed from $[360 - (\alpha + \beta)]/141$, where α and β are the two widest angles subtended at the thallium(I) atom [32]. In this case, $\tau_4 = 0.63$, which corresponds to a seesaw geometry ($\tau_4 = 0.64$); the widest angle corresponds to S2–Tl1–S3^{i}, i.e., 156.32(3)°. The Tm^{tBu} ligand is therefore tetra-coordinating, bridging two thallium(I) centers. A curious feature of the molecular structure is a close intramolecular B–H···ring centroid (Tl_2S_2) separation of 2.17 Å. This is a consequence of the coordination mode of the Tm^{tBu} ligand and has been observed in literature analogs, e.g., in the benzene mono-solvate of $[Tl(Tm^{tBu})]_2$ [15] and in $[Tl(Tm^{Ph})]_2$, as its chloroform mono-solvate [16], each of which features the same κ^3 coordination mode as described above but with variations in the magnitudes of the Tl–S1 bond lengths.

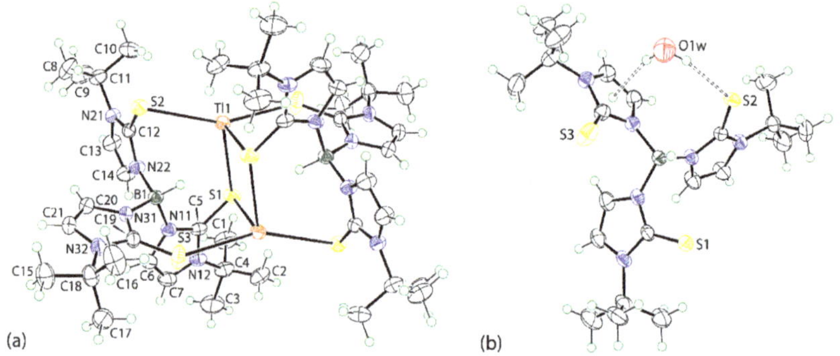

Figure 3. (**a**) Molecular structure of the complex molecule in $[Tl(Tm^{tBu})]_2{\cdot}2H_2O$, showing atom labeling scheme and displacement ellipsoids at the 50% probability level, and (**b**) detail of the supramolecular molecular association (dashed lines) involving the water molecule. Unlabeled atoms in (**a**) are related by the symmetry operation $2 - x, 1 - y, - z$.

Table 2. Selected geometric parameters (Å, °) for $[Tl(Tm^{tBu})]_2{\cdot}2H_2O$.

Parameter	Value	Parameter	Value
Tl1–S1	3.0408(9)	S1–Tl1–S3^{i}	82.04(3)
Tl1–S2	3.1696(9)	S2–Tl1–S1^{i}	91.89(2)
Tl1–S1^{i} [a]	2.9993(10)	S2–Tl1–S3^{i}	156.32(3)
Tl1–S3^{i} [a]	3.2764(11)	S1^{i}–Tl1–S3^{i}	104.23(3)
C5–S1	1.721(4)	Tl1–S1–Tl1^{i}	86.10(2)
C12–S2	1.700(3)	Tl1–S1–C5	120.26(12)
C19–S3	1.694(4)	Tl1–S2–C12	83.37(12)
S1–Tl1–S2	114.53(2)	Tl1–S1^{i}–C5^{i}	86.11(12)
S1–Tl1–S1^{i} [a]	93.90(2)	Tl1–S3^{i}–C19^{i}	127.75(14)
(N11,N12,C5-C7)/(N21,N22,C12-C14)	88.2(3)	(N21,N22,C12-C14)/(N31,N32,C19-C21)	78.7(3)
(N11,N12,C5-C7)/(N31,N32,C19-C21)	83.2(3)		

[a] symmetry operation: $2 - x, - y, - z$.

As highlighted in Figure 3b, the water molecule of crystallization is closely associated with the complex molecule, forming a hydrogen bond to the thione-S2 atom and close contact with the N31-pyrazolyl ring; see Table 3 for the geometric parameters defining these interactions. The H2w···ring centroid separation is 2.47 Å, with the closest contact to a specific atom within the ring being 2.45 Å, i.e., the C19 atom with the next closest interaction, i.e., 2.60 Å with N31, and the longest separation of 2.97 Å with C21. This pattern indicates that the interaction is best described as a localized H2w···π interaction [33].

Table 3. Geometric parameters (Å, °) characterizing the specified intermolecular contacts operating in the crystal of $[Tl(Tm^{tBu})]_2{\cdot}2H_2O$ [a].

A	H	B	H···B	A···B	A–H···B	Symmetry Operation
O1w	H1w	S2	2.47	3.2779(10)	160	x, y, z
O1w	H2w	Cg (1)	2.47	3.2816(18)	161	x, y, z
C13	H13	O1w	2.46	3.214(5)	136	$-1 + x, y, z$

[a] Cg(1) is the ring centroid of the (N31,N32,C19-C21) ring.

The asymmetric-unit of $[In^{III}(Tm^{tBu})_2](InCl_4)$ comprises two independent complex cations, each disposed about a center of inversion, and a $InCl_4^-$ anion; the molecular structures are shown in Figure 4a–c. Focusing on the In1-containing molecule, the In1 atom is coordinated by two tripodal Tm^{tBu} ligands to define a soft S_6 donor set, which defines an octahedral geometry. The geometry is slightly flattened, as the In1–S1 bond length of

2.5682(4) Å is systematically shorter than the In1–S2, S3 bond lengths of 2.6597 (4) and 2.6429(4) Å (Table 4). Despite this difference, the C5–S1 bond lengths are equal within experimental error. Otherwise, the deviations from the ideal octahedral geometry are small, as noted from the relevant angles included in Table 4. As highlighted by the dihedral angles between the pyrazolyl rings listed in Table 4 and the overlay diagram of Figure 4d, there is close agreement between the independent molecules. This observation notwithstanding, there are notable differences in the In–S bond lengths. Thus, in the first independent cation of $[In^{III}(Tm^{tBu})_2](InCl_4)$, one In–S bond length is 0.07–0.10 Å shorter than the others, whereas in the second independent cation, the differential between the In–S bond lengths is significantly smaller, i.e., the difference between the shortest and longer bond lengths is now 0.02–0.08 Å.

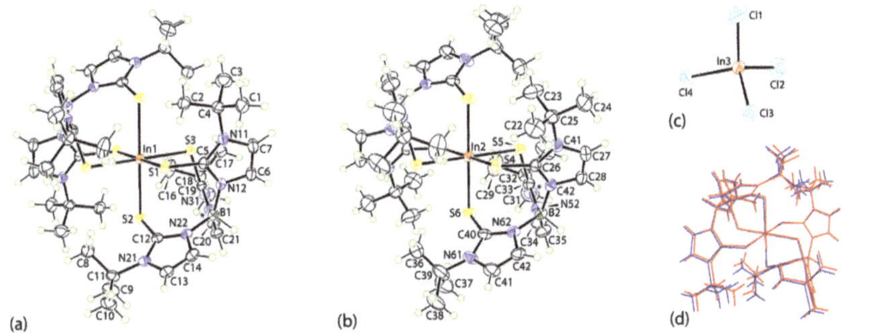

Figure 4. Molecular structure diagrams for $[In(Tm^{tBu})_2](InCl_4)$: (**a**) through (**c**) molecular structures of the independent In1- and In2-containing cations, and In3-anion, respectively, showing atom labeling schemes and displacement ellipsoids at the 50% probability level. (**a**) The N32 atom is indicated by an asterisk, and the C15 atom is not labeled. (**b**) The N51 atom is indicated by an asterisk, and the C30 atom is not labeled. The In atoms in (**a**) and (**b**) are located at crystallographic centers of inversion with the unlabeled atoms generated by the application of symmetry operations $1 - x, 1 - y, -z$ and $1 - x, 1 - y, 1 - z$. (**d**) overlay diagram of the In1- and In2-containing cations shown as red and blue images, respectively. The cations have been overlapped so that the S_3 faces are coincident.

Table 4. Selected geometric parameters (Å, °) for $[In(Tm^{tBu})_2](InCl_4)$.

Parameter	Value	Parameter	Value
In1–S1	2.5682(4)	In2–S4	2.6110(4)
In1–S2	2.6597(4)	In2–S5	2.6709(5)
In1–S3	2.6429(4)	In2–S6	2.5905(5)
C5–S1	1.7236(17)	C26–S4	1.7223(19)
C12–S2	1.7279(17)	C33–S5	1.7249(19)
C19–S3	1.7253(17)	C40–S6	1.725(2)
S1–In1–S2	91.715(13)	S4–In2–S5	92.567(14)
S1–In1–S3	93.179(13)	S4–In2–S6	95.416(15)
S2–In1–S3	92.142(13)	S5–In2–S6	89.139(15)
In1–S1–C5	105.90(6)	In2–S4–C26	109.61(6)
In1–S2–C12	107.64(6)	In2–S5–C33	106.11(7)
In1–S3–C19	106.57(6)	In2–S6–C40	106.41(7)
(N11,N12,C5-C7)/(N21,N22,C12-C14)	83.99(11)	(N41,N42,C26-C28)/(N21,N22,C12-C14)	82.84(12)
(N11,N12,C5-C7)/(N31,N32,C19-C21)	89.18(11)	(N41,N42,C26-C28)/(N51,N52,C33-C35)	88.91(13)
(N21,N22,C12-C14)/(N31,N32,C19-C21)	87.98(11)	(N51,N52,C33-C35)/(N61,N62,C40-C42)	84.43(13)

There are several literature precedents for $[In(Tm^{tBu})_2](InCl_4)$, including $[In(Tm^{tBu})_2]^+$ characterized as the Cl^- as an acetonitrile tri-solvate [15], and I^- as a benzene mono-solvate, salts [15], along with $[In(Tm^{Me})_2]^+$ characterized as the I^- salt as a diethyl ether mono- and dimethylformamide di-solvate [20]. While the indium(III) centers in all three literature complex cations exist within S_6 donor sets, the symmetries of these vary. Thus, in the chloride salt of $[In(Tm^{tBu})_2]^+$, each indium atom of the two independent molecules sits on an inversion center; in the iodide salt, the indium atom is located on a three-fold inversion center, implying that all In–S lengths are equivalent. The $[In(Tm^{Me})_2]^+$ cation is located on an inversion center.

3.2.2. Molecular Packing

The water molecule of solvation in $[Tl(Tm^{tBu})]_2 \cdot 2H_2O$ proves pivotal in assembling the dimeric molecules into a supramolecular chain along the *a*-axis, as shown in Figure 5a. Thus, the water molecule forms O–H···π(pyrazolyl) and O–H···S interactions, as shown in detail in Figure 3b, with the bay region of the Tm^{tBu} ligand. At the same time, the water-O atom accepts an interaction from a pyrazolyl-C–H atom; see Table 3 for geometric parameters. Chains pack in the crystal without directional interactions between them, as highlighted in the unit cell diagram of Figure 5b.

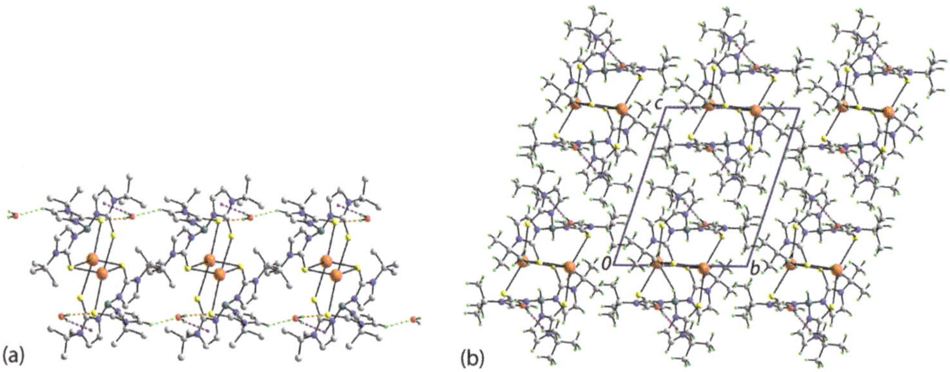

Figure 5. Molecular packing in the crystal of $[Tl(Tm^{tBu})]_2 \cdot 2H_2O$: (**a**) supramolecular chain aligned along the *a*-axis (non-participating H atoms are omitted) and (**b**) a view of the unit cell contents viewed in projection down the **a**-axis. The O–H···π(pyrazolyl), O–H···S and C–H···O contacts are highlighted as purple, orange and bright-green dashed lines, respectively.

A view of the unit cell contents for $[In(Tm^{tBu})_2](InCl_4)$ is shown in Figure 6. Globally, when viewed down the *b*-axis, molecules of In1- and In2-containing cations and $[InCl_4]^-$ anions stack in columns along this axis. The cations align in rows along the *a*-axis, and the rows alternate in an ···ABA··· fashion down the *c*-axis. Interspersed between rows are the $[InCl_4]^-$ anions, and these are pivotal in providing links between the constituent cations. The primary mode of interaction is via C–H···Cl contacts of the type pyrazolyl-C–H···Cl and methyl-C–H···Cl; for geometric parameters characterizing these interactions, refer to Table 5. The anion also forms an In–Cl···π(pyrazolyl) interaction, which is discussed in more detail in Section 3.3.

Table 5. Geometric parameters (Å, °) characterizing the specified intermolecular contacts operating in the crystal of $[In(Tm^{tBu})_2](InCl_4)$ [a].

A	H	B	H···B	A···B	A–H···B	Symmetry Operation
C21	H21	Cl1	2.85	3.723 (2)	154	$x, -1+y, z$
C41	H41	Cl3	2.83	3.773 (2)	174	$1-x, -y, 1-z$
C42	H42	Cl4	2.81	3.562 (2)	137	$1-x, -y, 1-z$
C1	H1b	Cl4	2.83	3.797 (2)	168	x, y, z
In3	Cl1	Cg(1)	3.7819 (11)	6.0631 (9)	162.17 (3)	$x, -1+y, z$

[a] Cg(1) is the ring centroid of the (N21,N22,C12-C14) ring.

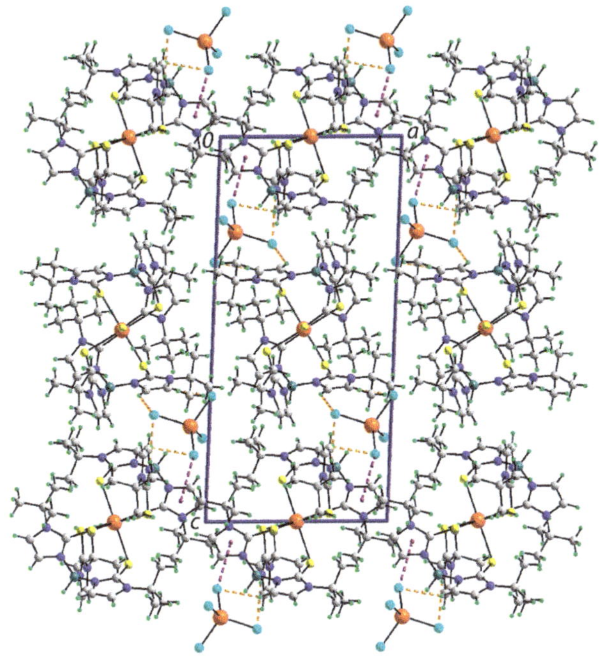

Figure 6. A view of the unit cell contents for $[In(Tm^{tBu})_2](InCl_4)$ viewed in projection down the *b*-axis. The In–Cl···π(pyrazolyl) and C–H···Cl contacts are highlighted as purple and orange dashed lines, respectively.

3.3. Hirshfeld Surface Analysis

To further understand the nature of the contacts between molecules in their respective crystals, an analysis of the calculated Hirshfeld surfaces was conducted employing *Crystal Explorer 17* [34] following established protocols [35]. The analysis shows that there are some close contacts present in each crystal, as evident from the red spots of various intensity observed on the respective d_{norm} maps indicating contact distances shorter than the sum of van der Waals radii [36].

For $[Tl(Tm^{tBu})]_2{\cdot}2H_2O$, the d_{norm} map of the $Tl(Tm^{tBu})$ fragment shows several red spots of moderate to strong intensity (Figure 7), which can be attributed to O1w–H1w···S2, O1w–H2w···C19(π), and C13–H13···O1w close contacts, with the respective contact distances being 2.37, 2.35, and 2.34 Å compared to the respective sums of the van der Waals radii (ΣvdW) of 2.89, 2.79, and 2.61 Å for H···S, H···C, and H···O (Table 6).

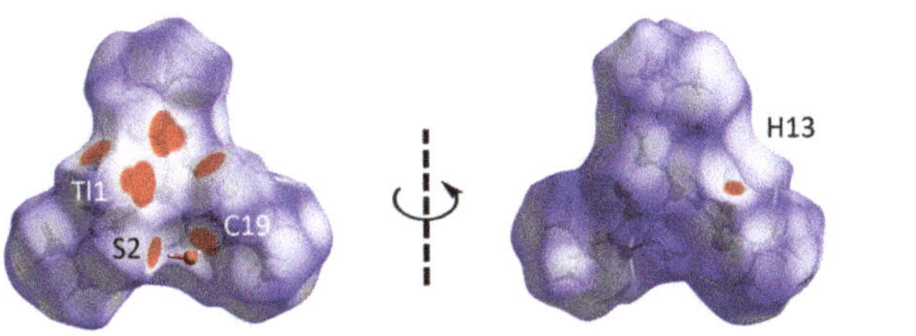

Figure 7. The two views of the d_{norm} Hirshfeld surface mapping for the $Tl(Tm^{tBu})$ fragment of $[Tl(Tm^{tBu})]_2{\cdot}2H_2O$ within the range −0.1027 to 1.3365 arbitrary units, highlighting close contacts as red regions on the surfaces with their intensity relative to the contact distance.

Table 6. d_{norm} contact distances (adjusted to neutron values) of intermolecular interactions identified in the crystals of $[Tl(Tm^{tBu})]_2 \cdot 2H_2O$ and $[In(Tm^{tBu})_2](InCl_4)$, as computed through the Hirshfeld surface analysis and in comparison to the corresponding sum of van der Waals radii (ΣvdW).

Contact	Distance (Å)	ΣvdW (Å)	Δ(ΣvdW—Distance)	Symmetry Operation
$[Tl(Tm^{tBu})]_2 \cdot 2H_2O$				
H1w···S2	2.37	2.89	0.52	x, y, z
H2w···C19	2.35	2.79	0.44	x, y, z
H13···O1w	2.34	2.61	0.27	$-1+x, y, z$
$[In(Tm^{tBu})_2](InCl_4)$				
H41···Cl3	2.40	2.84	0.44	$1-x, -y, 1-z$
H21···Cl1	2.49	2.84	0.35	$x, -1+y, z$
H37c···H30a	1.93	2.18	0.25	$1+x, y, z$
H42···Cl4	2.72	2.84	0.12	$1-x, -y, 1-z$
H1b···Cl4	2.73	2.84	0.11	x, y, z
H6···Cl4	2.78	2.84	0.06	x, y, z
H35···C19	2.76	2.79	0.03	x, y, z

Hirshfeld surface analysis was also performed for the central thallium atom (Figure 8). The d_{norm} map displays several weak to intense red spots indicating additional intramolecular interactions between B1–H1···Tl (d_{norm} distance = 2.69 Å vs. ΣvdW = 3.05 Å) and C10–H10a···Tl (d_{norm} distance = 3.00 Å vs. ΣvdW = 3.05 Å) on top of S1···Tl, S2···Tl and S3···Tl contacts, as described above.

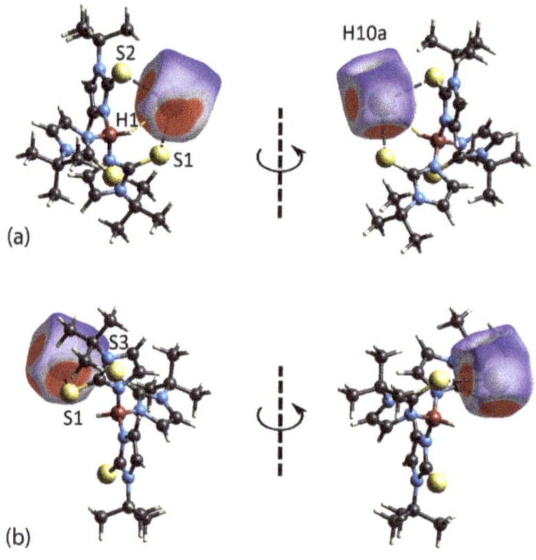

Figure 8. The two views of the d_{norm} Hirshfeld surface mapping within the range −0.1027 to 1.3365 arbitrary units for the thallium center in $[Tl(Tm^{tBu})]_2 \cdot 2H_2O$, showing the coordination modes of the metal with two Tm^{tBu} ligands of different symmetry: (**a**): x, y, z and (**b**) $2-x, 1-y, -z$.

The d_{norm} analysis was also conducted for each of the In1- or In2-containing cations and In3-anion in $[In(Tm^{tBu})_2](InCl_4)$ (Figure 9). A number of red spots with moderate intensity are observed for close contacts comprising C41–H41···Cl3, C21–H21···Cl1, and C42–H42···Cl4, in accordance with those interactions identified through the geometric analysis conducted with PLATON [31]. However, additional contacts were noted through the d_{norm} map arising from H37c···H30a, with moderate intensity, as well as from C1–H1b···Cl4, C6–H6···Cl4, and C35–H35···C19 with weak intensity. Table 6 summarizes these contacts and compares the separation distances with the respective van der Waals radii.

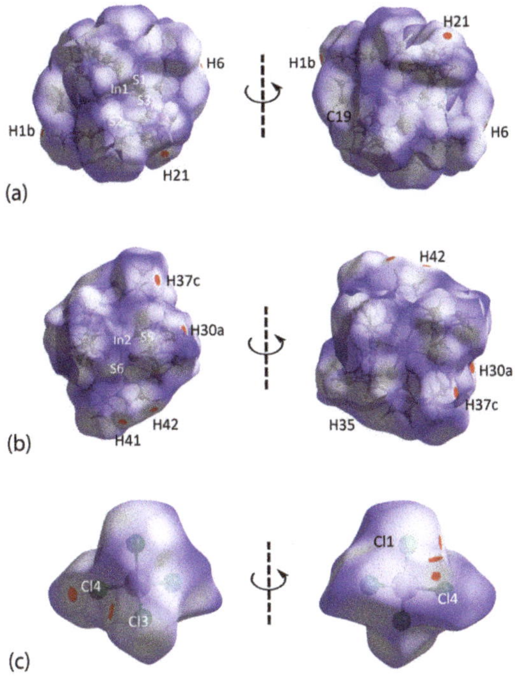

Figure 9. The two views of the d_{norm} Hirshfeld surface mapping within the range −0.0195 to 1.7591 arbitrary units for $[In(Tm^{tBu})_2](InCl_4)$: (**a**) In1-cation, (**b**) In2-cation, and (**c**) In3-anion, with the close contacts indicated by the corresponding red dots on the surfaces with their intensity relative to the contact distance.

In addition to the close contacts as identified through the direct observation of red spots on the Hirshfeld surface of $[In(Tm^{tBu})_2](InCl_4)$, two other important contacts, specifically In3–Cl1···π(pyrazolyl) and C1–H1b···In3, are detected based on the complementarity of shape as indicated by the hollows and bumps on the shape index mapped over the Hirshfeld surfaces between the independent In1 cation and In3 anion, despite the observation that the contact distances between Cl1 and the closest carbon atom of the five-membered imidazole ring, i.e., Cl1···C14, as well as H1b···In3, are longer than the corresponding ΣvdW, Figure 10.

Quantification of the close contacts in each individual component of $[Tl^{I}(Tm^{tBu})]_2{\cdot}2H_2O$ and $[In^{III}(Tm^{tBu})_2](InCl_4)$ was performed through two-dimensional fingerprint plot analysis. The overall and delineated fingerprint plots for the individual components profiled in $[Tl(Tm^{tBu})]_2{\cdot}2H_2O$, namely the Tm^{tBu} fragment, the water molecule, the thallium atom as well as the $[Tl(Tm^{tBu})]_2$ dimer, are illustrated in Figure 11. The fingerprint profiles for both Tm^{tBu} and $[Tl(Tm^{tBu})]_2$ resemble each other and have the shape of a flying fox, while those for H_2O and Tl display a cicada- and pincer-like profile, respectively. The decomposition of the corresponding overall profiles for Tm^{tBu} and $[Tl(Tm^{tBu})]_2$ show that their close contacts are dominated by H···H (69.5 and 76.6%, respectively), H···S/S···H (12.8 and 9.8%), H···C/C···H (6.2 and 6.6%), H···Tl/Tl···H (3.7 and 3.3%), S···Tl (for Tm^{tBu} only, 2.9%), and H···O/O···H (2.3 and 2.6%) as well as by other minor contacts that constitute less than 1.0% of all surface contacts, including H···N/N···H, S···O, and others. Among those contacts, only S···H, C···H, H···Tl/Tl···H, and O···H feature a distinctive tip with $d_i + d_e$ values corresponding to O1w–H1w···S2, O1w–H2w···C19, B1–H1···Tl/C10–H10a···Tl, and C13–H13···O1w, while the rest of the contacts present indistinct profiles.

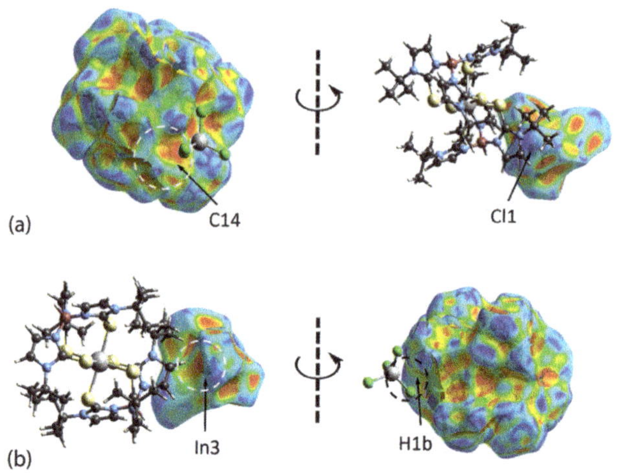

Figure 10. Two views of the shape-index mapped over the Hirshfeld surfaces for $[In(Tm^{tBu})_2](InCl_4)$ within the property range −1.0 to +1.0 arbitrary units highlighting (**a**) In3–Cl1···π(C14) and (**b**) C1–H1b···In3 contacts, showing the shape complementarity as indicated by the hollow (orange) and bump (blue) on the surfaces.

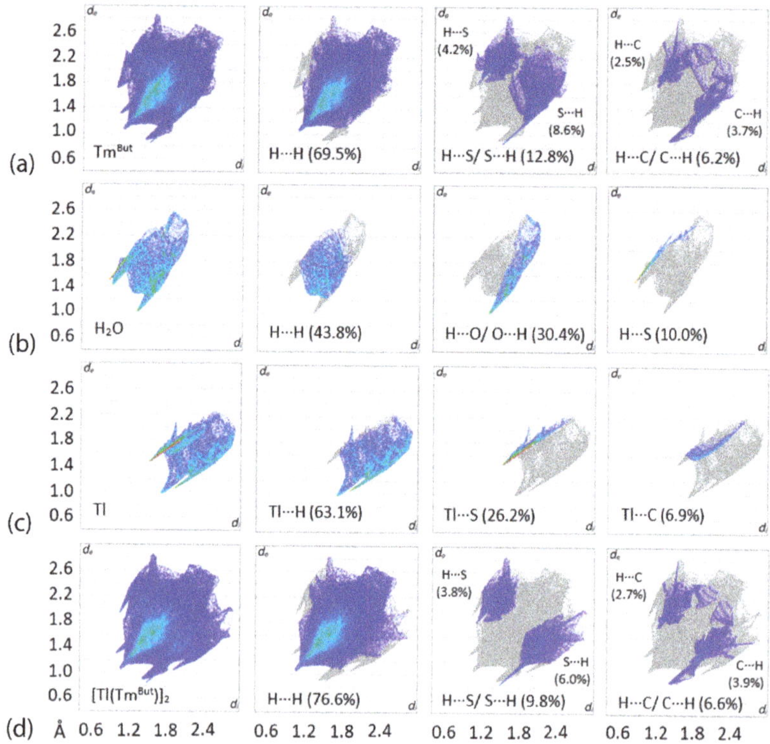

Figure 11. The overall two-dimensional fingerprint and decomposed plots delineated into the major contacts along with the percentage distributions for (**a**) Tm^{tBu}, (**b**) H_2O molecule, (**c**) thallium center, and (**d**) $[Tl(Tm^{tBu})]_2$ dimer for $[Tl(Tm^{tBu})]_2{\cdot}2H_2O$.

As for the H_2O molecule and thallium atom, the major contacts for the former appear in the order H···H (43.8%) > O···H (30.4%) > H···S (10.0%) > H···C (7.2%) > H···N (6.2%) > O···S (2.5%), while the latter is dominated by H···Tl (63.1%) > S···Tl (26.2%) > C···Tl (6.9%) > N···Tl (3.9%). The d_i + d_e distances associated with the significant peaks in these profiles correspond to the reciprocal contacts as identified in the Tm^{tBu} and $[Tl(Tm^{tBu})]_2$ profiles.

The overall and selected delineated two-dimensional fingerprint plots for the In1- and In2-cations as well as the In3-anion in $[In(Tm^{tBu})_2](InCl_4)$ are presented in Figure 12. As a general observation, the In1- and In2-cations exhibit a paw-like overall profile, which can be delineated mainly into H···H, H···Cl, and H···C/C···H contacts that constitute more than 93% of the Hirshfeld surfaces in each case. Specifically, the plots of the H···Cl contacts for the In1- and In2-cations display a distinctive spike in their decomposed fingerprint plots comprising 13–14% of all surface contacts with d_i + d_e tipped at about 2.70 Å, which can be, respectively, attributed to H1b···Cl4 and H42···Cl4. The (inner)-C···H-(outer) and (inner)-H···C-(outer) contacts, represented by the pair of pincer-like profiles in the decomposed H···C/C···H fingerprint plots for the In1- and In2-cations, are due to H35···C19, with both constituting about 5% of the Hirshfeld surfaces with a d_i + d_e distance of 2.76 Å, while the reciprocal (inner)-H···C-(outer) of the In1-cation and (inner)-C···H-(outer) of the In2-cation contribute 4.3 and 5.4%, respectively, to the contact surfaces but with a less significant d_i + d_e distance, being longer than ΣvdW. A distinct feature is observed in the decomposed H···H fingerprint plot for the In2-cation compared to that for the In1-cation, in that for the former, a relatively prominent tip at d_i + d_e = 1.93 Å is assignable to H37c···H30a.

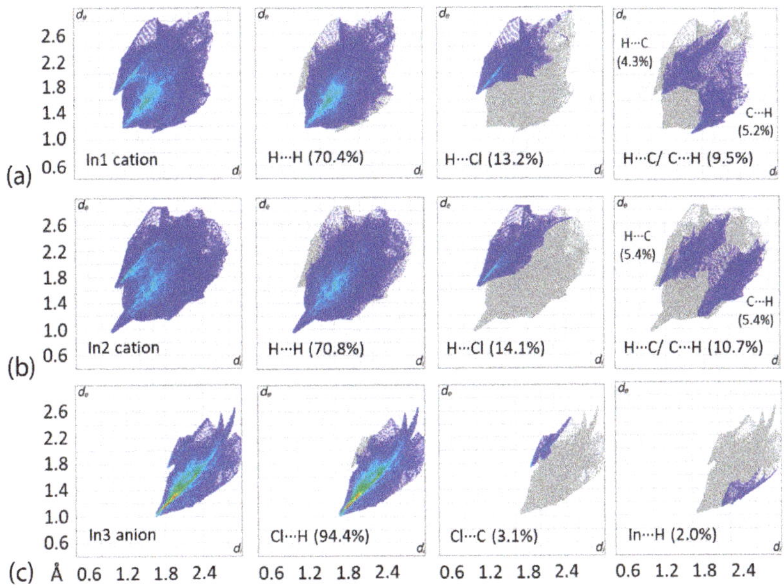

Figure 12. The overall two-dimensional fingerprint and decomposed plots delineated into the major contacts along with the percentage distributions for (**a**) In1-cation, (**b**) In2-cation, and (**c**) In3-anion for $[In(Tm^{tBu})_2](InCl_4)$.

On the other hand, the decomposition of the squid-like overall fingerprint plot for the In3-anion shows that about 94.4% of the contact surfaces are dominated by Cl···H with a d_i + d_e distance tipped at 2.66 Å ascribed to the close C41–H41···Cl3 contact, while other minor contacts include Cl···C and In···H, which contribute about 3.1 and 2.0%, respectively, to the total surface, both with d_i + d_e tips arising from the In3–Cl1···π(C14) and C1–H1b···In3 contacts as discussed above.

4. Conclusions

The synthesis, spectroscopic and X-ray crystallographic characterizations of two main group element complexes of a soft S_3-tripod-type ligand have been described. S_4-seesaw and S_6-flatenened octahedral geometries were found for the central atoms in $Tl(Tm^{tBu})]_2{\cdot}2H_2O$ and $[In(Tm^{tBu})_2](InCl_4)$, respectively. The analyses of the calculated Hirshfeld surfaces confirmed the geometric analysis of the molecular packing. These results suggest soft S_3-tripod-type ligands related to that discussed herein are a potentially useful class of ligands for coordination to *p*-block elements. The coordination of sulfur atoms gives rise to distinct electronic characteristics compared with the well-known hard N_3-type ligands. Therefore, the present chemistry paves the way for new coordination chemistry based on soft S_3-tripod-type ligands.

Supplementary Materials: The following supporting information can be downloaded at: https://www.mdpi.com/article/10.3390/cryst13050745/s1. Figure S1: IR spectra for $[Tl(Tm^{tBu})]_2{\cdot}2H_2O$ and $[In(Tm^{tBu})_2](InCl_4)$; Figure S2: FT-Raman spectra for $[Tl(Tm^{tBu})]_2{\cdot}2H_2O$ and $[In(Tm^{tBu})_2](InCl_4)$; Figure S3: ^{1}H-NMR spectrum of $[Tl^I(Tm^{tBu})]_2{\cdot}2H_2O$; Figure S4: ^{1}H-NMR spectrum for $[In^{III}(Tm^{tBu})_2](InCl_4)$; Figure S5: ^{13}C-NMR spectrum of $[In^{III}(Tm^{tBu})_2](InCl_4)$; Figure S6: UV–Vis spectra for $[Tl(Tm^{tBu})]_2{\cdot}2H_2O$ and $[In(Tm^{tBu})_2](InCl_4)$.

Author Contributions: Investigation, formal analysis, A.K. Investigation, formal analysis, writing—original draft preparation, writing—review and editing, K.F., S.L.T. and E.R.T.T. All authors have read and agreed to the published version of the manuscript.

Funding: This research was funded by an Ibaraki University Priority Research Grant and by the Joint Usage/Research Center for Catalysis (proposals 22DS0143 and 23DS0198).

Data Availability Statement: Crystallographic datasets for the structures $[Tl(Tm^{tBu})]_2{\cdot}2H_2O$ and $[In(Tm^{tBu})_2](InCl_4)$ are available through the Cambridge Structural Database with deposition numbers CCDC 2253015 ($[Tl(Tm^{tBu})]_2{\cdot}2H_2O$) and 2253016 ($[In(Tm^{tBu})_2](InCl_4)$). These data can be obtained free of charge via https://www.ccdc.cam.ac.uk/structures/ (accessed on 31 March 2023).

Conflicts of Interest: The authors declare no conflict of interest.

References

1. Trofimenko, S. *Scorpionates: The Coordination Chemistry of Polypyrazolylborate Ligands*; Imperial College Press: London, UK, 1999.
2. Pettinari, C. *Scorpionates II: Chelating Borate Ligands—Dedicated to Swiatoslaw Trofimenko*; Imperial College Press: London, UK, 2008.
3. Fujisawa, K.; Tobita, K.; Sakuma, S.; Savard, D.; Leznoff, D.L. Binuclear and mononuclear copper(II) chlorido complexes with hindered neutral N3 type ligands: Influence of ligand framework and charge on their structure and physicochemical properties. *Inorg. Chim. Acta* **2019**, *486*, 582–588. [CrossRef]
4. Lehnert, N.; Fujisawa, K.; Camarena, S.; Dong, H.T.; White, C.J. Activation of non-heme iron-nitrosyl complexes: Turning up the heat. *ACS Catal.* **2019**, *9*, 10499–10518. [CrossRef]
5. Fujisawa, K.; Ono, T.; Okamura, M. Synthesis and characterization of catecholato copper(II) complexes with sterically hindered neutral and anionic N3 type ligands: Tris(3,5-diisopropyl-1-pyrazolyl)methane and hydrotris(3,5-diisopropyl-1-pyrazolyl)borate. *Inorganics* **2020**, *8*, 37. [CrossRef]
6. Fujisawa, K.; Sakuma, S.; Ikarugi, R.; Jose, A.; Solomon, E.I. Thermally stable manganese(III) peroxido complexes with hindered N3 tripodal ligands: Structures and their physicochemical properties. *J. Inorg. Biochem.* **2021**, *225*, 111597. [CrossRef]
7. Trofimenko, S. Scorpionates: Genesis, milestones, prognosis. *Polyhedron* **2004**, *23*, 197–203. [CrossRef]
8. Lennartson, A. Toxic thallium. *Nat. Chem.* **2015**, *7*, 610–611. [CrossRef]
9. Renouf, C. A touch of indium. *Nat. Chem.* **2012**, *4*, 862. [CrossRef]
10. Fujisawa, K.; Kuboniwa, A.; Kiss, M.; Szilagyi, R.K. Mono- and binuclear tris(3-tert-butyl-2-sulfanylidene-1H-imidazol-1-yl)hydroborate bismuth(III) dichloride complexes: A soft scorpionate ligand can coordinate to p-block elements. *Acta Crystallogr. Sect. C Struct. Chem.* **2016**, *72*, 768–776. [CrossRef]
11. Steel, G.; Rajasekharan-Nair, R.; Stepek, I.A.; Kennedy, A.R.; Reglinski, J.; Spicer, M.D. Observations on the steric impact of N- and S-donor scorpionate ligands. *Eur. J. Inorg. Chem.* **2016**, *2016*, 2409–2412. [CrossRef]
12. Reglinski, J.; Spicer, M.D. Chemistry of the p-block elements with anionic scorpionate ligands. *Coord. Chem. Rev.* **2015**, *297–298*, 181–207. [CrossRef]
13. Spicer, M.D.; Reglinski, J. Soft scorpionate ligands based on imidazole-2-thione donors. *Eur. J. Inorg. Chem.* **2009**, *2009*, 1553–1574. [CrossRef]

14. Rong, Y.; Palmer, J.H.; Parkin, G. Benzannulated tris(2-mercapto-1-imidazolyl)hydroborato ligands: Tetradentate κ^4-S_3H binding and access to monomeric monovalent thallium in an [S_3] coordination environment. *Dalton Trans.* **2014**, *43*, 1397–1407. [CrossRef]
15. Yurkerwich, K.; Buccella, D.; Melnick, J.G.; Parkin, G. Monovalent indium in a sulfur-rich coordination environment: Synthesis, structure and reactivity of tris(2-mercapto-1-tert-butylimidazolyl)hydroborato indium, [Tm^{Bu^t}]In. *Chem. Commun.* **2008**, *28*, 3305–3307. [CrossRef]
16. Kimblin, C.; Bridgewater, B.M.; Hascall, T.; Parkin, G. The synthesis and structural characterization of bis(mercaptoimidazolyl)(pyrazolyl)hydroborato and tris(mercaptoimidazolyl)hydroborato complexes of thallium(I) and thallium(III). *J. Chem. Soc. Dalton Trans.* **2000**, 1267–1274. [CrossRef]
17. Ojo, J.F.; Slavin, P.A.; Reglinski, J.; Garner, M.; Spicer, M.D.; Kennedy, A.R.; Teat, S.J. The synthesis of soft tripodal ligands: Restrictions on the preparation of hydrotris(thiazolyl)borate anions from borohydride melts. *Inorg. Chim. Acta* **2001**, *313*, 15–20. [CrossRef]
18. Slavin, P.A.; Reglinski, J.; Spicera, M.D.; Kennedy, A.R. Preparation and crystal structure of the [bis{hydrotris(methimazolyl)borato} thallium(III)] cation: Modulated chemistry resulting from the use of soft and hard tripodal ligands. *J. Chem. Soc. Dalton Trans.* **2000**, 239–240. [CrossRef]
19. Yurkerwich, K.; Yurkerwich, M.; Parkin, G. Synthesis and structural characterization of tris(2-mercapto-1-adamantylimidazolyl) hydroborato complexes: A sterically demanding tripodal [S_3] donor ligand. *Inorg. Chem.* **2011**, *50*, 12284–12295. [CrossRef]
20. Dodds, C.A.; Reglinski, J.; Spicer, M.D. Lower main-group element complexes with a soft scorpionate ligand: The structural influence of stereochemically active lone pairs. *Chem. Eur. J.* **2006**, *12*, 931–939. [CrossRef]
21. Armarego, W.L.F.; Chai, C.L.L. *Purification of Laboratory Chemicals*, 7th ed.; Butterworth-Heinemann: Oxford, UK, 2012.
22. Kreider-Mueller, A.; Rong, Y.; Owen, J.S.; Parkin, G. Molecular structures of tris(2-mercapto-1-tert-butylimidazolyl)hydroborato and tris(2-mercapto-1-adamantylimidazolyl)hydroborato sodium complexes: Analysis of [Tm^R] ligand coordination modes and conformations. *Dalton Trans.* **2014**, *43*, 10852–10865. [CrossRef]
23. Mihalcik, D.J.; White, J.L.; Tanski, J.M.; Zakharov, L.N.; Yap, G.P.A.; Incarvito, C.D.; Rheingold, A.L.; Rabinovich, D. Cobalt tris(mercaptoimidazolyl)borate complexes: Synthetic studies and the structure of the first cobaltaboratrane. *Dalton Trans.* **2004**, 1626–1634. [CrossRef]
24. Kimblin, C.; Bridgewater, B.M.; Churchill, D.G.; Parkin, G. Mononuclear tris(2-mercapto-1-arylimidazolyl)hydroborato complexes of zinc, [Tm^{Ar}]ZnX: Structural evidence that a sulfur rich coordination environment promotes the formation of a tetrahedral alcohol complex in a synthetic analogue of LADH. *Chem. Commun.* **1999**, 2301–2302. [CrossRef]
25. Reglinski, J.; Garner, M.; Cassidy, I.D.; Slavin, P.A.; Spicer, M.D.; Armstrong, D.R. Sodium hydrotris(methimazolyl)borate, a novel soft, tridentate ligand: Preparation, structure and comparisons with sodium hydrotris(pyrazolyl)borate. *J. Chem. Soc. Dalton Trans.* **1999**, 2119–2126. [CrossRef]
26. *Rigaku Oxford Diffraction, CrysAlis PRO*; Oxford Diffraction Ltd.: Oxfordshire, UK, 2015.
27. Sheldrick, G.M. A short history of SHELX. *Acta Crystallogr. Sect. A Found. Crystallogr.* **2008**, *64*, 112–122. [CrossRef] [PubMed]
28. Sheldrick, G.M. Crystal structure refinement with SHELXL. *Acta Crystallogr. Sect. C Struct. Chem.* **2015**, *71*, 3–8. [CrossRef] [PubMed]
29. Farrugia, L.J. WinGX and ORTEP for Windows: An update. *J. Appl. Crystallogr.* **2012**, *45*, 849–854. [CrossRef]
30. Brandenburg, K. *DIAMOND, Crystal Impact GbR*; Crystal Impact GbR: Bonn, Germany, 2006.
31. Spek, A.L. checkCIF validation ALERTS: What they mean and how to respond. *Acta Crystallogr. Sect. E Crystallogr. Commun.* **2020**, *76*, 1–11. [CrossRef]
32. Yang, L.; Powell, D.R.; Houser, R.P. Structural variation in copper(I) complexes with pyridylmethylamide ligands: Structural analysis with a new four-coordinate geometry index, τ_4. *Dalton Trans.* **2007**, 955–964. [CrossRef]
33. Schollmeyer, D.; Shishkin, O.V.; Rühl, T.; Vysotsky, M.O. OH–π and halogen–π interactions as driving forces in the crystal organisations of tri-bromo and tri-iodo trityl alcohols. *CrystEngComm* **2008**, *10*, 715–723. [CrossRef]
34. Spackman, P.R.; Turner, M.J.; McKinnon, J.J.; Wolff, S.K.; Grimwood, D.J.; Jayatilaka, D.; Spackman, M.A. CrystalExplorer: A program for Hirshfeld surface analysis, visualization and quantitative analysis of molecular crystal. *J. Appl. Crystallogr.* **2021**, *54*, 1006–1011. [CrossRef]
35. Tan, S.L.; Jotani, M.M.; Tiekink, E.R.T. Utilizing Hirshfeld surface calculations, non-covalent inter action (NCI) plots and the calculation of inter action energies in the analysis of molecular packing. *Acta Crystallogr. Sect. E Crystallogr. Commun.* **2019**, *75*, 308–318. [CrossRef]
36. Spackman, M.A.; Jayatilaka, D. Hirshfeld surface analysis. *CrystEngComm* **2009**, *11*, 19–32. [CrossRef]

MDPI
St. Alban-Anlage 66
4052 Basel
Switzerland
Tel. +41 61 683 77 34
Fax +41 61 302 89 18
www.mdpi.com

Crystals Editorial Office
E-mail: crystals@mdpi.com
www.mdpi.com/journal/crystals

www.ingramcontent.com/pod-product-compliance
Lightning Source LLC
Chambersburg PA
CBHW061442210926
51850CB00008B/1589
9783036579283